AF875773

Managing Critical Infrastructure Risks

NATO Science for Peace and Security Series

This Series presents the results of scientific meetings supported under the NATO Programme: Science for Peace and Security (SPS).

The NATO SPS Programme supports meetings in the following Key Priority areas: (1) Defence Against Terrorism; (2) Countering other Threats to Security and (3) NATO, Partner and Mediterranean Dialogue Country Priorities. The types of meeting supported are generally "Advanced Study Institutes" and "Advanced Research Workshops". The NATO SPS Series collects together the results of these meetings. The meetings are coorganized by scientists from NATO countries and scientists from NATO's "Partner" or "Mediterranean Dialogue" countries. The observations and recommendations made at the meetings, as well as the contents of the volumes in the Series, reflect those of participants and contributors only; they should not necessarily be regarded as reflecting NATO views or policy.

Advanced Study Institutes (ASI) are high-level tutorial courses intended to convey the latest developments in a subject to an advanced-level audience

Advanced Research Workshops (ARW) are expert meetings where an intense but informal exchange of views at the frontiers of a subject aims at identifying directions for future action

Following a transformation of the programme in 2006 the Series has been re-named and re-organised. Recent volumes on topics not related to security, which result from meetings supported under the programme earlier, may be found in the NATO Science Series.

The Series is published by IOS Press, Amsterdam, and Springer, Dordrecht, in conjunction with the NATO Public Diplomacy Division.

Sub-Series

A.	Chemistry and Biology	Springer
B.	Physics and Biophysics	Springer
C.	Environmental Security	Springer
D.	Information and Communication Security	IOS Press
E.	Human and Societal Dynamics	IOS Press

http://www.nato.int/science
http://www.springer.com
http://www.iospress.nl

Series C: Environmental Security

Managing Critical Infrastructure Risks

Decision Tools and Applications for Port Security

Edited by

Igor Linkov
INTERTOX, Inc.
Brookline, Massachusetts, USA

Richard J. Wenning
ENVIRON International Corporation
San Francisco, California USA

Gregory A. Kiker
University of Florida at Gainesville, USA

Springer

Based on papers presented at the NATO Advanced Research
Workshop on Management Tools for Port Security,
Critical Infrastructure and Sustainability,
Venice, Italy March 2006

A C.I.P. Catalogue record for this book is available from the Library of Congress.

ISBN 978-1-4020-6383-1 (PB)
ISBN 978-1-4020-6384-4 (HB)
ISBN 978-1-4020-6385-5 (e-book)

Published by Springer,
P.O. Box 17, 3300 AA Dordrecht, The Netherlands.

www.springer.com

Printed on acid-free paper

All Rights Reserved
© 2007 Springer
No part of this work may be reproduced, stored in a retrieval system, or transmitted in any form-or by any means, electronic, mechanical, photocopying, microfilming, recording or otherwise, without written permission from the Publisher, with the exception of any material supplied specifically for the purpose of being entered and executed on a computer system, for exclusive use by the purchaser of the work.

PREFACE

At the beginning of each year, there is a deluge of top-10 lists on just about every subject you can imagine. A top-10 list of biggest news stories, best-selling books, most popular music and movies, richest companies, and best places to visit or live. It seems everyone has his or her own top-10 list, reflecting, perhaps, differences in regional, national, and cultural values. Companies and governments most often tend to focus their top-10 lists on economic priorities, or priorities related to national defense, security, public health, and new infrastructure. This year, 2007, was no exception. Yet, increasingly, we see governments, private organizations, and companies advocating a new type of prioritization.

The complexity of societal change requires an enhanced capacity for scientific assessment, monitoring, and emergency response. New uncertain and multifaceted risks and stressors as well as globalization and public pressure for decision transparency drive the need for a new framework for thinking about prioritization. This framework needs to reach beyond the realms of economics, world trade, and corporate management to include the environment, stakeholders, public preferences, and social goals. Moreover, corporations and individuals are not only interested in generic 10-best lists; they want lists tailored to their values, goals, and current economic and social state. For example, the U.S. Government Performance and Results Act (GPRA) of 1993 [2] requires federal agencies to establish objective, quantifiable goals describing the outcomes of agency programs to improve confidence, performance, accountability, feedback, decision making, and management. Clearly, prioritization in this case should be customized for the needs of each individual agency. On the global scale, about 100 countries have now prepared national environmental strategies, and tangible changes in addressing environmental issues are apparent in about half of these countries [1]. Implied in all of these top-10 lists, whether Steer's new environmentalisms [1], the Ceres[1] Principles, the UN's Global Compact [4], the Equator Principles [5], or the U.S. Forest Stewardship Council (FSC) forest management principles [6], is recognition of the globalization of environmental issues. In all cases, scientific input is crucial to inform governments and public opinion and to minimize long-term impacts and risks.

It is within this context that we believe decision-making tools are urgently needed to support environmental management, particularly in the world's

[1] Ceres (pronounced *series*) is a network of U.S. investors, environmental organizations, and public interest groups working to address sustainable environmental challenges [2].

ports and coastal areas where a delicate balance exists between human demands for clean air, land, and water and the demands imposed by an increasingly global economy. Addressing threats and identifying actions to mitigate those threats necessitate an understanding of the basic risk assessment paradigm and the tools of risk analysis to assess, interpret, and communicate risks. It also requires modification of the risk paradigm itself to incorporate a complex array of quantitative and qualitative information that shapes the unique political and ecological challenges of different countries and regions around the world.

Over the past seven years, the authors of this book and their colleagues have convened a series of meetings to explore the topic of making policy decisions under uncertainty in general and environmental management specifically. Each workshop has tried to demonstrate the power of risk assessment and decision analysis as tools that decision makers should use to understand complex environmental, economic, legal, social, and technological information and to make the best informed decisions. Risk analysis, these experts have argued, offers a relatively objective, unbiased, and rational approach to framing and solving complex problems. It imposes a certain degree of quantitative rigor, as opposed to the all-too-common tendency to make environmental decisions based on assumptions and anecdotal evidence. Decision analysis could supplement risk analysis by providing a means for integrating heterogeneous scientific information and value calls by both scientists involved in the assessment process and decision makers facing decision challenges.

The first workshop in this series, entitled "Assessment and Management of Environmental Risks: Cost-efficient Methods and Applications" (Lisbon, Portugal, October 2000) [7], confirmed the role risk assessment could play as a platform for providing a scientific basis for environmentally sound and cost-efficient management policies, strategies, and solutions to various environmental problems. The second workshop, entitled "Comparative Risk Assessment and Environmental Management" (Anzio, Italy, May 2002) [8], explored the development and application of comparative risk assessment (CRA) and other risk-based decision-analysis tools in environmental management. The use of CRA was exceptional for facilitating decision making when various social, political, and economic activities compete for limited environmental resources. The third workshop, entitled "The Role of Risk Assessment in Environmental Security and Emergency Preparedness in Mediterranean Region" (Eilat, Israel, April 2004) [9], focused on environmental security challenges in the Middle East and how risk assessment could resolve some of the region's pressing environmental needs. The fourth workshop, entitled "Environmental Security in Harbors and Coastal Areas: Management Using Comparative Risk Assessment and Multi-Criteria Decision Analysis Framework" (Thessaloniki, Greece, April 2005) [10], explored environmental security issues in ports, harbors, and coastal areas and how the use of multi-criteria decision analysis (MCDA), in conjunction with risk assessment, could evaluate environmental security threats, formulate responses to those threats, and assess the efficacy of different responses on threat reduction.

This book is based on discussions and papers presented at a fifth workshop, entitled "Management Tools for Port Security, Critical Infrastructure, and Sustainability. " This meeting was the first to bring together concepts of environmental risk and critical infrastructure vulnerability with the objective of developing harmonized risk management and decision support methods and tools. The meeting was held in Venice, Italy, in March 2006. More than 40 science, risk assessment, decision-making, and security analysts from nine countries discussed the current state-of-knowledge on environmental security issues at industrial ports and harbors and the adequacy of available systematic, quantitative tools to guide vulnerability and threat assessments, evaluate the consequences of different events and responses, and support decision making. This workshop, like those previous, was sponsored jointly by the Society for Risk Analysis and NATO.

The organization of the book reflects sessions and discussions during the meeting in Venice. The goal of the meeting was to review the needs, methods, and tools of critical infrastructure managers and to offer a solution framework and tools. Even though ports were chosen as the focus of discussions and the prime application area, many papers and ideas are easily applicable to critical infrastructure in general. The papers in the introductory section provide a broad overview of the concepts and frameworks important for environmental security and risk management. Part 1 focuses primarily on critical infrastructure and related stressors, including management of hazardous materials and installations, transportation of dangerous goods including supply chain risk management, vulnerability to earthquakes, and network attacks. Part 2 focuses on ecological stressors affecting ports and coastal area infrastructure, including contaminants, algal blooms, and invasive species. Part 3 provides methodology and tools that are useful for assessing risks and vulnerabilities and supporting management decisions. These include multiple risk assessment tools, multi-criteria decision analysis, advanced modeling, and economic tools. The book concludes with a series of case studies that illustrate application and needs across multiple cultures and regions. This section is quite heterogeneous (both with topical coverage and depth of information assessment), but it preserves the nature of the meeting as a forum for scientists from developing and developed countries to discuss topics, methods, and tools from their own top-10 lists.

The collection of papers presented in this book reflects the workshop conclusion that our ability to fully evaluate environmental security and critical infrastructure vulnerability remains inexact. Integrated tools are urgently needed if we are to respond to the exponentially growing environmental and technological challenges that, until only a few years ago, seemed to be several decades away. While scientists and policymakers may argue about whether and when mankind will reach the tipping point for global change, there is little doubt that we are running out of time. We must begin to include in future environmentally and socially focused top-10 lists the consideration of new decision-making tools and new frameworks for investigation, assessment, and response to environmental challenges. Establishing, maintaining, or enhancing a sense of environmental security and vulnerability in coastal regions will require (i) matching human demands with available environmental resources; (ii) recognizing environmental security threats and infrastructure vulnerabilities; and (iii) identifying the range of available options for preventing and minimizing natural disasters, technological failures, and terror actions. These three considerations will require input from different stakeholder perspectives, and a broad range of quantitative and qualitative sociopolitical, environmental, and economic information.

Igor Linkov, Richard J. Wenning, and Gregory Kiker
April 2007

References

1. Steer A. Ten principles of the new environmentalism. Finance & Development. 1996; 33(4):4.
2. Government Performance and Results Act. Available at: http://www.whitehouse.gov/omb/mgmt-gpra/gplaw2m.html.
3. Ceres. Available at: http://www.ceres.org. Accessed 10 February 2007.
4. U.N. Global Compact. Available at: http://www.unglobalcompact.org/AboutTheGC/TheTenPrinciples/index.html. Accessed 10 February 2007.
5. Equator Principles. Available at: http://www.equator-principles.com/afp1.shtml. Accessed 10 February 2007.
6. Forest Management Principles. Available at: http://www.fscus.org/standards_criteria. Accessed 10 February 2007.
7. Linkov I, Palma Oliveira JM, eds. Assessment and management of environmental risks. Amsterdam: Kluwer; 2001.
8. Linkov I, Ramadan AB, eds. Comparative risk assessment and environmental decision making. Amsterdam: Kluwer; 2004.
9. Morel B, Linkov I, eds. Environmental security: the role of risk assessment. Amsterdam: Springer; 2006.
10. Linkov I, Kiker G, Wenning R, eds. Management tools for port security, critical infrastructure, and sustainability. Amsterdam: Springer; 2007.

ACKNOWLEDGEMENTS

The editors would like to acknowledge Dr. Abou Bakr Ramadan (NATO workshop co-director) and organizing committee members (Drs. Della Sala, Todd Bridges, and James Valverde) for their help in the organization of the event that resulted in this book. Our special thanks to Dr. Stefano Della Sala for serving as our host in Venice and arranging the workshop reception and use of the excellent facilities at Venice International University. We also wish to thank the workshop participants and invited authors for their contributions to the book and peer-review of manuscripts. We are deeply grateful to Deb Oestreicher for her excellent editorial assistance and management of the production of this book. We also wish to recognize the staff at Venice International University for the superb accommodations and teaching facilities, and Marta Citron and Francesca Peruzzo at the Autorità Portuale di Venezia for their assistance during the workshop. Additional technical assistance in the workshop organization was provided by Elena Belinkaia and Eugene Linkov. The workshop agenda was prepared in collaboration with the Society of Risk Analysis Decision Analysis and Risk Specialty Group. Financial support for the workshop was provided mainly by NATO. Additional support was provided by the ENVIRON International Corporation, Autorità Portuale di Venezia, Cambridge Environmental Inc., and the U.S. Army Corps of Engineers.

TABLE OF CONTENTS

PART 1

OVERVIEW OF CRITICAL INFRASTRUCTURE AND ENVIRONMENTAL SECURITY

CHAPTER 1

UNDERSTANDING ENVIRONMENTAL SECURITY AT PORTS AND HARBORS

R.J. WENNING

ENVIRON International Corporation, 6001 Shellmound Street, Ste. 700, Emeryville, CA 94608 USA

S.E. APITZ

SEA Environmental Decisions, Ltd., 1 South Cottages, The Ford, Little Hadham, Hertfordshire SG 11 2AT UK

A. BABA

Canakkale Onsekiz Mart University, Department of Geological Engineering, Terzioglu Campus, Canakkale 17100 Turkey

M. CITRON

Venice Port Authority, Direzione Sicurezza & Ambiente, Zattere 1401, Venezia 30123 Italy

K. ELLIOTT

U.S. Defense Threat Reduction Agency, 47-701 Hui Kelu Street, #6, Kaneohe, HI 96744 USA

N. AL-HALASAH

Royal Scientific Society, P.O. Box 1438, Al-Jubeiha, Amman 11941 Jordan

L. KAPUSTKA

Golder Associates Ltd., 1000 940-6th Avenue SW, Calgary T2P 3T1 Canada

I. Linkov et al. (eds.), Managing Critical Infrastructure Risks, 3–15.
© 2007 *Springer*.

M. LAZARESCU

National Research & Development Institute for Environment, Spl. Independent 294, Sector 6, Bucharest 78 060031 Romania

M. MOHAMED

National Centre for Nuclear Safety & Radiation Control, 3 Ahmed El Zomor Street, Nasr City 11762 Egypt

R. RUTJES
Environ-NL, Ziestereover 17, Zeist 3704 GB The Netherlands

Abstract: Port and harbors have expanded dramatically in the last few decades at a pace driven by rapid globalization of commercial trade and industrial activity, population expansion in marine regions, and the transfer of public and private wealth to coastal recreation. The impacts of ports on coastal environments can be considerable. The development of industrial facilities at ports and associated demands for water, land, and shoreline is unavoidable given their potential national and global economic significance. Just as the international community has increasingly become aware of the importance of ports and harbors, there has been a growing awareness of the need to ensure that health, environment, and safety are adequately protected and considered as an integral function of these facilities. This increased awareness has led to a closer examination of environmental security. At the NATO Advanced Research Workshop, *Risk Management Tools for Port Security, Critical Infrastructure, and Sustainability,* held 16–19 March 2006 in Venice, Italy, a working group of international risk assessment, disaster response, environmental modeling, and engineering experts from nine countries explored the importance of environmental security and challenges faced by coastal ports and harbors. The discussions drew heavily from experiences at ports located around the world. This chapter summarizes the findings of the working group and concludes with a summary of technical challenges and recommendations for future research.

1. Introduction

Global demand for coastal facilities has risen dramatically in the last decade due to the rapid globalization of the world's economy. Industrial and commercial shipping ports require large tracts of coastal land with deep water access, well-developed inland transportation networks to convey goods to the world's markets, and large, sustainable supplies of labor, raw materials, and fresh water to support their construction, expansion, and operation. Society's attention is increasingly focused on the potential hazards at these facilities due to inadvertent or purposeful disruption of port functions and major accidents near populated or environmentally sensitive areas. Such events can have immediate, profound, and potentially long-term consequences on environmental quality, ecosystems, local and international economies, and the safety and prosperity of nearby communities.

Though international organizations are aware of and may even understand these issues, the concepts of environmental security and attendant risk management and decision-making processes are defined differently by port organizations and political institutions in different parts of the world. This awareness inspired a working group of international risk assessment, disaster response, environmental modeling, and engineering experts from nine countries to explore more closely the importance of environmental security and the challenges faced by coastal ports and harbors. The discussions took place at the NATO Advanced Research Workshop, *Risk Management Tools for Port Security, Critical Infrastructure, and Sustainability,* held 16–19 March 2006 in Venice, Italy. The discussions drew heavily from different perspectives on environmental security and experiences at coastal ports and harbors located around the world.

2. Defining Environmental Security

As discussed by others, there is currently no consensus among scientists, governments, and organizations on the definition of environmental security, particularly as it applies to ports and harbors [23,24,13,12]. With only slight modification referencing the need for actions that are sustainable, the definition of environmental security provided by Belluck et al. [4] remains the most appropriate:

> ...environmental security involves actions that guard against environmental degradation in order to preserve or protect human, material, and natural resources at scales ranging from global to local *in a sustainable manner* [new text and emphasis added].

As discussed by Wenning et al. [24] and others, this modified definition of environmental security (or insecurity) implies a multitude of human, social, economic, and environmental connections. It is recognized increasingly that environmental security can be distinguished from environmental protection and environmental safety. If environmental security defines the balance between human, environmental, and socioeconomic needs, then environmental protection addresses the application of technical solutions to preserve that balance in a sustainable manner such that long-term stability is achieved between the manmade and natural worlds. The level of protection intended by various environmental quality benchmarks reflects society's goal to preserve or restore a certain level of ecological diversity, public safety, and well-being. In the same manner, the notion of environmental safety addresses the function of machine equipment and tools, industrial facilities, power stations, agricultural areas, and procedures for handling resources and waste materials in a manner that poses minimal risks to humans and does not negatively affect the environment through air, water, and soil pollution.

Since the terrorist events of 11 September 2001 in the United States, terrorism has been the most frequently identified primary purpose for the implementation of environmental security programs at ports and harbors and other critical infrastructure [25]. Regulations and guidelines embodied in international treaties and communiqués and in national legislation promulgated in the U.S. and elsewhere shortly after September 11th focused on preventing manmade catastrophes. While there is unanimous agreement in the international community that preventing terrorism is important, protection from terrorist violence is only one of many aspects of environmental security; and, in some cases, may be less important than securing potable drinking water supplies, preserving arable land for agriculture, or establishing a stable economy.

In this regard, the current focus on environmental security in some countries can be viewed as the response to reaching a tipping point on September 11th, whereby a single event inspired a call to action. Tipping points may also occur in response to a series of small, gradual changes. For example, the spate of river pollution incidents in China during the winter of 2005/06 affected drinking water for millions of people and is viewed as the tipping point leading to an overhaul of China's water protection regulations [22]. Recent concerns about the possibility of pandemic human and animal diseases such as avian flu, mad cow disease, and foot and mouth disease have triggered renewed recognition of the interconnectedness of humans and ecology, constituting a tipping point for global health care and environmental monitoring.

By aggregating the multitude of environmental, social, and economic stresses that currently plague human society, environmental security can best be viewed as a response to one or more of three categories of events:

1. Manmade gradual changes that slowly erode economic and environmental sustainability, and, in some cases, may even be irreversible.
2. Natural catastrophic events that, to some extent, may be predictable, so it is possible to plan response and protection measures.

3. Manmade catastrophic events, which are typically sudden and unpredictable.

These categories, and their characteristics, are discussed in greater detail in Apitz [26].

3. Environmental Security Framework for Ports and Harbors

In view of the multitude of environmental and economic demands and the varied manmade and natural challenges that are possible, a framework for addressing environmental security activities at ports and harbors is clearly needed. Specifically, an effective, sustainable, integrated, multi-media environmental security strategy is needed that encompasses a decision-making hierarchy, ranks different challenges, sets priorities, and supports practical actions.

The Driving Force - Pressure - State - Impact - Response (DPSIR) Framework used by the Organization for Economic Co-operation and Development (OECD), the European Environment Agency (EEA), and other European organizations appears to provide an overall mechanism for analyzing and responding to security challenges. The DPSIR approach defines the interactions between these various parameters as:

- **Driving forces**, such as industry and transport, produce;
- **Pressures** on the environment, such as polluting emissions, which then degrade the
- **State** of the environment; which then lead to
- **Impacts** on human health and eco-systems, causing society to
- **Respond** with various policy measures, such as regulations, information and taxes, which can be directed at any other part of the system [10].

Particularly useful for policy makers, DPSIR offers a basis for analyzing the multitude of inter-related factors that can impact on the environment. This conceptual approach is illustrated in Figure 1.

While Figure 1 appears straightforward, showing intuitive links between processes, the complexity lies in how D, P, S, I, and R are linked. For example, in contaminated sediment management [2,1], a direct comparison between sediment contaminant levels and target ecotoxicity values may be used to infer hazard. The selection of appropriate responses to remove or isolate the contamination may involve a complex assessment of financial, regulatory, scientific, and technical information. The most appropriate response may be affected by the scale of the problem, the possibilities for source control and/or natural attenuation, cost, technology limitations, and other factors. The success or failure of the selected response must also be evaluated over

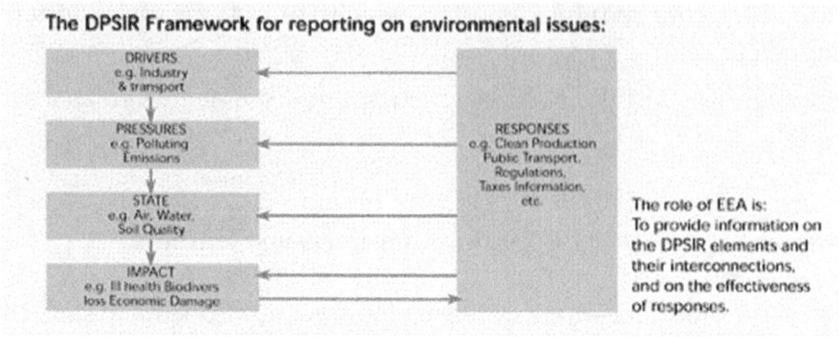

Figure 1. The DPSIR approach used by the European Environment Agency [10].

time, to monitor the possibility for changes in drivers and pressures. The use of a DPSIR framework to address environmental security in ports and harbors will require a systematic analysis of these links in terms of how various natural and anthropogenic drivers create pressures that can impact the environmental security issues of concern in ports and harbors. As can be seen in the next section, although there are some common concerns, the primary issues that drive environmental security concerns differ throughout the world.

4. Security Challenges at Ports and Harbors

With few exceptions, ports and harbors share a common interest in promoting or enhancing environmental security at their facilities. The guidance available from international organizations such as the International Maritime Organization [15,14] and treaties such as the London Convention promote a unified approach to protecting health, environment, and safety. Closer examination, however, suggests that, although there are some common concerns, different ports and harbors face different challenges, as shown in Table 1.

The challenges faced and lessons learned at different port facilities are of utmost value to the entire ports and harbor community, which includes facilities with military, commercial/industrial, and recreational focus. For example, concerns about public health, air and water quality, land use and development patterns, plus an improved understanding of environmental science, have resulted in a greater emphasis on freight transportation projects in the U.S. and in Europe that are both environmentally sound and economically sustainable [21]. Some of the key environmental issues facing the freight industry today include air pollution, sustainable dredging, the introduction of nonindigenous species, and noise.

TABLE 1. Examples of environmental security challenges that have caused concern at some of the world's active ports and harbors

Port / Harbor	One or More Environmental Security Challenges
Port of Venice, Italy	Managing the treatment and release of storm water; supporting large-scale industrial activity in a sensitive ecological habitat
Port of Rotterdam, The Netherlands	Maintaining operations at the world's largest industrial port facility
Port of Aqaba, Jordan	Protection of water resources
Port of Istanbul, Turkey	Responding to and minimizing the impacts of earthquakes
Port of Constantza, Romania	Protection from the damage caused by sudden storm events
Port of New Orleans, USA	Recovery after a major natural disaster
Western USA & Canadian ports	Reducing the impacts and avoiding the introduction of invasive species
Port of Alexandria & the Suez Canal	Managing a resource gateway for interior nations with no open ocean access

In western USA and Canadian ports (as well as ports in Australia, New Zealand, and elsewhere), the introduction of nonindigenous plant and animal species by air and water transportation from all over the world represents a significant ecological challenge. In the 1980s and early 1990s, more than 205 known nonnative species were either introduced or first detected in the U.S. Entering the U.S. Great Lakes in the 1980s when ship ballast water was discharged from European freighters, the zebra mussel has spread to 20 states, causing considerable damage to critical infrastructure such water intake pipes, filtration equipment, and electric generating plants [20]. According to the Commonwealth Scientific and Industrial Research Organization [5], invasive invertebrate and vertebrate species and nonnative plants cost Australia at least A$7 billion a year; globally, the damage caused by invasive species is estimated at around US$350 billion.

Earthquakes in Japan and Turkey have had serious social, economic and environmental repercussions on port operations. The Southern Hyogo Earthquake (The Great Hanshin-Awaji Earthquake) on 17 January 1995 paralyzed the Port of Kobe, interrupting the distribution of goods and raw materials throughout Japan and significantly affecting international distribution. According to the Ports and Harbors Bureau in the Ministry of Land, Infrastructure and Transport [17], the costs for reconstruction of the port reached approximately 570 billion yen. The disaster prompted development of the "Basic Policy on Construction and Reinforcement of Facilities to Cope with Large-Scale Earthquakes in Ports and Harbors" in 1996.

In Turkey, port and harbor facilities have been found to be particularly susceptible to submarine landslides or ground settlement due to liquefaction that may occur during earthquakes [8]. Tang [19] reported that ground shaking, settlement, and lateral displacement caused damage to port facilities on both the south and north shores of Izmit Bay following the 1999 Kocaeli earthquake. Erdik [9] reported similar widespread liquefaction and permanent ground deformation devastated the Port of Kobe, damaging more than 90% of the port's ship berths. Earthquake damage to port terminals at industrial facilities resulted in several naphtha and LPG spills into Izmit Bay from broken loading arms at Turkey's largest oil refinery following the Kocaeli earthquake. Economic losses in Turkey were estimated at US$16 billion [19].

5. Case Study: The Challenges of Protecting Water Resources at the Port of Aqaba, Jordan

The problem of water shortage in the Hashemite Kingdom of Jordan is exacerbated by high natural population growth, influxes of refugees and returnees to the country in response to the political situation in the Middle East, rural to urban migration, and increased modernization and higher standards of living. The average population growth rate stands at about 3.5% due to natural and involuntary migration, and about 78% of the population is located in urban areas. Consequently, Jordan faces a future of potentially severely limited water resources, among the lowest in the world on a per capita basis. Available water resources are projected to decline from more than 160 m^3 per capita per year for all uses at present to only 91 m^3 per capita per year by 2025, putting Jordan at risk of an absolute water shortage.

As a result, the Gulf of Aqaba (GOA), one of the two northward extensions to the Red Sea, is viewed as an essential resource and a priority environmental security concern. It is surrounded by Egypt in the east, Saudi Arabia to the west, and Jordan and Israel to the north. It ranges from 5 km wide at its northern tip, to a maximum of about 25 km wide, and then narrows back to 6 km wide at the straits of Tiran, where the GOA connects to the Red Sea. The GOA has an average water depth of 800 m. Its semi-enclosed nature—along with the semi-enclosed nature of the Red Sea—has created a unique ecosystem. The Red Sea supports higher levels of species diversity than any Eastern Indian Ocean region, and—in spite of limited water exchange—approximately 1,000 fish species, 110 species of reef building coral, and 120 species of soft corals inhabit the GOA. This diversity, along with the relatively small surface area of the GOA, results in vulnerability to pollutants, one of the most important risks to the sustainability of inland and offshore industrial activities.

At the present time, the number of visitors to Aqaba is estimated at 600,000, and is expected to increase sharply in the future as a result of major development and investment plans. Successful tourism and environmental protection of the GOA are interdependent. Successful tourism will depend on the availability of clean, nonpolluted water and healthy coral reef ecosystems.

However, recreational activities are a major source of solid waste and litter. Jordan's GOA beaches and near-shore reef and sea grass areas are heavily impacted by discarded plastic and other synthetic materials. Solid waste and litter have negative ecological effects on sea fauna. These effects include strangulation and entanglement, starvation and death resulting from ingesting plastics, and concentration of synthetic chemicals up the food chain. The Port of Aqaba places solid waste receptacles in port areas and additionally provides a daily garbage collection service via barge to ships anchored offshore. No systematic recycling is currently practiced in Aqaba.

The problem of solid waste and litter should be solved on a regional scale, as much of the litter found on the Jordanian side of the GOA has regional origin. A solid waste strategy that integrates land-based solid waste management issues with those associated with ship and boat-generated marine litter should be developed. In addition, sound solid waste management practices in Aqaba town, Aqaba ports, and the whole region of the GOA and the Red Sea should be promoted.

Numerous shipping and industrial activities have led to oil spills. In order to reduce the possibilities for large oil spills and reduce their effects, Jordan, Egypt, and Israel have initiated discussions to allocate responsibilities and financial burdens, conduct joint training exercises, and set mandatory design and operational standards for cargo vessels and oil tankers operating within the GOA.

Urban and industrial development as well as tourism have generated considerable environmental pressure. During the past three decades, 30 to 40% of the Jordanian coastline has changed from a pristine natural environment to a heavily used port and industrial area. With the declaration of Aqaba Special Economic Zone (ASEZ), rapid growth of port and industrial activities is expected, risking further degradation of sensitive coral reefs and the loss of biological diversity.

The total area of the GOA is comparatively small in comparison to its considerable population and associated economic activities that produce wastewater along the coast. In order to preserve the ecosystem for future generations, stringent regulations, monitoring, and enforcement are required. Further economic development (triggered by declaration of the ASEZ) is expected to dramatically increase water demand and wastewater generation. Advanced wastewater treatment or irrigation projects far from the coastline may ameliorate wastewater issues. New wastewater reuse applications are under consideration.

Industrial activities have put fringing reefs under pressure and increased coral diseases nearly tenfold. Phosphate concentrations, heavy metals, and algal cover are higher at or near industrial sites. In addition, a 50% reduction in fish abundance and changes in the relative abundance of different fish communities have been observed since the escalation of industrial activities began. It has been suggested that future industrial development should be restricted to already industrialized areas, in order to preserve the remaining GOA coral reefs and seagrass meadows.

Environmental hazards associated with land transportation result from the high truck volume to and from the Aqaba Port and the industrial facilities. Inadequate provisions for waste oil collection and recovery pose a hazard to the Aqaba marine environment. The use of waste oil as a dust suppressant in unpaved truck repair yards is a further environmental hazard. Air pollution, especially from leaded gasoline—still the main gasoline type used in the Aqaba region—is an additional environmental hazard. In addition, the absence of a clear hazardous waste management policy has caused used asbestos brake linings, exhausted automotive batteries, and worn tires to be discarded in a haphazard and—in some cases—health-endangering manner.

All the above problems also exist on the Israeli and Egyptian sides of the GOA. A new and unique source of pollution, however, comes from the Israeli side: pollution from fish farms. These farms are continuing to expand and cover a large area. Limits should be imposed on these farms in order to preserve the GOA's corals and sea creatures.

6. Future Focus and Path Forward

According to Swatuk [18], almost as soon as the words *environment* and *security* were linked, critics questioned the appropriateness of linking environmental issues to (national) security practices [7, 3,11,16]. Traditionally, the study of environmental security revolved around the central idea that environmental challenges—in particular, resource scarcity and environmental degradation—may lead to violent conflict among countries and regions. Proponents of environmental security argue that if environmental change is a potential source of social conflict, and if societies face dangers from environmental change, then security policies must be redefined to account for these threats [18].

Indeed, the field of environmental security is changing rapidly. Government and academic research in the U.S. appears to be reshaped practically on an annual basis by public health scares and sudden tragic world events. Environmental security will continue to change and evolve as new threats—manmade or natural—reveal themselves locally, nationally, and/or internationally. Frameworks for organizing environmental security programs at ports and harbors, therefore, must be flexible and must adapt as either current or new challenges and responses to those challenges manifest in additional or unforeseen consequences.

In addition to decision frameworks, new or improved technologies and environmental monitoring programs are needed to enhance prevention, response, and mitigation strategies and to anticipate or forecast when threats might occur.

In the future, environmental monitoring must:

1. Support real-time decision making.
2. Provide accurate and impartial data to avoid human interference.
3. Provide for stable, long-term safekeeping of data.
4. Support other environmental applications.
5. Support long-term planning schemes such as emergency response, and early warning systems.

According to Cruz et al. [6], engineering has contributed to our overall understanding of natural hazards and their impacts and the vulnerability of the built environment to these hazards. Improved understanding of natural hazards results in better forecasting of such hazards, and more effective disaster prevention and mitigation practices and preparedness planning. However, engineered hazard mitigation to protect ports and harbors and critical infrastructure will not guarantee protection from natural and other unforeseen hazards (e.g., terrorist actions). Therefore, a holistic multi-hazard perspective that integrates social, economic, and environmental issues into hazard reduction is desirable. The engineering, risk assessment, and decision analysis professionals who contribute to environmental security assessment will be increasingly required to work with many different stakeholders. Collaboration across disciplines, jurisdictions, and national boundaries will increasingly be required to meet the daunting challenges posed by environmental security issues. Workshops such as this one are a first step toward establishing effective partnerships for resolving such challenges.

7. References

1. Apitz SE, White S. A conceptual framework for river-basin-scale sediment management. Journal of Soils and Sediments. 2003; 3(3):132-138.
2. Apitz SE, Davis JW, Finkelstein K, Hohreiter DL, Hoke R, Jensen RH, Kirtay VJ, Jersac J, Mack EE, Magar VS, Moore DW, Reible D, Stahl R. Assessing and managing contaminated sediments: part I, developing an effective investigation and risk evaluation strategy. Integrated Environmental Assessment and Management. 2005; 1:2-8.
3. Barnett J. The meaning of environmental security: ecological politics and policy in the new security era. London: Zed Books; 2001.

4. Belluck DA, Hull R, Benjamin SL, Alcorn J, Linkov I. Environmental security, critical infrastructure and risk assessment: definitions and current trends. In: Kiker GA, Linkov I, editors. Environmental security and environmental decision-making. Dordrecht: Springer; 2005.
5. Commonwealth Scientific and Industrial Research Organization. Emerging biological threats and invasive species. 2007. Available at: http://www.csiro.au/science/ps1ow.html. Accessed on 28 March 2007.
6. Cruz AM, Steinberg LJ, Vetere-Arellano AL, Nordvik JP, Pisano F. State of the art in Natech (natural hazard triggering technological disasters) risk assessment in Europe, report EUR 21292 EN, DG. Ispra: Joint Research Centre, European Commission and United Nations International Strategy for Disaster Reduction; 2004.
7. Dalby S. Security and ecology in the age of globalization. Environmental Change and Security Project Report. 2002; 8:95-108.
8. Durukal E, Erdik M, Sanli B. Damage to and vulnerability of industry in earthquakes in Turkey. Geophys Res Abstracts. 2006; 8:10041.
9. Erdik M. Seismic vulnerability of megacities. In: Booth E, editor. Seismic design practice into the next century: research and application. Rotterdam: Balkema; 1998.
10. European Environment Agency. Conceptual framework: how we reason. Copenhagen: European Environment Agency; 2005.
11. Gleditsch NP. Armed conflict and the environment: A critique of the literature. Journal of Peace Research. 1998; 35(3): 381-400.
12. Glenn JC, Gordon TJ, Perelet R. Defining environmental security: implications for the U.S. Army, AEPI-IFP-1298. 1998.
13. Glen JC, Gordon TJ. State of the future, ISBN:0-9657362-8-8. AC/UNU Millennium Project Publications; 2001.
14. International Maritime Organization. International ship and port facility security code. Conference of Contracting Governments to the International Convention for the Safety of Life at Sea. 2002; 1974:9-13. Available at: http://www.imo.org/Newsroom/mainframe.asp?topic_id=583&doc_id=2689. Accessed on 25 July 2005.
15. International Maritime Organization. IMO's global programme on maritime and port security - the work continues. 2004. Available at: http://www.imo.org/Newsroom/mainframe.asp?topic_id=848&doc_id=3656. Accessed on 22 July 2005.
16. Levy MA. Is the environment a national security issue? International Security. 1995; 20(2):35-62.
17. Ports and Harbors Bureau. Ports and harbours in Japan. 2006. Available at: http://www.mlit.go.jp/kowan/english/pdms/2_6_4.html. Accessed on 28 March 2007.
18. Swatuk LA. Environmental security in practice: transboundary natural resource management in Southern Africa. Paper presented in Section 31 of the Pan-European Conference on International Relations, The Hague, 9-11 September 2004.
19. Tang AK. Izmit (Kocaeli) earthquake of August 17, 1999 including Duzce earthquake of November 12, 1999: lifeline performance. Technical Council on Lifeline Earthquake Engineering, Monograph No. 17.2000; Virginia: ASCE.
20. United States Environmental Protection Agency. Petition to EPA to regulate ballast water. 2001. Available at: http://www.epa.gov/OWOW/invasive_species. Accessed on 28 March 2007.
21. United States Federal Highway Administration. Key freight transportation challenges. 2007. Available at: http://ops.fhwa.dot.gov/freight/freight_analysis/freight_story/environ.htm. Accessed on 28 March 2007.
22. Wenning RJ, Leung K. Editorial: protecting China's rivers. Integrated Environmental Assessment and Management. 2006; 2(2):1-2.
23. Wenning RJ, Della Sala S, Magar VS. Role of risk assessment in environmental security planning and decision-making. In: Kiker GA, Linkov I, editors. Environmental security and environmental decision-making. Dordrecht: Springer. In press.

24. Wenning RJ, Apitz SE, Belluck DA, Chiesa S, Figueria J, Filip Z, Ignatavieius G, Kiker MK, Lahdelma R, MacDonell M, Mannini A, Raissouni B, Sinka D, Xenidis Y. Environmental security: approaches and tools. In: Linkov I, Kiker G, editors. Environmental security and environmental decision making. Dordrecht: Springer. In press.
25. Zaidi MK. Risk assessment in detecting and preventing of terrorist attacks in harbors and coastal areas. In: Linkov I, Kiker G, editors. Environmental security and environmental decision making. Dordrecht: Springer. In press.
26. Apitz, SE. Conceptual frameworks to balance ecosystem and security goals. In: Linkov I, Wenning RJ, Kiker G, editors. Managing critical infrastructure risks: decisions tools and applications for port security. Dordrecht: Springer. In press.

CHAPTER 2

A BRIEF ANALYSIS OF THREATS AND VULNERABILITIES IN THE MARITIME DOMAIN

N.O. BAKIR

University of Southern California
Center for Risk and Economic Analysis of Terrorism Events (CREATE)
3710 McClintock Avenue, RTH 322
Los Angeles, CA 90089-2902 USA

Disclaimer: This research was supported by the United States Department of Homeland Security through the Center for Risk and Economic Analysis of Terrorism Events (CREATE) under grant number N00014-05-0630. However, any opinions, findings, and conclusions or recommendations in this document are those of the authors and do not necessarily reflect views of the United States Department of Homeland Security.

Abstract: The attacks of September 11 exposed the vulnerability of the American homeland to terrorism. Terrorists have already expressed their intentions to continue their aggression towards United States. Their goal is to incur maximum economic damage, inflict mass casualties, spread unprecedented fear among citizens, and thus destabilize the nation to further their agenda. Many critical sites lie across U.S. maritime borders, all of which could be potential targets to accomplish these goals. All these sites are simple elements of a complex body where the vulnerability of the whole system is a function of the vulnerability of the weakest element against an adaptive adversary. Ports, nuclear facilities, liquid natural gas (LNG) facilities, urban areas, bridges, chemical plants, and other critical infrastructure are all elements of this complex system. In this paper, we review the current status of security in the American maritime realm and discuss the programs and initiatives that seek to minimize terrorism risk. Our goal is to direct attention to various possible avenues that could be used to illegally introduce weapons, explosives, and other contraband as well as to penetrate terrorists into the American homeland.

I. Linkov et al. (eds.), Managing Critical Infrastructure Risks, 17–49.
© 2007 *Springer*.

1. Introduction

U.S. sea borders include 95,000 miles of coastline and 3.4 million square miles [44] of exclusive economic zone. The huge economic value of trade, number of jobs provided, and a multitude of stakeholders involved render security along coasts and waterways critical for the American homeland. Port security is the underpinning of the U.S. economy and a terrorist attack may deliver a serious blow to supply chain operations and continuity of business. Besides, terrorists have already expressed their intentions to target the economic lifeline of the U.S., which raises concerns that an attack in the maritime domain may be in the making. Despite efforts after September 11 (9/11) to improve security, U.S. waterways and critical infrastructure along the borders remain vulnerable to a terrorist attack.

Maritime security is closely related to border security, which requires a systems approach to protect the American homeland from a terrorist infiltration or attack. The main objective of border security should be to minimize casualties, injuries, and economic losses due to terrorism, while ensuring the flow of commerce and continuity of business, conserving the environment, as well as supporting international partnerships for research, development, and education. As terrorist strategies are dynamic and terrorists have shown their capability to develop tactics that are adaptive to new environments, gaps in border security should be addressed following a comprehensive approach seeking to reduce the risk at all potential points of illegal entry or a terrorist attack. As such, to reduce the risk of terrorism along borders, a systems-based risk management approach that captures the complex relationship between multiple elements and their exposure to interdependent risks should be implemented. Such an approach will produce comparative risk assessments that will help deploy necessary resources to sites facing the greatest risk, and mitigate exposure to a nationally tolerable level.

The U.S. Department of Homeland Security (DHS) has adopted a risk-based approach to counterterrorism. In this context, risk is defined as the cross-product of threat, vulnerability, and consequence [47]. Threat is defined as the probability of a terrorist attack. One can extend this definition to specify the timeframe, the location, type of weapon used, and the terrorist groups involved in the attack. Nevertheless, this is the component of risk over which the U.S. administration has relatively less control. Vulnerability is the probability of damage if a terrorist attack occurs. Damage can take any form stated in the border security objective statement, as well as other consequences for which a widely accepted measure does not exist. Consequences are simply the expected damage inflicted from a successful terrorist attack. Following this definition of risk, the goal in this paper is to discuss what constitutes threat in the maritime domain and vulnerabilities that could potentially be exploited by a sophisticated adversary. Consequence assessments are beyond the scope of this paper.

2. Threats in the Maritime Domain

Terrorism threats from the maritime domain may come in various forms. For example, weapons or explosives may be concealed in containers, ships may be used as weapons to destroy critical infrastructure, or terrorists may illegally cross borders to launch attacks in the homeland. Terrorists have a wide array of options, which could culminate in a catastrophic attack. They have already shown their capacity to operate successfully on the open seas for non-terrorism purposes: piracy, illegal smuggling of contraband, and illegal human trafficking across borders. Therefore, they may utilize expertise that has accumulated over the years in other forms of maritime crime to launch more damaging attacks exploiting vulnerabilities in the global maritime system. Accordingly, increased capability to respond to each terrorist activity may prove to be very valuable, and require cooperation between nations.

2.1. PIRACY

Although historically not intertwined with terrorism, piracy is reemerging as a serious threat impeding global business. In 2003, there were 445 attacks in which 21 crew members were killed, 71 reported missing, and 359 taken hostage[1]. The number of attacks dropped to 325 in 2004 with an increase in the death toll from 21 the previous year to 30. Actual figures may be far more disturbing. Shipping companies tend to underreport incidents due to fears of increasing insurance premiums and lengthy investigations that may result in loss of reputation. As Singapore's Deputy Prime Minister, Tony Tan, said, "Piracy is entering a new phase; recent attacks have been conducted with almost military precision. The perpetrators are well-trained, and have well laid out plans" [29]. Annual cost of lost cargo has risen to $16 billion, mainly due to piracy, truck hijacking, and theft around the ports.

Pirates have excelled in hijacking ships over the years. Once the ship is hijacked, "turning it into a phantom ship, erasing its original identity, is relatively easy" [5]. The ships are then known to be painted at remote docks and given a completely new identity. A relatively simple way to do this is getting a new registry by changing flags in "flag-of-convenience" countries[2]. The fleets of these countries are growing. Registration standards in these countries are relatively lax and there are no requirements on the nationality of crew members. Most "flag-of-convenience" ships are relatively unprotected

[1] Annual statistics released by International Maritime Bureau (IMB).

[2] A "flag-of-convenience" ship is defined as the ship that flies the flag of a country other than the country of ownership. International Transport Workers' Federation (ITF) maintains the list of "flag-of-convenience" countries.

against piracy. In 2003, 63% of all losses in absolute tonnage were accounted for by just 13 flag-of-convenience registers [24]. These characteristics of flag-of-convenience ships render them and their cargo high risk.

There is minimal law enforcement in international waters where pirates are known to be operating. Most piracy incidents take place in the Far East, in regions such as Indonesia, Malacca Straits, Malaysia, Singapore Straits, and the South China Sea. Other geographical locations with reported cases include India, Philippines, Bangladesh, Gulf of Aden, Colombia, Venezuela, Vietnam, Red Sea, and the Dominican Republic. Most of these countries have minimal resources for maritime patrolling and long coastlines, granting the pirates the freedom to loot. Corruption among maritime officials adds to the complexity of law enforcement in these waters. Modern pirates use technology for vessel surveillance, automatic weapons, and motorized boats to hijack ships with valuable cargo. Investment in this technology is easily justified with potential returns ranging from $8 million to $200 million per vessel [26]. This is also a good financing source for terrorist operations and a catalyst in developing the interconnection between piracy and terrorism.

There is minimal cooperation between nations to combat piracy, and each country is responsible for enforcing the law in its territorial waters. Pirates have a good understanding of their operational environment. They usually elude maritime officials by crossing national sea boundaries and exploit vulnerabilities due to lack of information sharing and international cooperation. The Malaysian Maritime Enforcement Center stated:

> Under no circumstances would we intrude into each other's territory. If we chase a ship and it runs into the other side, we let the authorities there handle it [45].

Therefore, penalizing maritime criminals is quite difficult. It requires arrest authority unlimited by national boundaries and the willingness of authorities to enforce law in the maritime domain.

2.2. ILLEGAL HUMAN AND CONTRABAND TRAFFIC

Another concern is human and drug smuggling across maritime borders. This is particularly important because terrorists may use similar pathways to sneak illegal weapons through the border or to deliver a dirty bomb to seaports. Most maritime human smuggling operations into the U.S. are launched from Caribbean waters. However, in recent years an increasing number of Chinese migrants are caught in container ships at seaports. For instance, in 1999, 259 Chinese migrants aboard the freighter Wing Fung Lung were intercepted while traveling with no bed or sanitation facilities [39]. In January 2005, 32 Chinese nationals were found in two shipping containers at the Port of Los Angeles [40]. Those traveling in containers are usually found in miserable conditions. In 2001, Irish police found a cargo container with eight dead and five sick immigrants [27]. Similarly, in 2000, British police discovered 58 Chinese nationals, suffocating and at the brink of death, in a truck that traveled to England on a ferry. High risk of death does not seem to deter aliens from seeking illegal entry into developed countries because of the prospect of economic prosperity.

Terrorists may use illegal human trafficking tactics to cross U.S. borders without bearing the same inhumane conditions as economic migrants. In 2001, a stowaway was discovered at the Italian port of Gioia Tauro, traveling from Egypt to Canada [32]. An Egyptian, Rizik Amid Farid, converted a container into a hotel room with a bed, restroom, food supplies, a laptop computer, two mobile phones, and cameras. Among his belongings were a Canadian passport and airport security passes, which aroused the suspicion that he was involved in a terrorist plot to copycat the 9/11 attacks. While this was never confirmed, the incident was a clear indication that terrorists may enter the U.S. hiding in a container. In 2004, the Israeli port of Ashdod was the location of a suicide bomb attack by two Palestinians who were able to hide in a container[3].

Like piracy, stowaways are becoming a common phenomenon in the international maritime domain. In many cases, stowaways board the ship without being recognized, exploiting lax security at seaports, and may actually attempt to hijack the ship[4]. The crew may not have effective means to confront the problem once a stowaway is found on the ship. In some cases, stowaways are completely ignored to eliminate the possibility of an armed confrontation. Most countries refuse to accept stowaways if they are citizens of another country. Hence, handing stowaways over to port officials may not be an alternative solution to the problem. The best solution from the perspective of ship crew seeking to minimize trouble on board could be simply ignoring the stowaway.

Drug smuggling is also a part of the problem. Smugglers introduce drugs and other illegal contraband in small boats or in containerized cargo. It is widely believed that the multibillion dollar drug trade has historically exploited low inspection rates on containers at various ports. Detecting such illegal contraband has been likened to finding a needle in a haystack. In recent years, approximately 75% of cocaine seizures by the U.S. Coast Guard (USCG) took place in the Eastern Pacific [43]. While there has not been any confirmed explosive or weapon smuggling into the U.S. through maritime borders using containerized cargo or small boats for terrorism

[3] The suicide bombers were reportedly members of Al-Aqsa Martyr's Brigade.

[4] One such incident reportedly took place in 2000, when 14 Iranian stowaways hijacked an Italian ship.

purposes, similar tactics can clearly be used by terrorists in the future. In 2003, ABC News deliberately sent depleted uranium in a container from Indonesia, and government screeners failed to detect the nuclear content. More recently, Armen Barseghyan of Armenia was reportedly charged in an alleged scheme to smuggle grenade launchers, shoulder-fired missiles, and other Russian military weapons into the U.S. A more disturbing aspect of the case is its discovery by an FBI informant posing as an Al Qaeda representative.

2.3. TERRORISM IN THE MARITIME DOMAIN

While terrorists have largely targeted land sites, it has been evident over the course of recent history that seaborne terrorism poses an unignorable threat. International waters have long been penetrated by terrorists, and there is virtually no protection for commercial ships against this growing threat. U.S. vessels are not immune from the rise of maritime terrorism as shown by the boat attack on naval destroyer, Cole, in 2000. It is even more disturbing to realize that terrorists have a continuum of options to sabotage the flow of international trade with relatively less effort than in 9/11. As terrorists develop their maritime terrorism skills, the probability of launching an attack with catastrophic consequences on U.S. economic interests will increase. Al Qaeda has already stated its interest in this objective. After the attack on French tanker, Limburg, in 2002, Osama Bin Laden released an audio tape in which he stated:

> By God, the youths of God are preparing for you things that would fill your hearts with terror and target your economic lifeline until you stop your oppression and aggression.

Further intelligence seems to confirm that Al Qaeda may still be planning attacks on maritime targets. In 2002, Al Qaeda's former chief of naval operations confessed plans to attack ships passing through the Strait of Gibraltar [38]. The scheme was later foiled by Moroccan officials. The alleged mastermind of the terrorist attacks on 9/11, Khalid Shaikh Mohammed, was reportedly involved in a plot to export weapons and explosives into the U.S [25]. He reportedly made an offer to an import/export firm to use their containers for shipping illegal contraband to the port of New York and New Jersey. For years, drug dealers have been known to employ a similar tactic of buying out a trustworthy shipping company to disguise their shipments [13]. Drug smuggling chains can be discovered after observing patterns of shipments. However, maritime security officials have no luxury to observe such patterns to counter terrorism threats because one successful attempt to evade detection at seaports may bring catastrophe.

Al Qaeda is believed to control approximately 15 ships which fly Yemeni and Somalian flags. Other terrorist organizations have been active in the maritime domain for almost half a century. Most seaborne terrorism attacks were carried out by local groups as rather isolated incidents seeking to gain independence and oppose regional governments. However, relatively high-profile incidents in the past have exposed the capability of their perpetrators to successfully operate in the maritime domain and had political ramifications in the global arena. Some of the terrorist organizations and other militant groups involved in these incidents include:

- *Hezbollah*: While Hezbollah is less known for its attacks in this domain, it was active in laying mines in the Red Sea in the mid-1980s to impede access to Israeli ports. In 1984, Hezbollah organized a mine attack that hit 19 ships at the southern entrance of the Suez Canal.
- *Polisario Front*: This militant group, whose goal was the independence of the Western Sahara, was quite active in 1970s and 1980s. Its main targets were Spanish and Portuguese vessels operating off the northwestern coast of Africa. The group was known to carry out direct attacks on their targets with mortar and machine-gun fire.
- IRA: The Irish Republican Army targeted cruise liners and cargo ships in the 1970s and 1980s and was involved in the illegal transport of weapons and munitions.
- *Palestinian Liberation Front*: The PLO hijacked the Italian cruise liner, Achille Lauro, in 1985 off the Egyptian coast and took 511 passengers hostage, demanding the release of Palestinian prisoners in Israel. One American passenger was killed in this assault.
- *Abu Nidal*: This organization was active in the late 1980s in hijacking incidents. In 1987, the group captured a French yacht off the coast of Gaza Strip to warn Arab leaders not to assign the late King Hussein of Jordan to represent Palestinians in peace talks. All hostages were released at the request of Qaddafi.
- *Chechen Rebels*: In 1996, this group hijacked a ferry sailing from a northern port of Turkey (Trabzon), and demanded the withdrawal of Russian troops from a Daghestani village. After a series of negotiations, the hijackers were captured.
- *Liberation Tigers of Tamil Eelam (a.k.a. Tamil Tigers)*: This terrorist organization has been quite active in recent years, attacking ships off the coast of Sri Lanka. While its primary target has been Sri Lankan Navy ships, the group was also involved in attacking Chinese and North Korean ships with the intent of disrupting regional maritime traffic. The

Tamil Tigers are unusual among groups engaged in seaborne terrorism because of their suicide tactics.

- *Free Aceh Movement (GAM)*: GAM has discontinued its terrorist activities following a peace accord with the Indonesian government. Before the agreement, the group was active in Malacca Straits, carrying out relatively small-scale attacks.

Other organizations besides Al Qaeda are believed to be considering launching seaborne attacks targeting the U.S. and its allies. In Southeast Asia, where piracy is rampant, Jemaah Islamiya, Lashkar Jundullah, and Kampulan Mujahidin Malaysia are among the active terrorist organizations that could direct their attention to maritime terrorism. Elsewhere, as mentioned in the previous section, Al-Aqsa Martyr's Brigade shipped two suicide bombers to an Israeli port hidden in a container. It is expected that terrorism on the open sea will continue be a threat unless radical measures that foster coordination and intelligence sharing between nations are taken.

3. Vulnerabilities along Maritime Borders and Countermeasures for Risk Mitigation

The current status of maritime security in the U.S. can be analyzed under two main categories: port security and security in U.S. waters. The Maritime Transportation Security Act (MTSA), signed by President Bush on November 25, 2002, was prepared to address the security of ports and waterways. As a result of this act, maritime security enforcement responsibility has been mainly assigned to the USCG, the Bureau of Customs and Border Protection (CBP), and the Transportation Security Administration (TSA) under the DHS as well as the Maritime Administration under the Department of Transportation. The USCG has the lead responsibility in most MTSA assignments, as well as the security of U.S. waters and coastal targets, while the CBP assumes a key role in improving security of inbound cargo at maritime ports.

3.1. PORT SECURITY

Seaports are arguably the most critical nodes in the global supply chain and hence have a central role in the business continuity. They make a huge contribution to the U.S. economy by facilitating trade and tourism, providing jobs, and supplementing energy needs. U.S. seaports handle 95% of overseas trade by weight and 75% by value. Many critical coastal targets, including petroleum tank farms, hazardous material storage facilities, and factories are located around the ports. In addition, several major U.S. cities lie in close proximity to waterways and seaports. Therefore, a well-organized attack at a major U.S. port is likely to inflict high number of casualties and cause grave nationwide economic damage.

Estimates of the economic impact of a major terrorist attack at a U.S. seaport vary. Earlier studies predicted that a port closing could cost the economy as high as $1 trillion [33]. A more recent study by Gordon et al. [22] puts the figure at around $45 billion for a dirty bomb attack. For a potential nuclear attack, partially due to the higher level of uncertainty, experts provide a wider range of potential consequences. In a 2003 study [1], direct trade losses were calculated to be around $100–200 billion, with property damage expected to fall between $50 and $500 billion. A more disturbing figure is the estimated $300 billion to $1.4 trillion in indirect costs to the economy. On the casualty side, the study group estimates between 50,000 and 1,000,000.

Port security has been increasingly intertwined with the security of containerized cargo in the public and the media. While containers are arguably the "Trojan Horses" of the modern era, failure to recognize other vulnerabilities may leave the U.S. homeland unguarded for another surprise attack. In this paper, port security is analyzed under four headings: cargo security, access to secure areas, cruise lines, and security around the port perimeters. Most of the discussion centers on cargo security because monitoring the contents of containers without interrupting the flow of trade poses tremendous challenges to both public and private stakeholders. However, as will be evident in the discussion, other weaknesses in port security are equally important because any attack on the port facility may trigger a slowdown of cargo processing and inflict casualties.

3.1.1. Cargo Security

An estimated 10 million containers arrive in the U.S. from overseas each year [6]. With the current level of technology available at U.S. seaports, 5% of these incoming containers are screened. Containers—the key medium of goods movement—have made intermodal transportation possible since the 1950s. Containers enable the smooth transfer of cargo between various modes of transportation. Since the introduction of containers to the global supply chain arena, goods movement has become more efficient. However, containers reduce the transparency of cargo. This presents serious challenges to trade security in the modern era. Concerns over the potential shipment of radiological, nuclear, chemical, or biological weapons in containers rose after 9/11 as nations realized the nature of the terrorism threat confronting them.

Companies today are challenged to simultaneously achieve efficiency and security in containerized cargo transportation. While some security improvements implemented by the private sector have the potential advantage of increasing cargo transparency, incentives to install a wide variety of technologies that could provide continuous time monitoring and hence minimize the risk of container tampering are rather limited. Therefore, public stakeholders have to step in to screen containers at certain points in the supply chain network to reduce the likelihood of harmful cargo reaching its final destination in the U.S. However, with thousands of containers moving across the globe each day and intense competition that puts enormous pressure on companies for timely delivery, fully guaranteeing the security of cargo is virtually impossible. The problem becomes even more complicated as we realize that not all cargo is containerized and technologies to screen other forms of cargo are very limited.

Containerized Cargo. U.S.-bound containerized overseas cargo moves through various phases, all of which present unique security challenges. These phases include:

- Loading at the warehouse
- Land transportation
- Port of origin
- Sea transportation
- Port of destination

Loading at the Warehouse. Terrorists may load illegal weapons and explosives at the warehouse or distribution center from which the cargo is dispatched. To achieve this, terrorists may exploit loopholes in physical, personnel, and procedural security of these facilities. Access controls, background checks of employees, security awareness training, storage of containers, procedures for visitor admission, security of warehouse perimeters, and standardization of paperwork security should be addressed properly to reduce the risk of terrorist tampering with containers. Most of these warehouses are operated by foreign business partners of importers in the U.S. and located in foreign countries.

The security of international supply chains has become one of the priorities of the U.S. federal government after 9/11. To enlist the cooperation of private stakeholders, the U.S. unveiled a new initiative, Customs-Trade Partnership against Terrorism (C-TPAT) in November 2001. C-TPAT is a voluntary program of partnership between public organizations and private companies to improve supply chain security. Participating companies are required to perform self-assessments of their whole supply chain and to develop shipping guidelines for their suppliers. The U.S. government seeks to employ the private sector's leverage on their global suppliers to address one of the most vulnerable phases of container shipment: the loading phase. Governmental bodies in the U.S. have no power to make foreign companies follow proper guidelines in loading and transporting goods to the U.S. Companies can use their buying power to have their downstream suppliers enforce relevant security measures.

Companies that are interested in becoming C-TPAT members need to prepare and submit a supply chain security profile. The profile should include security assessments of foreign facilities, cargo movement, and background checks of people across the supply chain. Currently, C-TPAT members are granted benefits after CBP reviews the profile information provided by the company, and the history of its compliance with laws and regulations. For importers, benefits may be granted after this review. If CBP is convinced that the information provided is reliable and the company has a clean historical record, then the company starts enjoying benefits. Otherwise, CBP performs an additional examination that includes on-site visits and meetings with the company representatives. A final decision is made after this review.

The private sector enjoys reduced and expedited inspections under this program as their cargo is low-risk. Companies are also encouraged to use smart containers as their standard medium of trade. A potential benefit of smart containers is to eliminate routine inspections. Reduced delays for shipments are vital for companies who are already operating under tight profit margins. Members are promised priority treatment from the federal agencies to process their shipments in the case of an attack. According to a recent study by Peleg-Gillai et al. [35], some participants in C-TPAT have realized the benefits of membership. The study, which was based on input from 11 manufacturers and three logistics service providers, reports that security improvements resulted in 38% reduction in theft/loss/pilferage and 37% reduction in tampering. In addition to improvements in security, participants noted related benefits that contribute to business value, such as: 49% reduction in cargo delays, 48% reduction in cargo inspections, 29% reduction in transit time, and 28% reduction in delivery time windows. This study clearly demonstrates the potential of C-TPAT membership in increasing supply chain visibility, resilience, and customer satisfaction.

However, security improvements under the C-TPAT program have been limited in combating terrorism threat. The U.S. Government Accountability Office (GAO) released a report [41] in 2005 that discusses the weaknesses in its implementation. GAO review of the program shows that the validation phase is not based on an independent audit. Validation is the last phase in CBP review where selected domestic and foreign sites are visited. The goal is to ensure that the security profile accurately reflects the level of countermeasures taken at each site to reduce the terrorism risk exposure. GAO criticizes CBP for not performing a comprehensive review of the security profile. Besides, the criteria for site selection are not clear. Validations will be most effective when conducted at critical nodes of the supply chain with high import volume and at strategic geographic locations where suspected terrorist activity is of concern.

The C-TPAT program can be further improved by standardizing security countermeasures across the international supply chain networks. CBP issues recommendations under the C-TPAT program. However, these recommendations are not binding and each member may choose different sets of security solutions to follow the guidelines. Best practices to secure supply chains should be determined and applied uniformly to make sure that the relative vulnerability of various supply chains is minimal. The system's exposure to terrorism risk is a function of the vulnerability of the weakest link when facing an adaptive adversary. Reducing the risk of tampering on each supply chain to a level at a par with what the best practices entail will be a powerful deterrent to those who have aggressive designs on U.S. seaports.

Land Transportation. This phase starts with the dispatch of the container from the originating warehouse and ends at the port of origin. Cargo theft during this phase of transportation is a huge problem that companies need to confront in the modern era. Cargo insurers in industrialized nations (Italy, Australia, Germany, and France, with which the U.S. has extensive overseas trade ties) face enormous claims from cargo theft every year [7]. More recent statistics suggest that cargo theft is on the rise in Belgium, the Netherlands, France, and the United Kingdom [11][5]. In particular, high-value cargo—e.g., pharmaceuticals, luxury clothing, electronics, and computer hardware—are also high risk.

During this phase, cargo moves by truck or rail. However, due to economies of scale, the mode of transportation may not remain the same during travel to the port. Cargo may be transferred from one mode to another (intermodal transfer), or simply within the same mode (intramodal transfer). Security breaches at the transfer points and frequent stops in transit may leave cargo vulnerable to tampering while waiting unattended for the next pickup. However, due to caps on daily work hours, truck drivers have to make multiple stops for long-distance hauls. Many cargo theft incidents take place with insider help. Therefore, background checks on the truck and locomotive drivers are vital to transportation security. Other challenges to cargo security in this phase include funding sources and the tremendous variety of freight hauled on railroads and highways.

Extending U.S. borders beyond the homeland is part of a layered defensive strategy against terrorism. C-TPAT extends the borders by cooperating with U.S. companies to use their leverage on their global trade partners. While C-TPAT helped improve security at the container-loading phase, there is relatively little security improvement in the transportation phase. Supply chains may not enjoy a high level of security during the transportation of

[5] Other reports published by EUROWATCH earlier in 2005 suggest a similar trend in Spain, Russia, Ireland, and Italy.

U.S.-bound containers in foreign countries as the U.S. government cannot reach beyond American seaports and foreign private companies may underinvest in transportation security. Those countries with minimal historical exposure to terrorism may overlook some of the key security issues. Cargo theft is a serious issue that companies have been countering for years. The most vulnerable points for cargo theft are intermodal transfer locations. Local governments have to take the initiative to beef up security at these locations, so that terrorists are deterred from tampering with cargo in transit.

Anti-tamper seals have been developed to ensure integrity of cargo in transit. Because most anti-tamper seals are cheap, they have minimal impact on security. In some cases, terrorists can easily break the seal, and replace it with an identical copy after tampering [12]. Electronic seals can flag an intrusion once broken. Thieves have already developed the expertise to open containers without breaking ordinary anti-tamper seals [42]. Electronic seals are relatively new, but it is still possible to open a container without tampering with these seals. A dirty bomb can be located in a container by removing the doors completely without breaking the seal [9]. It is claimed that a Sandia technician can easily pass an electronic seal barrier in a couple of minutes [10].

Likewise, global positioning systems and radio frequency identification (RFID) technologies do not ensure the integrity of containers. Both provide good surveillance to determine where the container is and to improve transparency, but this will not be a deterrent to a sophisticated terrorist. Placing the bomb by partially removing the contents of the container, it is possible to avoid detection. The current technology does not allow detection of items that are not tagged. Unless both technologies are made smarter, terrorists may tamper with containers unnoticed. Installation of electronic sensors in containers is another option, but this technology is still in the development phase and the rate of false alarms is high. CBP seeks to reduce the false positive rate to 1% or less. A stumbling block to widespread use of technology is the cost. For example, some companies today are hesitant to replace bar codes which cost next to nothing with RFID tags that cost less than $1 per tag.

In an effort to promote the development of smart containers that external parties cannot unseal, the Office of Domestic Preparedness is currently administering a program called Operation Safe Commerce (OSC). The goals of the program are to increase the transparency of cargo, employ state-of-the-art technology for inspection, prevent cargo tampering, and reduce theft. While CSI and C-TPAT seek to reach out foreign ports and loading facilities to standardize the guidelines for securing container shipments, OSC seeks to set the standards that should be followed during transportation and when the cargo reaches U.S. seaports. Use of sensors, seals, and other cargo tracking and security technology are encouraged under this program to identify the best technologies and practices to prevent cargo tampering.

This program initially began as a partnership between government bodies and private sectors to address vulnerabilities in cargo security. OSC consists of three phases. The first two phases were completed by the end of 2004, and involved security assessment of the entire supply chain. In particular, security assessments at the Los Angeles and Long Beach, Seattle and Tacoma, and New York and New Jersey ports were completed along with a detailed analysis of the vulnerabilities of 19 separate supply chains. The best technology and best practices identified in the first two phases are currently subject to further evaluation in the third phase. If the best practices in various supply chain risk assessments are set as industry standards, OSC can have a significant impact on a layered defense against terrorism.

Port of Origin. After land transportation, cargo reaches the port of origin. Cargo containers may be stored around the port perimeter before loading on the ship. Physical security around port perimeters is a concern. In particular, empty containers may be enticing targets for terrorists planning to plant and ship bombs across the ocean. Risk of unauthorized access to secure areas in the port and its perimeters with counterfeit documents or fake ID is another complicating factor in cargo security at foreign ports. The human element is key to preventing such attempts. Trained port workers may be able to detect anomalies and respond accordingly. However, in some countries corruption at customs is a problem.

Security problems may be further compounded by the lack of state-of-the-art inspection equipment. Due to increased levels of competition between seaports, there may not be enough incentives to inspect outbound cargo. The Container Security Initiative (CSI) was unveiled in 2002 to address this problem. It is an effort to extend U.S. borders to confront external threats outside the homeland. The main objective is to identify and prescreen containers that pose terrorism risks at the port of origin. By implementing this initiative, CBP seeks to reduce the inspections at U.S. ports, thereby ensuring efficient flow of trade. As of September 2006, 50 operational ports in Europe, Africa, Asia, and the Americas participate in CSI.

The criteria for expansion to new foreign ports are based on trade volume, location, and strategic importance. Eligibility of a foreign port for this program requires installation of non-intrusive inspection (NII) equipment and deployment of trained customs officials who can perform inspections. U.S. customs officials visit participating ports under the initiative to target and pre-screen the containerized cargo. They also visit loading facilities to check the security standards. CBP then assembles a CSI team with members from either CBP or Immigration and Customs Enforcement. It is expected that participating foreign ports are willing to share critical information with the CBP officials to help target high-risk cargo. CBP recognizes that advance information is key to security operations. In order to address this issue, the

"24-Hour Rule" was initiated. The rule requires carriers to report their cargo manifests 24 hours before the cargo leaves the port of origin. This provides a timeframe for risk assessment of cargo.

High-risk cargo is selected using the Automated Targeting System (ATS) and various intelligence reports. ATS uses cargo manifest information, targeting rules, shipper information, and customer/importer information to assign each container to a risk category (e.g., low, medium, high). Intelligence reports and research assistance provided by the National Targeting Center are also checked to make the final decision regarding inspection of a specific container. The containers with a high risk score are to be inspected. The medium risk containers are subject to further research. Inspections are performed by local officials, and a CSI team is given the option to monitor inspections.

Should the officials choose to inspect the cargo, a gamma-ray or X-ray shot of the container is taken using the Vehicle and Cargo Inspection System (VACIS) to perform visual detection. Inspectors may decide to open the container and conduct physical inspections in case they detect anomalies. The physical inspection decision is technically made by both CBP and local officials. For nuclear content, radiation detection devices such as radiation portal monitors (RPMs) are used. The inspection process as described here may slow down the container flow through seaports. In order to attract foreign seaports to the initiative, both parties agree that CSI port shipments will be given priority in processing in the case of a terrorist attack that halts port operations. Likewise, U.S. port authorities may choose to limit their operations to CSI ports in the case of a minor attack. Accordingly, shipments that originate from non-CSI ports may experience long delays if a terrorist attack triggers full or partial shutdown of maritime ports.

Is it possible to locate explosives or illegal weapons in a container owned by a trusted shipper originating from a relatively secure CSI port? Stephen Flynn, a senior fellow in national security studies at the Council of Foreign Relations, describes one such scenario: terrorists can exploit security gaps during transportation of containers to a CSI port by loading a dirty bomb at a railroad facility [12]. They may target a container from a well-established manufacturer to reduce the probability that it's selected for inspection. According to this scenario, the shipment originates from Rotterdam. The security breach around this particular port has been confirmed by a recent drug interdiction incident in Australia. Australian customs officials seized 370,000 ecstasy tablets in a Melbourne freight warehouse from a shipment originating in Rotterdam[6].

Other CSI ports have been vulnerable to drug trafficking in recent years. In June 2002, a shipment of cannabis that originated from the Port

---

[6] 91.7 kg of drugs were hidden in plastic pipes inside eight metal German-made barbecues [52].

of Antwerp was interdicted in Ireland[7]. On December 12, 2003, an ecstasy shipment from the port of Tilbury was discovered in Sydney[8]. According to a recent report prepared by the National Criminal Intelligence Service of the United Kingdom:

...the use of feeder vessels to transport cocaine from Rotterdam, Antwerp, and Hamburg to container ports in southern and eastern England is known to take place.

Terrorists have also utilized CSI ports for small arms smuggling in the recent past. According to Jane's Intelligence Review, an arms cache shipped from the port of Hong Kong via Singapore was interdicted in Bangladesh[9]. Terrorists may exploit vulnerabilities on inland or maritime shipping routes to transport weapons and explosives to the U.S.

There are some other challenges to implementation of the CSI program. The success of this initiative relies on the level of cooperation provided by foreign port officials. According to the current procedure, when high-risk cargo is identified, the foreign customs officials are responsible for inspection and U.S. officials are entitled to observe the inspection process. According to recent reports, U.S. officials' roles in this inspection process have been relegated to the review of cargo manifests [23]. Some foreign ports perceive monitoring by U.S. officials as violation of sovereignty. However, observing the inspection process is an important element of CSI as the U.S. government is unable to perform background checks for workers at participating ports.

As of September 2004, 65% [20] of the containers originating from CSI ports (which constitute 43% of all shipments to the U.S.) have been subject to targeting using ATS, which implies that the remaining 35% were not subject to any risk assessment or inspections overseas. GAO attributes this to staffing imbalances at the CSI ports. 72% of the high risk containers were inspected overseas, and the rest were denied inspection for a variety of reasons. Of the remaining 28%, 93% were inspected upon arrival at a U.S. port. The remaining 7% were not subject to any inspection due to a lowered risk score based on further incoming intelligence, or because the port of destination was other than a U.S. port. It should be noted that further cooperation from officials at participating ports is required because it may be too late to respond to the terrorism threat when the container reaches a U.S. seaport.

Implementation of ATS has not been problem free either. The human element in understanding and evaluating risk factors in transportation of various forms of cargo is vital in raising the effectiveness and reliability of the targeting system. Expertise in commercial shipping and supply chain

---

[7] Drugs were reportedly discovered in a 40-foot container [53].

[8] Total value of the shipment was estimated to be around $16 million [8].

[9] Insurgent groups northeast India are reportedly tied to the shipment [4].

operations gives the ability to point to anomalies in shipment routes to detect possible intrusion into containers or the loading of illegal contraband during visits to ports en route. Without providing such training to customs personnel and understanding the human element in successfully targeting containers, risk scores will have deficiencies. Another weakness of ATS is its reliance on cargo manifests, and the route information provided by the shippers. There is virtually no way to verify the accuracy of the route information, including the seaports visited between the port of origin and the U.S. port as well as the waterways the vessel navigated.

CSI is not the only program seeking to improve capabilities of non-U.S. seaports in detecting harmful cargo transported to the U.S. The Megaports Initiative, unveiled in 2003 by the U.S. Department of Energy (DOE), is a program to improve radiological material and nuclear weapon detection capabilities at non-U.S. ports. It complements CSI in that cargo is inspected for nuclear and radiological content at the port of origin. The ultimate goal is to deter illicit nuclear and radiological material trafficking. DOE officials negotiate with host governments and port officials and finalize an agreement to install portal monitors at critical seaports. The agreement addresses critical issues like port equipment needs, placement of equipment, and optimal calibration to detect nuclear and radiological content. A challenge that the DOE faces is personnel training. When the agreements are finalized, the host government takes control of all the equipment, but receives support from the DOE for maintenance. As of December 2006, the Megaports Initiative is operational in six countries.

The sensitivity of the equipment provided to non-U.S. seaports under the Megaports Initiative is under the control of foreign port officials. Therefore, port officials may choose to reduce the sensitivity of RPMs at will if the number of false alarms hampers the flow of trade. U.S. officials have no control over the equipment after their calibration and testing is completed. Another factor that could limit the effectiveness of the equipment is the environmental conditions. RPMs tend to be less effective in cold and windy climates. It is difficult to measure the effectiveness of radiological and nuclear content detection equipment installed at foreign imports under the Megaports Initiative. Other factors that limit the effectiveness of this initiative are its very limited coverage and slow installation progress due to operational difficulties [19].

Sea Transportation. In this phase, cargo travels between the port of origin and the port of destination in the U.S. It is a critical phase as there is minimal law enforcement in international waters and the vessel carrying cargo may visit other ports before arriving at a U.S. port. This leaves the vessel exposed to the threat of piracy and stowaways. The lack of security guidelines for combating piracy makes the problem worse. Most ships arriving at U.S.

seaports carry foreign flags and foreign crew. Therefore, backgrounds of the crew are not verifiable, and there is no way to detect anybody who committed crimes in foreign countries. As mentioned earlier, security breaches at seaports visited en route is another loophole that could be exploited. In an effort to reduce crime in open waters and seaports, international bodies promote the use of technology that could track vessels, improve port perimeter security, and help enforce the law. The USCG recognizes the severity of the problem and has taken some steps to reduce terrorism risk in U.S. waters. We will elaborate more on these efforts later in this paper when general security in the U.S. and open waters is discussed.

Port of Destination. Containerized cargo arrives at a U.S. port after sea transit. Based on the risk score assigned by ATS, each container may be subject to inspection. Of the nearly 10 million containers arriving at U.S. ports each year, an average of 5% is subject to inspections every day. The capability to detect harmful cargo largely depends on the effectiveness of NII technologies. Devices with NII technology encompass X-ray and gamma imaging systems, portable and hand-held radiation detectors, remote monitoring equipment, and portal sensors. The VACIS, which includes mobile VACIS, truck X-ray, mobile truck X-ray, and rail VACIS, is a system of devices used to take radiographic snapshots of containers. CBP has deployed nuclear and radiological equipment as well. Among these devices are Personal Radiation Detectors, Radiation Isotope Identifier Devices and RPMs.

Current RPM technology does not guarantee an acceptable detection rate. Highly enriched uranium is not necessarily detected by the current technology if concealed under sufficiently thick shielding. A study by Wein et al. [46] concluded that with the current system of equipment, a shielded nuclear weapon can be detected with less than 0.1 probability. In this sense, radiation portals offer limited detection capability. Another concern about the technology is the high rate of false positives. Items such as granite, porcelain toilets, and bananas can set off a radiation alert. High numbers of false positives could disrupt the flow of trade at seaports.

DHS plans to deploy a next-generation RPM, Advanced Spectroscopic Portal (ASP), by 2009 to improve detection capability at ports of entry. Detection likelihood of unshielded or lightly shielded radiological and nuclear material is expected to be higher with ASP. Also, ASPs offer reduction of false positives. However, these portals use passive detection capability, like current technology, and thus heavily shielded nuclear material is likely to evade detection with ASP. Another drawback is the cost. GAO reports that expert estimates for the cost of ASP range between six and eight times the cost of RPMs currently installed at U.S. seaports [2]. The Domestic Nuclear Detection Office is currently developing a next-generation detection system, Cargo Advanced Automated Radiography System, with the ability to detect heavily shielded nuclear material. However, this revolutionary system is still in its initial stages of development.

There is no effective detection equipment for biological and chemical weapons. There are many avenues to introduce contagious diseases and infectious bioagents into the U.S. Should terrorists choose to bring harm to the nation via bioterrorism, containers will probably be a less preferred medium of transport. However, containers could be quite instrumental in some forms of chemical attacks. Tons of ammonium nitrate, an explosive chemical substance used as fertilizer, are shipped in U.S. waterways every year as bulk cargo. Terrorists may choose to ship a container with ammonium nitrate and detonate the load at a U.S. port. Such an explosion could inflict a high number of casualties. Terrorists may also consider container explosions that could release poisonous nerve agents around a port complex and disrupt the port operations. Development of technology that could detect containers with undeclared hazardous cargo is vital to reduce chemical threat at U.S. ports.

Another piece of the container security paradigm that has been largely overlooked in the past is export cargo. With the implementation of CSI and the 24-hour rule, inbound containers with manifests such as "freight of all kinds" were no longer allowed at U.S. ports. However, containers with such descriptions are still allowed for export cargo. This leaves the system vulnerable to an attack. A container explosion incident at the Port of Los Angeles in 2004 was a warning sign [30]. An outbound container carrying hazardous material exploded when it was left unattended for three days at the port without any special precautions taken, because the cargo was supposedly "freight of all kinds." In addition, the security of outbound empty containers has loopholes. 40% of containers present at a West Coast port on a given day are empty [30]. A particular vulnerability with empty containers is the lack of any protocol or requirement to lock them in transit. Thus, it is relatively easy to place a bomb in these containers. Empty containers may lie unattended at a port facility for days. Gaps in ensuring security of export cargo and empty containers should be minimized to reduce terrorism at U.S. ports.

General Cargo. General cargo categories include liquid bulk (petroleum), dry bulk (grain, paper), and iron ore or steel loads which are usually not shipped in containers. General cargo ships have been used historically to smuggle drugs and other contraband[10]. Cocaine smugglers are known to prefer shipping their cargo disguised in iron ore or charcoal shipments because of the relatively low probability of detection. However, general cargo shipments may not provide the same level of convenience in weapons

[10] For instance, the USCG found 10,000 pounds of cocaine hidden below iron ore pellets in 1999 [54].

smuggling as containerized shipments. Weapons may be smuggled in general cargo ships if the detection probability for weapons smuggling in containers is relatively high and thus terrorists are deterred from using containers as a medium of transport.

General cargo ships have reportedly been used for terrorism and other suspicious activities in the past. In 2002, Italian officials arrested 15 Pakistani men with false passports and suspicious documents on a Tonga-registered vessel carrying a cargo of lead[11]. In November 2001, a Cambodia-registered vessel, supposedly carrying a cargo of timber, was found smuggling cigarettes into Ireland to finance Real IRA operations [37]. In 2002, the general cargo vessel, Karine A, loaded with 50 tons of weapons, was captured by Israeli officials[12].

While considerable attention has been focused on container security, general cargo has received little interest, at least in the media. All vessels that transport goods to the U.S. have to report their cargo information 24 hours before they leave the port of origin and schedule their arrival 96 hours before they dock at a U.S. port. Those vessels whose schedules of arrival have not changed more than six hours are not required to submit an update. The arrival schedule is reported by completing advanced notice of arrival (ANOA) documents, which require information about the last five ports of entry. If a ship is bound to visit multiple ports in the U.S., it has to file a new ANOA for every single port. The goal is to determine the route of each vessel in order to assign a risk score. Vessels larger than 300 gross tons file this document with the National Vessel Movement Center (NVMC).

For vessels that are less than 300 gross tons, an ANOA is not required. However, some ports may choose to ask for ANOA information[13]. U.S.-flag recreational vessels are exempt from this requirement. Most commercial vessels that are less than 300 gross tons are fishing boats. Inspection will be triggered if given the information from ANOA or other intelligence information, a vessel is believed to engage in an illegal activity. Without ANOA, the route information for general cargo ships won't be known. This creates another pathway to introduce a dirty bomb or illegal contraband into the U.S. The risk is particularly high for ships carrying hazardous cargo. Terrorists may have an interest in exploding ships carrying hazardous cargo to inflict more damage.

---

[11] They were suspected to have links with Al Qaeda [55].

[12] Among the weapons were 122mm Katyusha rockets, 107mm rockets, 120mm mortars, Sager and LAW anti-tank rockets, mines, sniper rifles, Kalashnikov assault rifles, bullets, and explosives [56].

[13] For instance, Port of Los Angeles does not require this information, whereas the USCG in South Florida began to enforce the rule on May 21, 2004. The vessels have to file ANOA with the captain of the port [36].

General cargo ships may also be used as weapons to attack coastal targets, or as indicated above, to detonate a bomb. These scenarios are discussed in later sections on security of coastal facilities.

3.1.2. Security of Port Area and Perimeters

Seaports are vulnerable to threats that could originate from port perimeters or inside a port area. Critical infrastructure around seaports and the vast size of the perimeter area augment the fears that physical security gaps can be exploited. Realizing that a port facility is only as secure as its perimeter, there is a need to develop systems that would deter, detect, document, and deny any unauthorized entry into a port area and its perimeters.

Access to Secure Areas. The human element is critical to ensure security at a port area. The goal is to deny access to those who are not authorized or have criminal backgrounds to secure areas around seaports. The USA Patriot Act of 2001 requires background checks for all those individuals involved in transportation of hazardous materials in commerce. Ports are among the intermodal sites where tons of hazardous materials cross. This legislation was customized to the maritime environment by the Maritime Transportation Security Act of 2002. The act requires issuance of transportation security cards with biometric information and an overall background check for all workers employed at maritime ports. Otherwise, those with criminal records or those who have stolen identities can get access to secure areas at seaports.

In the current system, verification of identity is achieved by checking photo ID, which does not require any background check or check against national security databases. Obtaining a driver's license did not even require legal status in the U.S. until recently[14]. In the old system, consular cards issued by other countries such as Mexico would be sufficient to issue a driver's license. This opened the door for illegal aliens to get jobs in port areas. TSA is planning to overcome this identity problem by designing a Transportation Workers Identification Card (TWIC) that will be issued to all the 12 million transportation workers who need unescorted access to secure areas. These cards will have embedded fingerprint information and be tamper-resistant.

Progress toward the use of TWIC has been rather slow. TSA awarded a $12 million contract in August, 2004, and launched a prototype TWIC program at four sites in November, 2004. When TSA initiated the

---

[14] President George Bush signed legislation that standardizes the procedures across all U.S. states. The legislation became effective on May 11, 2005. Under the current legislation, the driver has to provide a birth certificate, proof of SSN, a photo ID, and a document showing name and principal address [31].

TWIC program in 2002, the plan was to issue the first cards in August, 2004 [16]. However, serious delays in critical policy decisions stood as stumbling blocks to progress. Delays were partly attributed to the late decision on the type of technology to use, the late approval by DHS to conduct prototype test, and data challenges [3]. In order to mitigate risks due to delays, TSA currently requires transportation workers to carry different identification cards for each facility they access. However, without full implementation of the TWIC program, access to secure areas at seaports will remain a vulnerable point of the system.

Each year, approximately 7,500 foreign-flagged ships carry the bulk of shipments into the U.S., including 175 billion gallons of oil and other fuels. 93% of U.S. trade sails in non-U.S.-owned or non-U.S.-flagged ships with foreign crews. Before 9/11, foreign crews could obtain visas without visiting an embassy or consular office in person as the shipment companies could obtain visas for all crew members by submitting a list of the crew to a U.S. embassy. This vulnerability was addressed by requiring all foreign crew members to acquire their own visas. However, the threat posed by stowaways remains unless port surveillance capabilities are improved and access to secure areas is granted with state-of-the-art identification cards. Terrorists may use this route to either sneak through U.S. borders, or launch an attack in the port area. They may seek the cooperation of foreign crew members in sympathy with their agenda to board the ship and penetrate a port facility when the ship arrives at a port. At the federal level, there does not seem to be any effort to incentivize U.S. shippers to enlarge their commercial fleet or to encourage U.S. vessel owners, who operate their ships under foreign flags, to switch to the American flag system.

Port officials have taken steps to install new technology to improve surveillance capabilities in the port area. So far, some ports have installed integrated security management systems that have video surveillance, automated access control, and perimeter intrusion detection capabilities[15]. These are positive steps to minimize the risk of unauthorized access to secure areas in a U.S. seaport. Nevertheless, these technologies also have limitations. As in the case of RPM, most sensor technology comes with the dilemma of sensitivity adjustment versus false alarm rate. For example, motion sensors can be adjusted to track a flying bird, wall vibration sensors may react to any vibrating object or machinery, and ultrasonic sensors may detect any noise in the ultrasonic range. In order to reduce the time to

[15] One example is Port of Oakland [57]. The ports of Galveston, Toledo, and Los Angeles/Long Beach are also among those that installed similar systems.

identify false alarms, cameras may be installed around sensors. However, lights may change the thermal environment and affect the operation of sensors. In sum, a layered monitoring approach will reduce the likelihood of success for an intrusion attempt.

Port Perimeters. Port police and the USCG are responsible for policing waterways inside and around every port. Many naval vessels, commercial vessels with hazardous material, nuclear power plants, densely populated areas, and critical infrastructure such as bridges are located on or near the open waterways. A discussion of security around coastal targets is provided later in this paper.

3.1.3. Cruise Lines

Terrorists have hijacked passenger vessels in the past to accomplish political goals. Like airplane hijacking incidents, they attracted a lot of attention, which helped spread the political message of terrorist groups. Cruise ships are enticing targets for terrorists, for multiple reasons. First, terrorists can hijack cruise liners for piracy and looting purposes, because there is a widespread belief that cruise ship passengers are wealthy. Second, some cruise ships have around a 5,000-passenger capacity, exposing them to a single terrorist attack with the potential to claim thousands of lives. Such an attack would have ripple effects on tourism, aviation, and the entertainment industry. In this regard, a well-organized attack on a cruise liner will fulfill terrorists' objectives: economic damage and a high number of casualties. If terrorists aim to kill the maximum number of people, cruise ships may suffer either seaborne or aerial attacks. It is less likely that cruise ships will be used as weapons themselves. A hijacking incident won't remain a secret for long, and port officials will have enough time to secure the critical targets on the coast.

After the 1985 hijacking of Italian cruise liner, Achille Lauro, the cruise industry implemented a wide array of new anti-terrorism measures. Since 1985, there has been only one cruise ship hijacking incident. In the current era, hijacking of cruise ships in the U.S. seems less likely, as the cruise liners have adhered to even more strict security measures since 9/11. Specific measures that are enforced by the USCG include:

- Screening of all passenger baggage, carry-on luggage, and ship cargo. Metal detectors, X-ray machines, human searches, and canine teams are used to do screening.
- Screening of passenger lists against criminal and terrorist watch lists.
- Restricting access to secure areas in the port and on the vessel.
- Maintaining a 100-yard security zone around cruise ships.
- Underwater surveillance at high risk ports.

There is no doubt that these security measures have been effective in reducing terrorism risk on cruise liners. The degree of vulnerability to a terrorist attack is more a function of technology failure risk which is and will be present in most security systems. Cruise lines are a relatively more secure component of the entire U.S. border security system.

3.2. SECURITY OF U.S. WATERS AND COAST

Most U.S. energy power plants, critical bridges, and densely populated urban areas lie close to waterways. For instance, 75% of oil refineries, a great majority of 103 nuclear reactors, and all liquid natural gas (LNG) terminals in the U.S. are located onshore. Nearly all major cities are accessible by waterways. Operation of the infrastructure onshore is crucial for the U.S. economy. A single attack on any of this infrastructure is likely to inflict a significant number of casualties and bring serious economic damage. Furthermore, they may be easier to penetrate from the shore, making them attractive targets for terrorists. The USCG has a central role in confronting such seaborne threats.

3.2.1. Discussion of USCG Capabilities

The USCG currently assumes homeland and non-homeland security responsibilities, such as enforcing security laws around ports, waterways, and coastlines, interdicting drug and human smugglers, monitoring fishing areas, responding to pollution, and conducting search and rescue operations. All these missions are accomplished with a fleet consisting of aircraft, cutters, patrol boats, and special-purpose vessels, such as icebreakers. Before 9/11, most USCG operational hours were dedicated to search and rescue missions, along with three categories of law enforcement: protecting fisheries, interdicting illegal migrants at sea, and controlling the flow of drugs. As 9/11 shifted focus to preventing terrorism, USCG resources were largely allocated to homeland security-related activities, while some of the traditional missions, such as search and rescue, underwent a significant reduction in operational hours[16].

The USCG was provided $1.5 billion between 2002 and 2004 for equipment replacement and modernization under the Deepwater Acquisition Program (DAP), which was initiated in 1996. The program seeks to modernize ships and aircraft used in missions that cannot be carried out by shore-based small boats. The equipment replacement and modernization phase, under the

[16] This reduction amounted to 22% [15]. Other non-homeland security missions with similar reduction in operational hours are law enforcement activities protecting living marine resources and foreign fish enforcement.

DAP, began in 2002 as the contract for acquisition and integration of necessary equipment was awarded to Integrated Coast Guard Systems. Funds appropriated for DAP are used to upgrade legacy assets, acquire new vessels, and maintain existing equipment [49]. Another modernization program in progress is Rescue 21, which will replace the equipment used for coastal communication needed for search and rescue operations. Total USCG funds requested for fiscal year 2006 amount to $8.1 billion, which constitute 20% of the DHS budget [48].

According to GAO, the USCG faces challenges in implementing both DAP and Rescue 21. GAO believes that DAP has not enjoyed a predictable and steady funding stream, which is key to acquisition and integration of new equipment to the system. Furthermore, they report that the acquisition program is behind schedule, and pouring more money into the program will not help get inline with the original schedule [15]. Due to these delays, GAO estimates that the cost projection of the program that was originally expected to be $15 billion at the end of 20 years first rose to $17.2 billion and then recently up to $24 billion. The approach adopted by the USCG in managing this huge acquisition project to hire a single systems integrator led to concerns about potential lack of competition in later phases of asset procurement that may eventually lead to an uncontrolled stream of acquisition costs. This prediction was realized in earlier years of the project, as the estimated cost of the project increased up to 60%. Another reason behind these escalating costs lies in deepwater aircraft and cutters "failing at an unsustainable rate" [21], which compelled the USCG to revise its implementation plan for this acquisition.

Similar problems have arisen in the Rescue 21 program that hampered search and rescue operations across U.S. waters. These problems raise concerns about the capability of the USCG to achieve some of its missions. GAO has also reported that most missions may also suffer from lack of station readiness due to staffing and training problems which lead to prolonged periods of work and shortage of experienced personnel [18]. While the USCG responded to these problems by increasing its operational efficiency through improved technology, port security assessments, stronger partnerships, and better information sharing, most of these improvements have remained rather local and should be spread all over the maritime security spectrum.

The USCG is currently developing the operational requirements for an Automatic Identification System (AIS). The International Maritime Organization's (IMO) package of security measures which extended Safety of Life at Sea (SOLAS) in 2002 included installation of AIS technology as a ship-related provision. These security measures that are known as International Ship and Port Facility Security Codes (ISPS) required AIS to improve monitoring of vessel movement. AIS is a technology that enables tracking of vessels by coastal stations and other USCG vessels using a portion of the radio frequency spectrum for communication. Information such as size of the ship, its course and speed, registration number, and other identifying characteristics of the vessel can be transmitted to the central USCG location using AIS. This technology is required for all vessels over 300 gross tons. These larger vessels are currently documented by the USCG. The USCG is still developing operational requirements of AIS technology and evaluating these requirements with all the stakeholders involved. AIS is crucial to identify all high tonnage vessels in U.S. waterways and accordingly will help the USCG extend waterway coverage. According to the USCG, approved AIS equipment ranges in price between $3,000 and $9,000, excluding installation costs. These prices make it cost-prohibitive to require expansion of AIS for vessels below 300 tons, as many of these vessels are not commercial. AIS will be implemented in 10 critical maritime areas, which is only a fraction of over 12,000 miles of coastline and 25,000 miles of river and inland shoreline[17]. These areas are currently monitored by radar based Vehicle Traffic Service (VTS) systems.

Smaller vessels that will not be required to install AIS technology are registered by individual states. Integration of this data into the USCG's vessel registry database is crucial to increase awareness of all vessels in the maritime domain. The USCG has faced problems in the past integrating state vessel registry data [14]. As of now, there is no legal requirement for individual states to share their vessel registry data with the USCG. These legal boundaries may reduce the capability of the USCG to monitor small vessels that may be involved in illicit arms and contraband traffic.

3.2.2. Vulnerabilities along U.S. Waterways and Countermeasures for Risk Mitigation

We discuss vulnerabilities along U.S. waterways in three categories:

- Critical coastal targets
- Pleasure and fishing boat security
- Waterway and underwater security

Critical Coastal Targets. Maritime terrorism has hit U.S. and non-U.S. coastal infrastructure in the past. Terrorists have a plethora of potential targets for launching deadly attacks on the U.S. coast. The history of maritime terrorism suggests that terrorists have already exercised a variety of options to execute such attacks. In recent years, many terrorist organizations have

[17] These areas cover 10% of the U.S. ports recognized by the Department of Transportation. Some of the major US ports are not included in these 10 maritime areas [17].

added various means of suicide attacks to their portfolios. This poses further challenges to those who seek to deter terrorists from attacking one of the most vulnerable points of the nation. LNG facilities, chemical plants, urban centers, bridges, and nuclear plants are among those critical targets that are exposed to waterborne terrorism threats.

There are various ways to attack critical coastal targets. Terrorists may acquire new vessels through piracy or smuggle humans on vessels that are destined to sail near the target of interest. Piracy is a rising threat across the globe, and terrorists are already known to use piracy for financing purposes. Having acquired a ship to execute an attack, terrorists may detonate explosives on the ship at a time or location of their choice, or ram the ship into the target. It may be difficult to execute an attack hijacking an LNG ship due to security guidelines followed by the USCG, but vessels with other forms of hazardous cargo may be vulnerable to hijacking. In particular, ships carrying bulk shipments of ammonium nitrate are potentially vulnerable. A huge volume of ammonium nitrate flows through U.S. inland waterways each year. For example, in 1997 over 400,000 tons of ammonium nitrate was shipped through the Mississippi river. These shipments pass near urban centers such as New Orleans, St. Louis, Memphis, and Pittsburgh. In order to monitor and reduce vulnerability against ammonium nitrate and other hazardous cargo shipments (commonly called certain dangers cargo or CDC), the USCG acted to introduce new regulations in 2004. These regulations include mandatory development of security plans at vessels and facilities handling ammonium nitrate, preparation of vessel maintenance and security records, training of a facility and vessel security officer, and installation of vessel and facility security systems. These new regulations increased the transportation cost of ammonium nitrate.

Inland waterways are vulnerable to attacks that could cripple freight routes and devastate waterfront cities, power plants, chemical facilities, and other critical and commercial targets. Some of the measures to address the terrorism threat in inland waterways include routine anti-terrorism patrols, establishment of maritime security zones covered by the AIS technology, increased inspections of domestic tankers, and technology-based surveillance around inland ports and critical facilities. The USCG partnered with the private sector to analyze the consequences of explosions on inland barges carrying CDC. Since 9/11, inland ports have been closed to non-U.S. flag vessels. The threat posed by barges and small watercraft is still present despite all the security improvements made so far. Large segments of inland waterways still do not have any AIS coverage, which increases the difficulty of surveillance [34]. It is urgent to expand AIS coverage as soon as possible along inland waterways to reduce waterborne threats in America's heartland.

The risk of suicidal attacks on coastal targets from sea shores can be mitigated by blocking access from the sea. Waterborne security barriers such as chains prevent terrorists from either ramming into the facility or exploding bombs in the vicinity. However, remote sensor technology at maritime ports, power plants, or other coastal targets is needed to detect these attacks in advance. In particular, sensors that can detect nuclear content from a distance could be a very valuable addition to the anti-terror armor.

Pleasure and Fishing Boats. As mentioned earlier, tracking of general cargo vessels under 300 gross tons is relatively poor as these vessels provide no information about their route. Since most fishing boats are under the 300 gross tons requirement, there is no way of gathering intelligence about routes fishing boats take in fisheries. Another point of concern is that fishing boats permitted to enter the port area could launch an attack on critical targets in and around a seaport. Likewise, other open targets on the coast are vulnerable. A similar threat is posed by pleasure boats. Without further intelligence that triggers boarding of a specific boat, there is little chance of intercepting terrorists.

Drug smugglers have been known to introduce illegal drugs on fishing boats for a long time. As the threat of terrorism intensifies, fishing boats may be a new means to introduce explosives or weapons in the American homeland. Most marinas in the U.S. have minimal protection from terrorists. Thus, as long as a fishing boat escapes the USCG's notice, success is very likely. According to GAO, resource hours allocated to drug interdiction has been reduced 44% [48]. Drug interdiction performance did not reflect this reduction on hours devoted to the mission. The USCG attributes this to improved efficiency in operations due to new technology. Recent evidence of increased criminal activity involving drug cartels beyond the southwestern land borders suggests that this could be attributed to increased drug traffic. In one Senate hearing, Michael O'Hanlon of Brookings Institution said:

> In the immediate aftermath of 9/11, more than half of the Coast Guard was devoted to port and waterway security against possible terrorist attacks. Even today, at least a quarter of Coast Guard assets are devoted to such missions. Other activities ranging from environmental protection to patrolling of U.S. economic maritime zones to counterdrug missions have suffered[18].

Reduction of hours devoted to drug interdiction may have negative impacts on weapons interdiction. An increased rate of random boat inspections may be an effective deterrence-based solution to the problem.

---

[18] Senate Governmental Affairs Committee Hearing, 20 March 2003.

Expansion of initiatives and technology investments increasing maritime domain awareness to include intelligence gathering on fisher and pleasure boats would be another effective risk mitigation effort.

Waterways and Underwater Security. Most of the discussion which is relevant to waterways security is already provided in previous sections, as there is significant overlap between waterways security and security of coastal targets. What distinguishes waterways security from the earlier discussion is the threat from commercial traffic flowing in U.S. waterways. Coordinated suicide and mine attacks on ships carrying hazardous or other commercial cargo are the most prominent scenarios that target economic prosperity by disrupting trade. Increased maritime domain awareness that could help the USCG observe, report, and respond to suspect vessel activities will mitigate both risks. Vessels should develop anti-piracy measures to detect any suspicious activity and report valuable information that could facilitate early response.

Terrorists may choose to disrupt trade by laying mines in a port, or around critical waterways. If mines hit a ship with hazardous cargo, then the damage can be compounded. Mines may be laid by vessels or frogmen. This threat raises questions about underwater security, which has been overlooked in the past. A countermeasure to this threat would be to equip USCG vessels with mine sensors. No extra resource hours would be allocated for mine search missions as regular patrolling activities would accomplish the task. The USCG is aware of the problem and unveiled an integrated anti-swimmer (IAS) system to address this threat [51]. The system integrates technology with human monitoring. Underwater and sonar cameras are used for continuous monitoring, whereas swimmers and divers can be used to thwart underwater bombers. Underwater weapons called "non-lethal interdiction acoustic impulse" devices were still in the testing phase early in 2005 [28]. The USCG plans to use these weapons after a verbal warning using underwater speaker systems. These countermeasures deter frogmen from laying mines around ports; they do not involve mine detection and sweeping.

The U.S. Navy has resources for mine sweeping and detection activities, but some of these resources are currently deployed for overseas operations, and all of the resources (Coastal Mine Hunters and Mine Countermeasures Ships) available domestically are based in Texas. This increases vulnerability to a mine attack at certain locations in U.S. waterways, as it may be difficult to deploy minesweepers in time. In particular, any West Coast port is particularly vulnerable as it may take a month for a minesweeper to cross the Panama Canal and arrive at the port under attack. It is mandatory to have highly trained divers at, or sufficiently close to, critical locations to clear mines upon an attack.

4. Conclusion

Terrorism is a multidimensional risk that will probably never be eliminated with limited resources as long as there are those willing to deliver harm. Therefore, the goal is to achieve maximum possible risk reduction by spending dollars to minimize security weaknesses at certain points of vulnerability. DHS has adopted a risk-based fund allocation principle to address the problem, but more than five years after the 9/11 attacks, U.S. borders remain vulnerable to terrorism threats. Maritime borders are particularly vulnerable due to the sheer size of waterways that need to be protected and plethora of targets that terrorists can attack. DHS employs a layered approach to seal maritime borders against terrorist plots and has introduced several initiatives to this end. While these initiatives and other DHS efforts have helped defend the American homeland against this persistent threat, they still leave gaps in maritime security which could be exploited by an adaptive adversary.

This paper is an attempt to summarize the status of homeland security in the maritime domain. More dollars will be spent and more resources will be mobilized in the future for better defense. Expected benefits of every dollar spent to mitigate terrorism risk should be measured after a comprehensive analysis of the status of security in the maritime domain. Different elements of maritime security have complex interdependencies that determine the marginal value of new countermeasures and initiatives. For example, new technology to improve container security may yield limited benefits if other avenues to deliver a dirty bomb are left open. Therefore, a systems-based risk analysis should be used to evaluate the benefits of new technologies, policies, and initiatives.

Intelligence is key to deploying resources to the right locations. Intelligence will help make correct predictions about the possible terrorist behavior, hence develop strategies to protect maritime borders. It will also help gauge the degree of terrorist sophistication and elicit the relative probability of various types of attacks. Terrorists may not have the immediate capability to launch all types of attacks, so intelligence will help characterize threats and also monitor terrorists' agendas. In addition, early interdiction rests on intelligence and a multilayered defense that engages all maritime domain stakeholders in security. Early interdiction is a powerful way to minimize economic damage, and in most cases eliminate human losses. It may be too late to respond if a dirty bomb or a weapon of mass destruction arrives at a U.S. seaport.

5. References

1. ABT Associates. The economic impact of nuclear terrorist attacks on freight transport systems in an age of seaport vulnerability. Executive summary available at: http://

www.abtassociates.com/reports/es-economic_impact_of_nuclear_terrorist_attacks.pdf. 2003.
2. Aloise G. Combating nuclear smuggling: challenges facing us efforts to deploy radiation detection equipment in other countries and in the United States, GAO testimony, GAO-06-558T. 28 March 2006.
3. Berrick CA. Transportation security: systematic planning needed to optimize resources. GAO Testimony, GAO-05-357T. 15 February 2005.
4. BurmaNet News. 3 March 2005.
5. Burnett J. Dangerous waters: modern piracy and terror on the high seas. New York: Penguin Group; 2002.
6. Cohen SS. Boom boxes: containers and terrorism. In: Haveman JD, Shatz HJ, editors. Protecting the nation's seaports: balancing security and cost. San Francisco: Public Policy Institute of California; 2006.
7. Coppack L. Hijackings haunt cargo underwriters. International Union of Marine Insurance convention report. 1993.
8. Cubic Analysis Center. Counterdrug press summary. 17 November 2004.
9. Durstenfeld B, Fuhr P, Haag WR, Hsi P, Ng J. Cargo container security. Occupational Health & Safety. August 2003.
10. Edmonson RG. Still trying on e-seals. Journal of Commerce. 16 May 2005.
11. EUROWATCH. September freight crime bulletin. Available at: http://www.cargosecurityinternational.com. Accessed on 10 October 2005.
12. Flynn S. America the vulnerable. New York: Harper Collins; 2004.
13. Fritelli JF. Port and maritime security: background and issues for Congress: CRS report to Congress. 27 May 2005.
14. GAO Report, GAO-02-477: Coast Guard: vessel identification system development needs to be assessed. 24 May 2002.
15. GAO Report, GAO-04-695: Coast Guard: deepwater program acquisition schedule update needed. 14 June 2004.
16. GAO Report, GAO-05-106: port security: better planning needed to develop and operate maritime worker identification card program. 10 December 2004.
17. GAO Report, GAO-04-868: maritime security: partnering could reduce federal costs and facilitate implementation of automatic vessel identification system. July 2004.
18. GAO Report, GAO-05-161: Coast Guard: station readiness improving, but resource challenges and management concerns remain. 31 January 2005.
19. GAO Report, GAO-05-375: preventing nuclear smuggling: DOE has made limited progress in installing radiation detection equipment at highest priority foreign seaports. 31 March 2005.
20. GAO Report, GAO-05-557: container security: a flexible staffing model and minimum equipment requirements would improve overseas targeting and inspection efforts. April 2005.
21. GAO Report, GAO-05-757: Coast Guard: progress being made on addressing deepwater legacy asset condition issues and program management, but acquisition challenges remain. 22 July 2005.
22. Gordon P, Moore JE, Richardson HW, Pan Q. The costs of a terrorist attack on Terminal Island at the twin ports of Los Angeles and Long Beach. In: Haveman JD, Shatz HJ, editors. Protecting the nation's seaports: balancing security and cost. San Francisco: Public Policy Institute of California; 2006.
23. Harrington C. Concern grows over gaps in foreign port screening. Congressional Quarterly. 17 May 2005.
24. International Transport Workers Federation. Website. Available at: http://www.itfglobal.org.
25. Klaidman D, Hosenball M. Terrorism: ties to a Qaeda chief. Newsweek. 18 August 2003.
26. Koknar AM. Terror on the high seas. Security Management. June 2004.
27. Lavery B. Irish police find 8 people dead and 5 sick in cargo container. New York Times. 9 December 2001.

28. Lipton E. Coast Guard turns its eyes underwater. New York Times. 2 February 2005.
29. Luft G, Korin A. Terrorism goes to sea. Foreign Affairs. Nov/Dec 2004.
30. Mitre M. Congressional testimony on maritime security to the House Committee on Transportation and Infrastructure by Michael Mitre, port security director of International Longshore and Warehouse Union (ILWU).
31. Murphy K. New rules will make your driver's license harder to get. Kansas City Star. 20 May 2005.
32. O'Brien D. Container stowaway raises new fears terrorist suspect tried to cross ocean in steel cargo box. The Virginian Pilot. 26 October 2001.
33. Orszag PR, Daalder IV, Destler IM, Gunter DL, Litan RE, Steinberg JB, O'Hanlon ME. Protecting the American homeland: a preliminary analysis. The Brookings Institution. 2002.
34. Pappalardo J. Federal agencies tackle maritime security, ports first. National Defense. June 2005.
35. Peleg-Gillai B, Bhat G, Sept L. Innovators in supply chain security. The Manufacturing Institute. 2006.
36. Reed LC. No ANOA needed for pleasure vessels coming to South Florida. Available at: http://www.the-triton.com
37. Richardson M. Raid at sea highlights flag abuses: Cambodia-listed ship was carrying cocaine. International Herald Tribune. 24 June 2002.
38. Richardson M. A time bomb for global trade: maritime-related terrorism in an age of weapons of mass destruction—address to the Victorian Branch of the Australian Institute of International Affairs. 2004.
39. Schneider DL, Steiner R, Romaine J. Human cargo: health conditions of Chinese migrants interdicted offshore by US authorities. Journal of Community Health. February 2003.
40. Slater E. Human smuggling operation probed. Los Angeles Times. 17 January 2005.
41. Stana RM. Homeland security: key cargo security programs can be improved: GAO testimony, GAO-05-466T. 26 May 2005.
42. Tirschwell P. No simple solutions for box security. Journal of Commerce. 30 October 2003.
43. U.S. Coast Guard. Fiscal year 2004 report.
44. U.S. State Department. Website. Available at: http://www.state.gov
45. Valencia MJ. Conflation of piracy and terrorism in Southeast Asia: rectitude and utility. Contemporary Southeast Asia. 01 August 2003.
46. Wein LM, Wilkins AH, Baveja M, Flynn SE. Preventing the importation of illicit nuclear materials in shipping containers. Unpublished manuscript. 2004.
47. Willis HH, Morral AR, Kelly TK, Melby JJ. Estimating terrorism risk. RAND Report. 2005.
48. Wrightson MT. Coast Guard: key management and budget challenges for fiscal year 2005 and beyond: GAO testimony, GAO-04-636T. 7 April 2004.
49. Wrightson MT. Coast Guard: observations on agency priorities in fiscal year 2006 budget request: GAO testimony, GAO-05-364T. 17 March 2005.
50. Wrightson MT. Coast Guard: preliminary observations on the condition of deepwater legacy assets and acquisition management challenges: GAO testimony, GAO-05-307T. 20 April 2005.
51. USCG Unveils 'anti-swimmer' system to thwart underwater terror attacks. Emergency Preparedness News. 22 March 2005.
52. $18 million drug seizure like a needle in a haystack. AAP General News Wire. 22 April 2005.
53. Gardai question Dutch man following drug seizure. Available at: http://www.irishtrucker.com. 22 July 2002.
54. Drug ship auction in Texas draws bidders worldwide. Knight Ridder Tribune. 26 May 1999.

55. Italy arrests 15 Pakistanis suspected of Al-Qaeda links, terror plans. Agence France Presse English. 12 September 2002.
56. Israel seizes Palestinian gun-running ship. United Press International. 4 January 2002.
57. Port of Oakland selects ADT to design and install a $4.75 million maritime security system. PR Newswire. 7 May 2003.

CHAPTER 3

ENVIRONMENTAL AND HUMAN SECURITY IN THE MEDITERRANEAN

New Threats and Policy for Reducing Vulnerability

J. GANOULIS

Department of Civil Engineering
Aristotle University of Thessaloniki
Greece

Abstract: The terms environmental and human security and vulnerability are defined in this paper in a broad sense. Human security is directly related to environmental preservation (water, air, soil, ecosystems, and biodiversity). Environmental and human security have been recognized as key factors for socioeconomic growth and prosperity and sustainable development. Regulation and policy to sustain environmental and human security in European Union (EU) countries, and more specifically in the Mediterranean region, are briefly reviewed in this presentation. Examples from environmental sectors like water resources may include the EU Water Framework Directive (EU-WFD) and institutional issues for facing water quality degradation and hydrological extremes, such as floods.

1. Introduction

Nowadays the concept of *human security* may be extended from its traditional meaning of local, state, regional, and worldwide civil and military security of citizens to also embrace the idea that every human being should be able to benefit from sustainable socioeconomic development. This paradigm shift is indicated schematically in Figure 1.

Environmental security has been recognized as the key factor for social security, economic growth, and prosperity. Human security can therefore be seen to be related to environmental preservation (water, air, soil, ecosystems, and biodiversity) and to socioeconomic stability and sustainable development. The concept of sustainable development and integrated management of environmental resources was first mentioned in Stockholm in 1972, during the United Nations World Conference, and then at the Rio Summit in 1992 with Agenda 21.

I. Linkov et al. (eds.), Managing Critical Infrastructure Risks, 51–61.
© 2007 *Springer*.

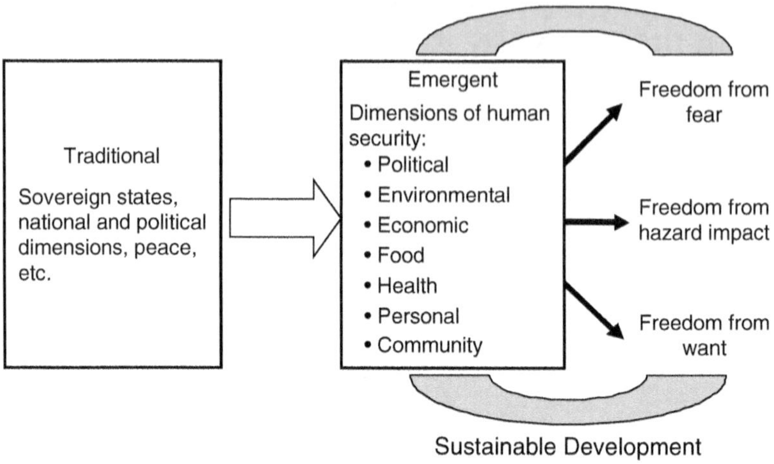

Figure 1. The paradigm shift for the concept of human security [9].

The term *vulnerability*—as applied to humans, ecosystems, or any environmental system—denotes the susceptibility of the system to be damaged under risk. It may be considered a performance index of the system, indicating the possible degree of a system's damage or the severity of consequences, due to an incident such as a flood or a drought [4, 5].

Regulation and policy to sustain environmental and human security in EU countries are briefly reviewed in this presentation. Examples from environmental sectors like water resources may include the EU-WFD [2] and institutional issues for facing hydrological extremes, such as floods.

Examples of recent floods and environmental incidents in Central and Southern Europe and elsewhere [3] illustrate that residual vulnerabilities and risks are always present and that an integrated risk-based management framework is needed in order to prevent and alleviate negative consequences to human society and the environment.

2. Main Threats to Environmental Security in the Mediterranean

The Mediterranean region is facing demographic, social, cultural, economic, and environmental changes. In the last four decades, rapid increase of population, ambitious agricultural policies in several countries, increase of economic activities, as well as unplanned utilization and mismanagement, have all led to natural resources being extensively depleted and even overexploited in many parts of the region. With withdrawal exceeding the internally renewable water resources, the resulting water scarcity is rapidly becoming a major concern in most countries of the Mediterranean. The varying climate in the north, south, and east of the region creates different conditions for water resources availability. Water resources are relatively plentiful in the countries in the north and scarce in the south and east. Thus today, of the 420 million Mediterranean people (United Nations estimate) more than 160 million live in countries with less than 1,000 m^3 water/year per inhabitant (annual average). Of these 160 million persons, 30 million in the Palestinian Territories, Israel, Jordan, Libya, Malta, and Tunisia live below the line of absolute water-poverty of 500 m^3/year per inhabitant [1,6]. See Figure 2.

In addition to their overexploitation, water resources in the region are being threatened and polluted by numerous point and nonpoint sources of pollution generated from anthropogenic activities, such as agricultural (e.g., saline and contaminated irrigation return flows with pesticides or fertilizers), industrial (e.g., discharge of hazardous and toxic industrial wastes, underground storage tanks, or surface and deep disposal of oil and gas brines), and domestic activities (e.g., discharge of inadequately treated domestic wastewater or municipal landfills).

2.1. AGRICULTURE

Agriculture is by far the most important water use activity in the Mediterranean region, and is also probably the least efficient sector in water use. Agricultural activities not only threaten the availability (quantity) but also the quality of

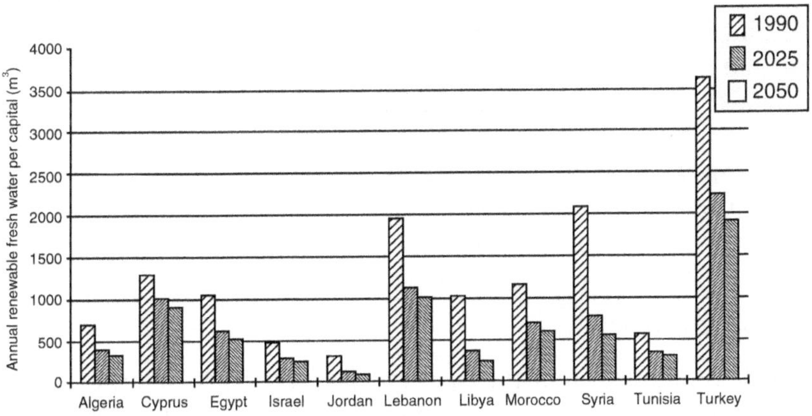

Figure 2. Available fresh water per capita in southern Mediterranean countries, 1990–2050 [10][1].

[1] 1700 m3 = periodic water stress; 1000 m3 = chronic water stress; 500 m3 = absolute water stress.

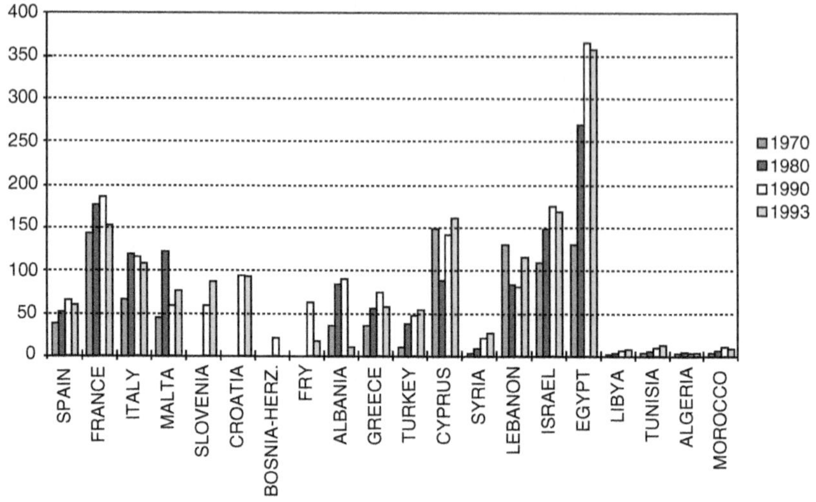

Figure 3. Fertilizer consumption in the Mediterranean countries from 1987–1993 (kg/ha) [12].

groundwater due to the extensive use of fertilizers, pesticides and release of olive-oil-mill wastes (Figure 3).

In spite of the rapid expansion of irrigated areas, irrigation and drainage have undergone little technological change over this period. Most irrigation systems in the Mediterranean countries perform far below their potential, mainly as a result of inadequate technologies, management practices, and policies.

Average losses of irrigation water in the Mediterranean are extremely high (55%), and they are divided between farm distribution (15%), field application (25%), and irrigation system losses (15%). Only about 45% of water diverted or extracted for irrigation actually reaches the crops. Losses vary widely, with those from the conveyance system varying between 5 and 50%.

Such low levels of efficiency in agricultural water use and the unsatisfactory features of irrigated agriculture in the region are undoubtedly the result of water resource mismanagement.

2.2. URBANIZATION

The resident population of the riparian states of the Mediterranean was 246 million in 1960 and is currently about 420 million. "Blue Plan" estimates that depending on the development scenarios applied, this figure will rise to 520–570 million in the year 2030, is expected to reach approximately 600 million in the year 2050, and possibly as much as 700 million at the end of the 21st century. The average annual population growth rate in the southern countries of the Mediterranean is estimated at 3% (Figure 4).

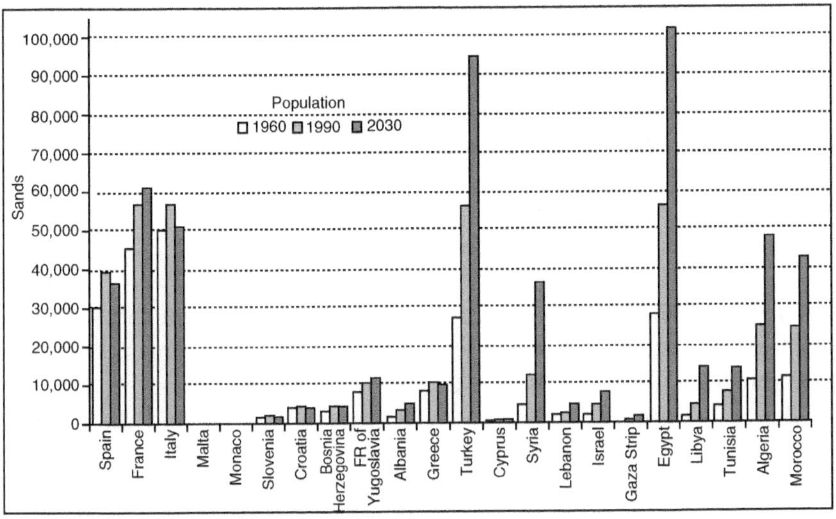

Figure 4. Population increase in the different Mediterranean countries [11].

2.3. TOURISM

The attractive climate and the historical and archaeological significance of the area make the Mediterranean countries the greatest tourist destination in the world, with 176 million visits in 1996 increasing at a rate of 2–3% per year, expected to be 250–275 million by the year 2010 and 290–355 million in the year 2025.

Tourism activity peaks in summer, coinciding with the time when natural water availability is at its lowest.

In certain areas and at certain times of the year the population can increase two, three, or even 10 or more times. This increase in population brings about a peak in water demand for domestic use.

Growing demand for drinking water in the localities that receive visitors is not the only effect of tourism. Tourism also involves services and leisure activities that use water extensively, resulting in the construction of huge distribution and purification facilities.

2.4. INDUSTRY

There is a large range of different industrial activities (from mining to manufactured products) scattered all around the Mediterranean basin, and a number of hotspots are concentrated mainly in the north, where there are heavy industr complexes. Discharges of contaminants from these industries pose a threat to water resources, especially in the area of the hotspots.

The impacts of industry on water resources can be direct or indirect. Direct impacts deriving from industrial effluents involve pollution problems at the site level that contribute to the creation of hotspots. Indirect impacts are related to the location of industries, ultimately leading to a concentration of activities and urban development in specific regions.

3. EU Environmental Regulation and the Water Framework Directive

EU environmental regulation aims at coordinating different measures taken at Community level to tackle particular environmental problems in order to meet established objectives. Key examples of such regulation are the Urban Waste Water Treatment Directive, the Nitrates Directive and the Integrated Pollution Prevention and Control Directive.

In 2000, the EU issued the Water Framework Directive (WFD) in order to ensure an analysis of the state of water bodies and "a review of the impact of human activity on the status of surface waters and on groundwater." The analysis and review are to be conducted so as to determine how far from the objectives each body of water is [2].

The purpose of the WFD is to establish a framework for the protection of inland surface waters, transitional waters, coastal waters and groundwater which:

1. Prevents further deterioration and protects and enhances the status of aquatic ecosystems.
2. Promotes sustainable water use based on a long-term protection of available water resources.
3. Aims at enhanced protection and improvement of the aquatic environment.
4. Ensures the progressive reduction of pollution of groundwater and prevents its further pollution.
5. Contributes to mitigating the effects of floods and droughts.

Key elements of the WFD include:

1. Technical considerations: monitoring, river basin planning, and management.
2. Institutional: adopt the river basin as a single system for water management.
3. Environmental: water quality and ecosystems.
4. Water economics.
5. Public participation.

3.1. MONITORING

For many years there have been two different approaches dividing European water quality monitoring practice:

- Control pollution sources through the application of available technologies.
- Focus on quality status of receiving environment.

There are potential shortcomings when only one of these approaches is applied. Source controls do not take into account the cumulative toxic effects of contaminants from a number of different sources of pollution. The diffuse impacts cannot be estimated. Quality standards applied to water bodies can underestimate the effects of particular substances on the ecosystem, due to lack of scientific knowledge regarding the final outcome of substances in the environment. This approach may also lead to gradual degradation of a water body, if its initial state was better than standard.

3.2. RIVER BASIN PLANNING AND MANAGEMENT

The WFD requires that River Basin Management Plans (RBMPs) are produced for each River Basin District (RBD) by 2009. These will be strategic management documents, developed via the river basin planning process, which will integrate the management of the water and land environment.

Preparation will involve a process of analysis, monitoring, objective setting, and consideration of the measures to maintain or improve water status. RBMPs will have a number of functions, but are primarily intended to:

- Establish a strategic plan for the long-term management of the RBD.
- Set out objectives for water bodies and state in broad terms what measures are planned to meet these objectives.
- Act as the main reporting mechanism to the European Commission.

3.3. RIVER BASIN DISTRICTS

Integrated water management within the WFD is based on RBDs. For each RBD there is a statutory requirement to produce and regularly review a RBMP. When the RBMPs have been produced, the river basins should be mapped and the quality of the water assessed.

3.4. ENVIRONMENTAL

Under the WFD, environmental monitoring programs are required and specific objectives for water quality are set up. The WFD operates using a cyclical management process. This process begins by identifying water bodies in each RBD and describing their natural characteristics. The second stage is to assess the pressures and impacts on the water environment. This assessment identifies those water bodies that are unlikely to achieve the environmental objectives set out in the Directive by 2015. This process is known as river basin characterization.

3.5. WATER ECONOMICS

The Directive calls for the application of economic principles (e.g., the recovery of the costs of water services and the polluter pays principle), approaches, and tools (e.g., cost effectiveness analysis), and for the consideration of economic instruments (e.g., water pricing) for achieving its environmental objective in the most effective manner; i.e., good water status for all waters. Although the different elements of the economic analysis appear in various parts of the WFD text, these should be well integrated in the policy decision and management cycle in order to aid decision making.

3.6. PUBLIC PARTICIPATION

The WFD recognizes the value and importance of involving all those with an interest in the water and land environment in how the WFD is put into practice. In certain areas (e.g., the development of RBMPs), stakeholder involvement is an inherent part of the Directive. In order to develop a coherent and consistent approach to information provision and consultation involvement it is important to provide:

- A framework for stakeholder engagement
- A communication strategy on how to engage the public
- A platform for consultation and involvement

4. Flood Management

Integrated flood risk management (IFRM) can mean very different things depending on the approach; e.g., engineering, social, or institutional. It is recommended that IFRM be defined as a multidimensional and multidisciplinary activity, which takes into account technical, institutional, economic, social, and environmental aspects of flood assessment, prevention, mitigation, and control, while promoting a more holistic view of the whole spectrum of human security and vulnerability under risk of flooding. With IFRM, the river basin is considered as a whole, with downstream/upstream solidarity. As part of integrated water management, IFRM contributes to rationalization of the use of river basin capacities and unifies the social, economical, hydrological, and environmental points of view in a global perspective. These considerations imply good communication and coordination between all the river basins' actors, perfect transparency, and dissemination of information to all the stakeholders, as well as public participation. The main objectives of IFRM are:

- Protection of human implantations and interests: reduction of flood damages to "acceptable" levels ensure the sustainability of human settlements and activities.
- Restoration of fluvial law, ecosystems, and water cycle: besides the pure environmental aspect, natural mechanisms and cycle rehabilitation are also guarantees of sustainability for human societies. They contribute to flood mitigation and to providing healthy drinking water.
- Promotion of risk culture: it should be understood that total protection is a myth, so that preventive measures can be seen to be necessary. This is a switch from dominating the risk to living with it.
- Promotion of basin wide solidarity and actions: mobilization of all stakeholders (water agencies, municipalities, inhabitants, companies), and impact studies for any initiatives.
- Preparation against extreme events partly due to climate change: a very long-term validity of IFRM actions is expected, even if climate change amplifies flood (and drought) scales.

In order to meet these objectives, a framework for IFRM (Figure 5) was proposed by Plate and Merz [8], among others. In contrast to other natural hazards like earthquakes, in IFRM it is possible to independently control both the load that is represented by the flood and the resistance of the endangered assets.

The four main elements shown in Figure 5 and described by Nachtnebel [2003] are:

1. Risk assessment, which includes the analysis of various failure modes together with the evaluation of the consequences in case of a given failure.

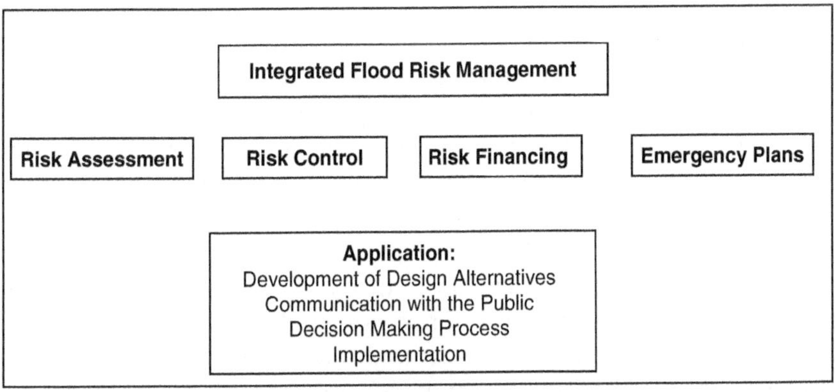

Figure 5. Main elements of integrated flood management.

2. Risk control, which includes risk prevention and risk mitigation measures. The first term refers to actions (either structural or nonstructural) to reduce the failure probability by reducing the flood peak. Nonstructural measures may include preservation of inundation areas, increase of infiltration rates by appropriate land use, and establishing river corridors by buying land along the river banks. Structural and nonstructural mitigation measures have to be considered and aim to reduce the vulnerability of the system by imposing regulations on land use and land development and enforcing technical regulations for any construction works in floodplains.
3. Risk financing has two aspects: risk acceptance by the people concerned and the transferal of risk to a broader community, either by agreements within different groups of the society or by an insurance mechanism.
4. Emergency plans. Due to the fact that some uncertainty will always remain in the system about the time and magnitude of an extreme event, precautionary measures have to be developed to be prepared for emergencies. These include the development of information systems, warning systems in case of emergency, evacuation plans, and response actions to efficiently avoid secondary losses.

The elements described above are mainly elaborated at the expert level, but the involvement of the public concerned is also needed; otherwise the measures will not be implemented or will not work in case of emergency. Public involvement in the selection of alternative strategies and in the communication of risk is indispensable.

5. Conclusions

The Mediterranean region currently faces many socioeconomic and environmental challenges due to various demographic, agricultural, industrial, climatic, and flood-related pressures. In order to deal with these challenges and risks, it is important to implement a consistent system of regulation and develop regional cooperation frameworks between riparian countries.

In this perspective, the EU environmental regulation is a very useful tool. It provides an integrated framework in order to take into account technical, environmental, institutional, economic, and public participation issues for protecting both water surface and groundwater resources.

Integration of all these dimensions is essential in order to ensure human and environmental security in the region.

6. References

1. Bemblidia M, Margat J, Vallée D, Glass B. Water in the Mediterranean region: blue plan for the Mediterranean. Sophia-Antipolis: Regional Activity Centre; 1996.
2. Directive 2000/60/EC of the European Parliament and of the Council of 23 October 2000 establishing a framework for community action in the field of water policy. Off. J. Eur. Communities. L 327, 22.12.2000.
3. Ganoulis J. Risk-based floodplain management: a case study from Greece. Int. J. River Basin Management. 2003; 1(1):41-47.
4. Ganoulis J. Integrated risk analysis for sustainable water resources management. In: Linkov I, Ramadan AB, editors. Comparative risk assessment and environmental decision making. Dordrecht: Kluwer Academic Press; 2004.
5. Ganoulis J. Risk analysis of water pollution: probabilities and fuzzy sets. Oxford: Wiley-VCH; 1994.
6. Margat J, Vallée D. Water resources and uses in the Mediterranean countries; figures and facts: the Mediterranean in figures–blue plan for the Mediterranean. Sophia-Antipolis: Regional Activity Centre; 2000.
7. Nachtnebel H.P. New strategies for flood risk management after the catastrophic flood in 2002 in Europe: integrated disaster risk management: coping with regional vulnerability. Kyoto: World Water Forum; 2003.
8. Plate E, Merz B. Naturkatastrophen. E. Schweizerbart Vlg. ISBN 3-510-65195-2 (in German). 2001.
9. Renaud F. Human and environmental security in the context of the CABRI-Volga project. Proc. Nizhny Novgorod: Great Rivers Forum (forthcoming); 2005.
10. UN-PD. 1994.
11. UN. World population prospect. Blue plan databases. 1994 revision.
12. World Bank. Social indicator of development. 1996.

PART 2

PORT CRITICAL INFRASTRUCTURE AND MANAGEMENT FRAMEWORKS

CHAPTER 4

PROTECTION OF HAZARDOUS INSTALLATIONS AND CRITICAL INFRASTRUCTURES - COMPLEMENTARITY OF SAFETY AND SECURITY APPROACHES

Application of the ARAMIS Methodology

F. FONTAINE, B. DEBRAY, O. SALVI

Institut National de l'Environement Industriel et des Risques (INERIS)
Accidental Risks Division
BP n°2—Parc Alata
60550 Verneuil en Halatte France

Abstract: Europe has recently faced several major industrial accidents generated by various causes. They have led to a lot of casualties and major damage to the environment (e.g., Enschede 2000, Toulouse 2001, and Ath 2004). Moreover, people have already stressed the influence of terrorism in the emergence of new risks. The geopolitical context has brought about risks which cannot be completely avoided. This is demonstrated by many events, including, for example, the sarin attack, anthrax letters, USS Cole and Limburg attacks, drinking water contamination, and actions on industrial sites. Regarding SEVESO establishments and similar critical infrastructures like ports and marshalling yards, these concerns are not included in the safety report required by the authorities. In this paper, the authors establish a brief inventory of the available methodologies and tools required to elaborate a threat assessment. Then they formulate some suggestions for integrating security aspects in a risk assessment, to globalize prevention and response approaches in case of accident or intentional acts on chemical sites. An example of the adaptation of the ARAMIS methodology is explained in the last part.

1. Introduction

During the last 10 years, Europe has faced several major industrial accidents generated by various causes, e.g.:

- Enschede 2000 – explosion of firework storage
- Toulouse 2001 – explosion of ammonium nitrate
- Ath 2004 – rupture, explosion, and fire of a gas pipeline

I. Linkov et al. (eds.), Managing Critical Infrastructure Risks, 65–78.
© 2007 *Springer*.

These incidents caused a lot of casualties and major damage to the environment, forced international authorities to examine these phenomena, and, moreover, led the European Commission to adopt legislation to prevent such events.

Two consecutive European Commission directives, in 1982 [1] and 1996 [2], gave guidance to member states to develop a prevention policy for major technological accidents. This guidance is transposed in each member state's regulations and provides global instructions for good risk management. The approach is summarized in the four-step scheme shown in Figure 1.

Moreover, in the third millennium people have already stressed the influence of terrorism in the emergence of new risks. The geopolitical context has brought about risks that cannot be completely avoided. This is demonstrated by many events, including the sarin attack, anthrax letters, USS Cole and Limburg attacks, drinking water contamination, and actions on industrial sites.

These threats are not taken into account in the safety report required for SEVESO establishment. In addition, the SEVESO Directive doesn't include marshalling yards, port areas, or multimodal transport platforms in its scope.

These observations lead to the proposal of a new approach that completes the safety report with a threat assessment, with the objective of developing a better understanding of the main natural and anthropic (e.g., manmade or terrorist) risks that the European Union faces.

2. Definition of the Work Field

This paper will focus on hazardous establishments, SEVESO plants, and areas dedicated to the transport of dangerous goods (e.g., marshalling yards and port areas).

In terms of safety, the majority of these are covered by regulations related to the control of major industrial hazards at a European or international level (e.g., SEVESO Directive, ADR, or IMDG code).

Figure 1. Major hazards control approach.

In terms of security, these establishments could be divided into several categories according to the existing regulatory framework (Figure 2):

- Some are considered critical infrastructure at the national level by each member state and regulations are applied to prevent the disruption of service (e.g., supply chain, electricity, or potable water). In this field, the European Commission affirms a need for agreement on criteria for the definition of critical infrastructure [3].
- Some are included in sensitive activities sectors in terms of economic intelligence.
- Regarding the transport of dangerous goods, international legislators, following 9/11, considered it necessary to develop security measures against possible terrorist acts; this will result in regulation changes (e.g., ISPS Code, ADR Chapter 1.10).

3. Existing Approaches and Methodologies for Threat Analysis for the Protection of Hazardous Installations against Deliberate Acts

In the U.S., the September 11 events have radically changed the risk level assessment for chemical facilities as it relates to intentional acts. The security of hazardous sites has now become a major concern to chemical process industries (e.g., oil refineries and petrochemical plants). Plants that handle hazardous chemicals are prime targets for terrorists and criminals [4].

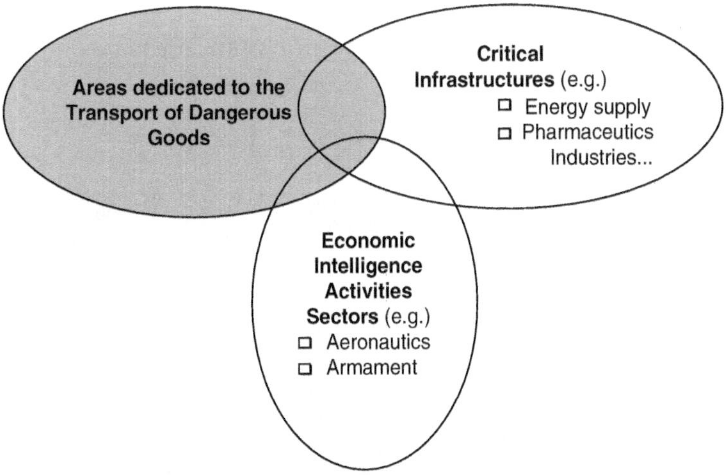

Figure 2. Activities related to security aspects.

American professional organizations [5,6], associations [7], and administrations [8] have developed guidelines to elaborate threat analysis methods for the protection of chemical sites from terrorism attacks.

The main methodologies are based on a four-step approach [4]:

- Threat analysis: Identify the sources and types of threats and their likelihood. Identify adversaries and conduct preliminary investigation of their intentions, capabilities, and prior history, if any.
- Vulnerability analysis: Identify ways in which the credible threats identified in threat analysis could be realized.
- Security countermeasures: Enhance the security level of chemical facilities in order to intercept and neutralize intentional acts.
- Mitigation and emergency response: Manage the emergency situations related to a terrorist attack and mitigate its consequences.

Moreover, European companies elaborated guidelines to apply the new international regulations related to transport of dangerous goods [9,10,11].

4. Synergy and Complementarity of Safety and Security Approaches

It could be interesting to elaborate a global approach for the safety and security aspects of chemical facilities with a European point of view. This new method would be developed on the basis of the SEVESO approach and would complete the risk assessment by a threat analysis. This step could be illustrated by Figure 3.

This approach will complete the major hazard control approach with:

1. A threat assessment phase.
2. A second phase aiming at defining relevant countermeasures in order to mitigate the threats.

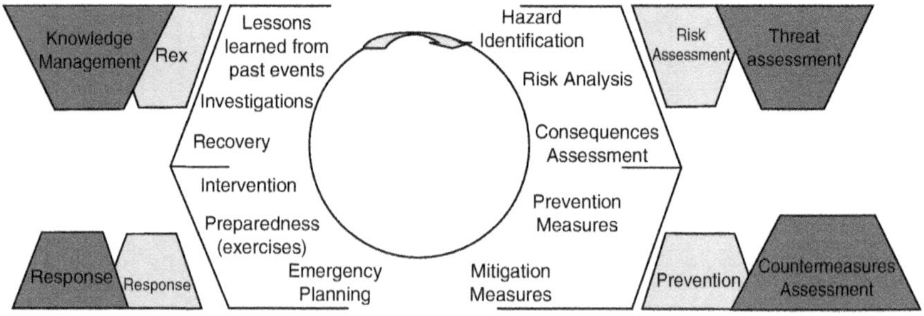

Figure 3. Major hazards control approach completed with threat step.

3. A third phase, where the relevance of the tools, means, and procedures used in an accidental situation, in terms of emergency response, will be examined.
4. A last phase will organize the knowledge of investigation issues and lessons learned from past terrorist or deliberate events.

4.1. THREAT ASSESSMENT

First of all, it is necessary to consider that a chemical site could be a target: terrorist assault against these plants could result in the release of hazardous materials [12] and cause fire, explosion, and toxic effects.

Moreover, chemical facilities could also be considered a potential resource for terrorists [13], where they could find hazardous materials (e.g., toxic substances) that could be disseminated in a public place (e.g., subway or railway station) or spread after a deliberate loss of containment.

A better identification of the hazard potential inside the facility is indispensable and is the basis of the risk assessment. It is essential to establish a precise material inventory including the intermediate and byproducts to identify exactly the potential of hazardous release from the plant. This inventory should take into account fixed installations as well as mobile storage (e.g., trucks and rail tanks parked on site).

Comparable to the risk analysis step in the classical risk assessment for hazardous installations, the second step of a threat assessment [14] consists of threat identification; in particular, a modus operandi analysis. Threats could be divided into two categories [15]:

- Internal threats, which include vandalism and sabotage by malevolent employees.
- External threats caused by terrorists.

The last step of this phase consists of assessing the consequences. In a first hypothesis, the safety barriers will not be taken into account to simulate the effects of an attack on industrial installations. They are considered unavailable. Consequently, the consequences of an attack will be more serious than in an accidental situation.

4.2. COUNTERMEASURES ASSESSMENT

Comparable to the definition of prevention and mitigation measures in the classical risk assessment approach, this phase consists of assessing the existing safety barriers in order to reduce or eliminate the identified threats.

If the safety barrier appears not to be adapted to an intentional context, a specific study has to be launched to define relevant security barriers in terms of technical and organizational measures.

The Centre for Chemical Process Safety precisely detailed the phase in specific guidelines [7].

4.3. RESPONSE

The SEVESO Directive[1] fixes rules and responsibilities for establishing emergency plans with the objectives of:

- Containing and controlling incidents so as to minimize the effects, and to limit damage to man, the environment, and property.
- Implementing the measures necessary to protect man and the environment from the effects of major accidents.
- Communicating the necessary information to the public and to the services or authorities concerned in the area.
- Providing for the recovery and cleanup of the environment following a major accident.

The operator has to draw up an internal emergency plan and the authorities are in charge of the offsite emergency plans. These plans, based on accidental scenarios, are regularly tested (every three years).

In the framework of deliberate acts, these contingency measures have to be completed by intentional hypothesis.

For example, in France, two types of plans cohabit without real interaction:

- Emergency preparedness plans, "*Plan Particulier d'Intervention,*" to manage accidental situations related to SEVESO establishments, nuclear plants, and gas underground storage.
- CBRN emergency plans (PIRATOX - BIOTOX) to manage emergency situations related to terrorist acts.

The last step consists of organizing the recovery after an event to come back to a nominal situation.

4.4. KNOWLEDGE MANAGEMENT

Regarding attacks on chemical facilities or industrial sites, it appears difficult to get information on investigations after the event, in particular on the modus operandi used by the terrorist.

[1] Council Directive 96/82 - Article 11.

In addition, there is no detailed database of lessons learned from past intentional acts on chemical plants or transport of dangerous goods.

This information is essential to establish a threat analysis.

Furthermore, it seems also necessary to consider Intelligence Services information as data inputs in order to select the different industrial sectors most concerned by terrorist acts; this watch has to be regularly established to update the threat analysis.

4.5. FIRST CONCLUSION

To conclude, regarding some of the SEVESO establishments, it seems necessary to complete the risk assessment (safety report) with a threat assessment.

Several methodologies and tools exist, in particular in the U.S., but they have to be adapted in a European context. Some attempts have been made in Germany, Austria, and in The Netherlands. In addition, it could be interesting to examine the methodologies and tools used in nuclear activities [16].

With this objective, it seems useful to develop a global approach which takes into account safety concerns and security concerns and globalizes risk and threat analysis and response (Figure 4).

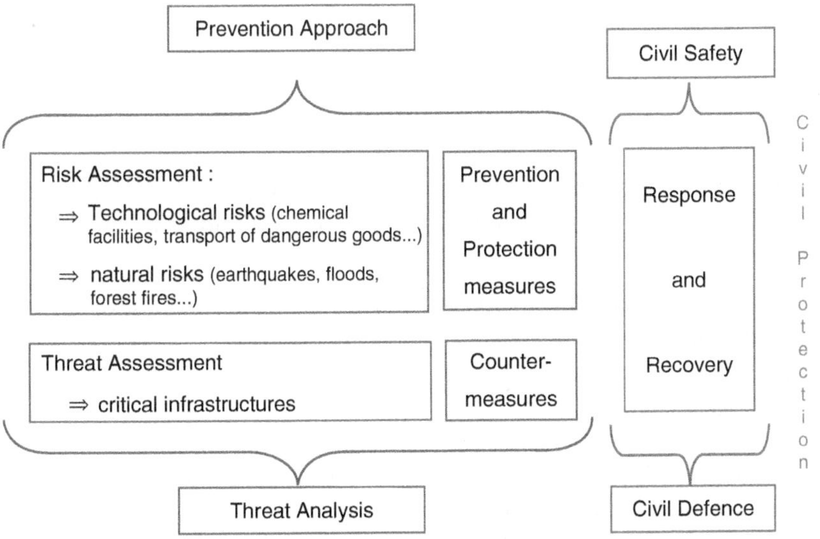

Figure 4. Global approach in risks and threat including prevention aspects and civil protection measures.

5. Application of the ARAMIS Methodology

A terrorism attack on a hazardous plant can have two types of consequences:

- Direct consequences of the attack, which can be economic, such as loss of production capacity due to consequences on the availability of a productive function; or injuries and fatalities among the workers and neighboring populations if the attack uses explosive or toxic substances.
- Indirect consequences on people and workers and/or buildings inside or outside the plant caused by the hazardous phenomena resulting from the release or the transformation of the hazardous substances present in the plant.

The present paper mostly concentrates on the second type of consequences, considered in the light of the ARAMIS methodology.

The ARAMIS methodology was developed in the fifth framework program of the European Commission with the objective of meeting the specific requirements of the SEVESO II Directive. It offers an alternative to purely deterministic and probabilistic approaches to carry out risk assessment for hazardous establishments.

ARAMIS is divided into six major steps (Figure 5):

- Identification of major accident hazards (MIMAH).
- Identification of the safety barriers and assessment of their performance.
- Evaluation of safety management efficiency to barrier reliability.
- Identification of reference accident scenarios (MIRAS).
- Assessment and mapping of the risk severity of reference scenarios.
- Evaluation and mapping of the vulnerability of the plant's surroundings.

When assessing the risks resulting from a terrorist attack, the two main questions are what could be the consequences of the release of the hazard potential due to the attack, and how vulnerable is the plant to such an attack. In other words:

- Identify equipment that is likely to be at the origin of a severely hazardous phenomenon.
- Describe how this release could take place.
- Assess what consequences it would have.
- Identify the safety systems that could protect the equipment against the attack or mitigate its consequences.
- Assess the reliability of these safety systems including their capacity to resist the attack itself.

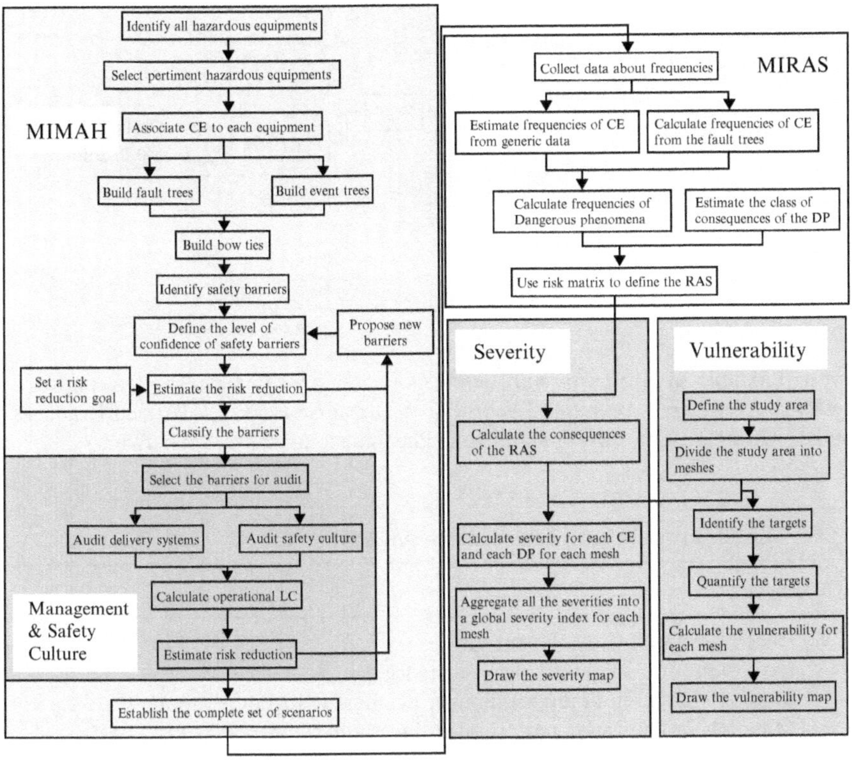

Figure 5. Main steps of the ARAMIS methodology.

ARAMIS provides useful tools for all these steps with a methodology for MIMAH, severity and vulnerability indexes, and a methodology for the identification of safety barriers.

5.1. IDENTIFICATION OF MAJOR ACCIDENT HAZARDS

MIMAH [17,18] is the method for the identification of major accident hazards. It is based mainly on the use of bow-tie diagrams (Figure 6), composed of a fault tree and an event tree. The major input of ARAMIS was to define a precise bow-tie structure and to define precisely and exhaustively a generic list of equipment, potential critical events, and their consequences. The critical events were defined to be either losses of containment for fluids or losses of physical integrity for solids. The complete list contains 12 critical events including breach, collapse, and explosion (Table 1).

From a description of the establishment including the chemical substances used, produced, or stored, the methodology allows for the selection

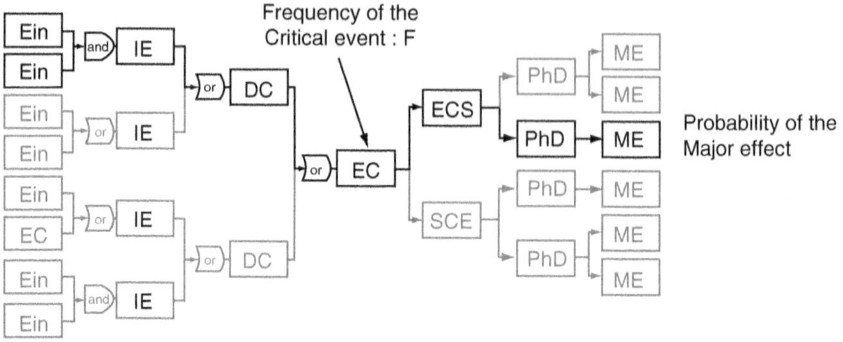

Figure 6. Example of a bow tie and description of a specific accidental scenario (Ein = Undesirable Event, IE = Initiating Event, DC = Direct Cause, EC = Critical Event, ECS = Secondary Critical Event, PhD = Dangerous Phenomenon, ME = Major Event).

TABLE 1. List of critical events considered in ARAMIS

Critical events	
CE1	Decomposition (of a solid substance)
CE2	Explosion (of a solid explosive substance)
CE3	Materials set in motion (entrainment by air of a dust or a vapor)
CE4	Materials set in motion (entrainment by a liquid)
CE5	Start of fire (LPI) (interaction between an oxidizing substance and a combustible)
CE6	Breach on the shell in vapor phase
CE7	Breach on the shell in liquid phase
CE8	Leak from liquid pipe
CE9	Leak from gas pipe
CE10	Catastrophic rupture
CE11	Vessel collapse
CE12	Collapse of the roof (of an atmospheric storage)

of hazardous equipment pieces to be considered in the risk analysis. It is then possible from MIMAH to list all the critical events likely to occur in the establishment. For each of these critical events, MIMAH allows the identification of all consequences in terms of secondary events and dangerous phenomena.

At this stage, MIMAH provides the user with a set of generic fault trees, which are based on the most frequently observed causes [19]. From these generic fault trees, the user will build specific fault trees that take into account the specificity of the establishment: types of process used, presence of equipment, etc.

The specific fault trees are obtained mainly by the suppression of causes and consequences which are not relevant to the context without any consideration of probability at this stage. It is important to note that both the fault and event trees are considered without safety barriers, which will be defined in the next step of the method. This has the advantage of allowing for an explicit identification of safety devices and operations.

When considering the risk of a terrorist attack, the bow ties can be used to assess consequences. The attack can be systematically introduced in the fault tree as a potential cause at the appropriate level. In fact the generic fault trees already contain a generic cause termed deliberate acts that could be the base for the introduction of a terrorist action.

The event trees are unchanged and directly provide the potential consequences of the attack.

5.2. IDENTIFYING SAFETY BARRIERS

The next step of MIMAH is the identification of safety barriers and the assessment of their efficiency. ARAMIS offers a list of generic safety functions and safety barriers for each event of the generic fault tree and generic event tree. For each event the following functions are considered:

- To avoid; i.e., to make the event impossible by—for example—changing the process or suppressing the hazard source.
- To prevent; i.e., reduce the probability of the event.
- To mitigate; i.e., to reduce the consequences of the event by reducing its intensity.
- To protect; i.e., to put obstacles between the hazard source and vulnerable elements (targets).

For each safety function, ARAMIS offers a list of safety barriers: devices or operations dedicated to the implementation of the safety function.

Whereas the safety barriers to prevent a terrorist attack will probably be specific, and could be called security barriers, it turns out that the other safety functions (to avoid, to mitigate, and to protect) can be—in principle—the same as for an accidental event. However, their efficiency can be affected by the terrorist attack itself. This aspect will have to be taken into account in the safety barrier assessment that follows. The safety barriers are characterized in terms of efficiency, response time, and reliability. This last criterion is covered in ARAMIS by the Level of Confidence (LC), directly inspired by the SIL level introduced by the IEC 61508 [20] and 61511 [21] standards. This Level of Confidence takes into account the redundancy and the common failure modes. We propose adding a criterion that would characterize the resistance to terrorist attack of a given type.

The concept of security barriers—i.e., barriers that can prevent a deliberate act from damaging the plant and thus resulting in a catastrophic event—could also be applied to scenarios where the plant is considered by terrorists as a source of dangerous substances for future intentional actions. Such scenarios are not taken into account in ARAMIS because their consequences are not local. But they could be easily added and treated by means of bow ties.

5.3. ASSESSING THE SEVERITY OF THE EVENT AND THE VULNERABILITY OF THE SURROUNDINGS

At this stage, using ARAMIS has provided a way to identify hazardous pieces of equipment, describe the consequences of a terrorist attack, identify the safety functions and safety barriers that can prevent such an event and allow for the mitigation of its consequences, and finally assess the efficiency and reliability of these safety barriers. It may be of interest now to assess the damages potentially caused by such an attack.

For that, it is necessary to characterize the intensity of the dangerous phenomena that would result from the attack and the vulnerability of the surroundings (e.g., toxic release or explosion). ARAMIS did not produce any new model for the estimation of the consequences of an accident, but it proposed the definition of two indexes used to represent these consequences on a map.

The first index, called the severity index, combines the frequency of accidents, the probability of incurring a dangerous phenomenon after a loss of containment, and a normalized representation of the intensity. To adapt such an index to the risk assessment of terrorist attacks, it would be necessary to replace the frequency of a critical phenomenon with the likelihood of a terrorist attack (taking into account the existence of security barriers). The other components of the index would remain unchanged.

The severity index is used to produce a map that is overlaid to a vulnerability index map to assess the probability of damage to the receptors (e.g., human targets, environment, or infrastructure). The vulnerability index is a linear combination of the number of vulnerable elements weighted by their relative vulnerability to specific dangerous effects. Various maps can be drawn—showing types of vulnerable elements (populations, buildings, natural environment) or types of dangerous effects (thermal, toxic, overpressure); or showing global vulnerability, integrating all potential hazardous effects and types of vulnerable target.

6. Conclusion

Recent technological accidents and the evolution of the international context following the 9/11 terrorist attack have stressed the necessity to improve and coordinate the safety report and the terrorist threat assessment. The

SEVESO II Directive offers a framework that must be supplemented to take into account deliberate acts. The parallel is made between threat assessment and more classical industrial risk assessment, showing that the adaptation of risk assessment methodologies is possible.

This conclusion is also illustrated by the example of the application of ARAMIS methodology to terrorist attack risk assessment. Most of the tools developed in the framework of the ARAMIS [22] project can be used and adapted to assess deliberate acts and their consequences, leading to the identification of security barriers aside from preventive safety barriers. A new criterion is proposed to assess the capacity of protective safety barriers to resist a terrorist attack. ARAMIS also provides tools to make a geographical representation of risks by combining a severity index, which can be adapted to take the probability of an attack into account, and a vulnerability index.

Combining safety and security approaches and concepts is probably a way to increase cost effectiveness and acceptance by the population and workers of measures that, unfortunately, turn out to be necessary to protect citizens.

7. Acknowledgements

The authors acknowledge the French Ministry of Ecology and Sustainable Development for its institutional and financial support.

Part of the work presented in this paper has been elaborated in the frame of the EU project ARAMIS "Accidental Risk Assessment Methodology for Industries," coordinated by INERIS (F).

8. References

1. Council Directive 82/501/EEC of 24 June 1982 on major accident hazards of some industrial activities (Seveso I).
2. Council Directive 96/82 of 9 December 1996 on the controlling of dangers connected with major accidents involving dangerous substances (Seveso II).
3. European Commission. Directorate General Environment report of the joint workshop on the protection of hazardous installations from intentional adversary acts. Budapest, Hungary. April 28-29 2005.
4. Bajpai S, Gupta JP. Site security for chemical process industries. Journal of Loss Prevention in the Process Industries (article in press). 2005.
5. American Chemistry Council. Site security guidelines for the US chemical industry. Washington, DC: American Chemistry Council; 2001.
6. American Petroleum Institute. Security guidelines for the petroleum industry. Washington, DC: American Petroleum Institute Publishing; 2003.
7. Centre for Chemical Process Safety. Guidelines for analysing and managing the security vulnerabilities of fixed chemical sites. 2002.
8. U.S. Department of Justice - National Institute of Justice. Method to assess the vulnerability of U.S. chemical facilities. 2002.

9. European Chemical Industry Council. Industry guidelines for the security of the transport of dangerous goods by road. 2005.
10. Union des Industries Chimiques. Guide d'élaboration du plan de sûreté relatif au transport des marchandises dangereuses à haut risque. 2005.
11. CIFMD. Guide interprofessionnel concernant le chapitre 1.10 de l'ADR, du RID et de l'ADNR – sûreté des transports terrestres des marchandises dangereuses. 2005.
12. Baybutt P. Assessing risks from threats to process plants: threats and vulnerability analysis. Process Safety Progress. 2002; 21(4):269-275.
13. Marlair G, Kordek MA. Safety and security issues relating to low capacity storage of AN-based fertilisers. Journal of Loss Prevention in the Process Industries. 2005.
14. Coster N, Hankin R. Risk assessment of antagonistic hazards for protecting process plants against terrorism, sabotage and other criminal acts. Journal of Loss Prevention in the Process Industries. 2003; 16:545-550.
15. Baybutt, P. Strategies for protecting process plants against terrorism, sabotage and other criminal acts. Homeland Defence Journal. 2003; 2:1-7.
16. International Atomic Energy Agency. The physical protection of nuclear material and nuclear facilities; international standards for physical protection revising INFCIRC/225. 1998.
17. Delvosalle C, Fiévez C, Pipart A. Deliverable D.1.C. - Report presenting the final version of the methodology for the identification of reference accident scenarios, ARAMIS Project - 5th Framework Program of the European Community, Energy, Environment and Sustainable development. Contract number: EVG1 – CT – 2001 – 00036. July 2004; Mons (Belgium). Available at http://aramis.jrc.it.
18. Delvosalle C, Fiévez C, Pipart A, Londiche H, Debray B, Hubert E. ARAMIS project: effect of safety systems on the definition of reference accident scenarios in SEVESO establishments. LP2004.
19. Debray B, Delvosalle C, Fiévez C, Pipart A, Londiche H, Hubert E. Defining safety functions and safety barriers from fault and event trees analysis of major industrial hazards. PSAM7-ESREL2004 conference. Berlin, June 2004.
20. IEC. IEC 61508: functional safety of electrical, electronic and programmable electronic safety-related systems, parts 1-7. Geneva: International Electrotechnical Commission; 1998.
21. IEC. IEC 61511: functional safety instrumented systems for the process industry sector, parts 1-3. Geneva: International Electrotechnical Commission; 2001.
22. Outcome of the ARAMIS project: accidental risk assessment methodology for industries in the framework of the SEVESO II directive - new stakes and opportunities in the control of major accident hazards in Europe: outputs from the ARAMIS project. Journal of Hazardous Materials. 2006; 130(3) 187-328.

CHAPTER 5

APPLYING RISK ASSESSMENT TO SECURE THE CONTAINERIZED SUPPLY CHAIN

M. VAN DE VOORT, H. WILLIS, D. ORTIZ, S. MARTONOSI

RAND-EUROPE
Newtonweg 1, 2333CP
Leiden, The Netherlands

Abstract: Over the past decades, container transport has been booming with an annual average growth of 9.3% increasing its modal share from 6.3% in 1980 to over 25% currently. The world's largest container ports now handle over 20 million 20-ft-equivalent units (TEUs) per annum and have shown spectacular growth over the past years, with the main pivot of growth being the Far East. The global flow of containers represents the lifeblood of modern economies, without which we would not be able to enjoy the way of life that we have. This unfortunately is exactly one of the criteria terrorists are known to apply when selecting their targets, rendering the global containerized supply chain a potentially attractive target.

However, which threats to the maritime domain are most important and how should resources be directed to reduce risks from these threats? Risk analysis (with risk being defined as a function of three components: threat, vulnerability, and consequence) provides a framework for prioritizing threats to the maritime sector. For this purpose we need to:

1. Assess the threats to determine what it is we are guarding ourselves against.
2. Assess vulnerabilities (e.g., determine the chances of damage from a successful hit).
3. Consider the consequences of attacks on and potential disruptions of supply chains.

In the context of container transport, two types of scenarios are relevant:

1. Maritime containers are misused to traffic illicit goods past our borders, especially when they involve potential nuclear detonation.
2. The freight transport system itself is targeted in order to create an extended disruption of operations at major ports.

I. Linkov et al. (eds.), Managing Critical Infrastructure Risks, 79–95.
© 2007 *Springer*.

Since only 5 to 6% of all containers handled at western ports are physically inspected, the chances of intercepting a container that carries illicit goods are small. Once a contraband-carrying container has eluded Customs checks, it travels fast, making hinterlands accessible within hours to days. These types of scenarios move up in the risk assessment ranking, particularly if nuclear weapons get into play. However, in spite of a desire to use these types of weapons, terrorists so far have shown little ability to execute large-scale unconventional attacks.

If terrorists attacked the supply chain itself, aiming to disrupt it, these disruptions could have a large economic impact. This is illustrated by the 2002 West Coast port lockout and the fall 2004 congestion at the ports of Los Angeles and Long Beach, the former estimated to have resulted in economic damage ranging from $0.5-19 billion. Another study estimates that any discovery of proposed attacks on U.S. ports leading to a shutdown of U.S. ports could result in economic damage to the tune of $58 billion.

Policy makers have taken note of these threats and enacted several measures to guard against them, such as implementing the International Ship and Port Facility Security (ISPS) code, Container Security Initiative, and Customs-Trade Partnership against Terrorism. The question however remains: have these measures made us any safer than we were? Risk analysis can also be used to assess measures on their merits. To illustrate how risk analysis can be used to assess the cost effectiveness of risk reduction resulting from a security measure, we consider a single security measure: scanning 100% of inbound containers at U.S. ports. This assessment shows that current approaches for inspecting cargo containers moving through the container shipping system are impractical and imperfect because of the large number of containers and the inherent errors (both false positives and false negatives) of inspection technologies.In addition, it demonstrates that the introduction of new technologies and innovative process solutions can overcome these problems if they result in fewer false positives.

The proposed risk assessment methodology has been applied to assess security risks to the containerized supply chain as well as the effectiveness of measures that aim to secure it. Additionally, the methodology allows the ranking of both risks and countermeasures to identify the most important and effective ones.

1. The Terrorist Threat to the Container Supply Chain

In recent years, policymakers have become increasingly concerned about the possibility of future maritime terrorist attacks. Although the historical occurrence of such attacks has been limited, concerns have nevertheless been galvanized by recognition that maritime vessels and facilities may be, in some respects, particularly vulnerable to terrorism. Some plausible maritime attacks could have very significant consequences, in the form of mass casualties, severe property damage, and attendant disruption of commerce. Since over the past decades transport including shipping has become more interwoven with companies' logistic processes, many threat scenarios to the maritime sector entail ripple effects into other sectors, which typically are much more significant than the mostly physical damages to the maritime sector incurred by an attack. For this reason, a risk analysis should address the entire supply chain and look beyond its maritime link. In this paper, we focus on the container supply chain, as this is by far the most common vehicle for transporting non-bulk cargo.

This paper presents a risk assessment methodology that can be applied to evaluate security measures that aim to secure the maritime sector (and the container supply chain in particular) against terrorist attacks. This risk assessment methodology can be applied to two types of scenarios:

1. Maritime containers are misused to traffic illicit goods past our borders such as radiological or nuclear devices.
2. The freight transport system itself is targeted to cause an extended disruption of operations at major ports.

The methodology also enables assessment of the merits of specific security measures. Applying the methodology to assess the merits of scanning all inbound containers illustrates that 100% inspection of arriving containers is not viable using current technologies. New technology, however, would improve the viability of 100% inspection.

2. Theoretic Framework: Risk Analysis Methodology

Understanding the nature of maritime terrorism risk requires an investigation of threats, vulnerabilities, and consequences associated with potential attacks, as grounded both by relevant historical data and by intelligence on the capabilities and intentions of known terrorist groups. Assessment of the risks associated with maritime terrorism can help policymakers and private firms to calibrate and prioritize security measures, prevention efforts, and mitigation plans.

In order to assess risk, the recently released National Infrastructure Protection Plan (NIPP) follows the best practices of risk analysis that are outlined in, among other places, the National Research Council in its foundational reports *Risk Assessment in the Federal Government: Managing the*

Process [10] and subsequently *Science and Judgment in Risk Assessment* [11]. These best practices require that risk assessments be:

1. Analytic, addressing threat, vulnerability and consequence preferably in a quantitative way.
2. Deliberative, as a way to incorporate values and risk perception, and make a tradeoff between financial and personal harm.
3. Practical, meaning the assumptions should be tenable and not be overly reliant on a single perspective [17].

As indicated above, risk is a function of three components: threat, vulnerability, and consequences. People and organizations represent threats when they have both the intent and capability to damage a target. The threats to a target can be measured as the probability that a specific target is attacked in a specific way during a specified period. Thus, a threat might be measured as the annual probability that a city's football stadium will be subject to attack with a radiological weapon. Vulnerability can be measured as the probability that damage occurs, given a threat. Damages could be fatalities, injuries, property damage, or other consequences; each would have its own vulnerability assessment. Consequences are the magnitude and type of damage resulting, given a successful terrorist attack. Figure 1 below illustrates how these components interact.

In the wake of September 11, 2001, good progress has been made in maturing the analysis behind security policy. The approach has moved from *consequence assessment* (using crude indicators, such as population, to approximate the consequences of terrorist events), to *vulnerability reduction* (analyzing the

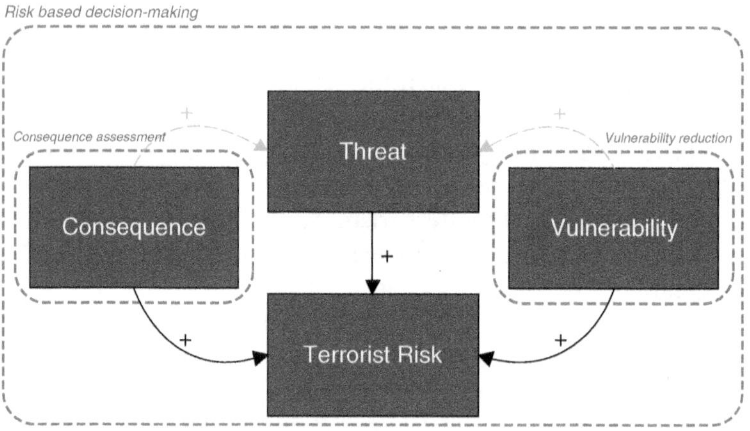

Figure 1. Maturing security policy development.

soft spots of potential targets and hardening these targets against potential threats), and, more recently, to the more comprehensive approach of risk-based decision making. The latter encompasses all of the elements mentioned before and takes these components to calculate the risk: the expected damage from a threat resulting in a successful hit. Figure 1 depicts how policymaking has evolved to include all components composing terrorist risk.

A next step in this process would be to include the cost effectiveness of measures that aim to reduce risk through cost-benefit analysis of these measures. However, in many cases the complexity of risk assessment and a lack of data to determine its components hinder transcendence to this level.

The above-mentioned NIPP reflects current progression in security policy development by defining a comprehensive approach to risk assessment across sectors that affect the U.S. economy [17], several of which are dependent on the container supply chain.

3. Risk Analysis Applied to the Global Container Supply Chain

Attacks to the supply chain aim to exploit weaknesses in the defenses of the supply chain or decrease supply chain performance. Therefore, we believe supply-chain performance and security are distinct but interconnected, meaning that all aspects of supply-chain performance must be considered when assessing security measures.

In order to assess terrorist risk to and posed through supply chains, we first need to understand how supply chains function. Second, we need to assess how we can measure supply chain performance, and third how this performance is affected by attacks. Finally, these effects on supply chain performance need to be monetized in order to assess the risks and balance them with the cost of mitigation measures.

Therefore, we describe the supply chain system and map the relations between the stakeholders within it. Typically when thinking of supply chains, an image of a physical structure, a network of vessels, port facilities, rail cars, trucks, and containers that transport goods in discrete units around the world comes to mind. However, that view pertains only to the physical components of a system that includes the cargo, information, and financial flows required for it to operate.

In fact, the supply chain can be viewed as three interdependent and interacting networks or layers, as shown in Figure 2, below:

1. A physical logistics system for transporting goods.
2. A transaction-based system that procures and distributes goods and that is driven primarily by information flows.

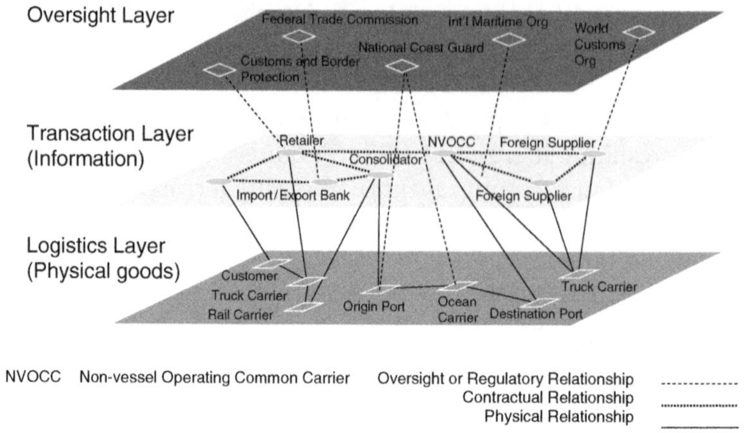

Figure 2. Actors and their interactions within supply chains.

3. An oversight system that implements and enforces rules of behavior within and among the subsystems through standards, fines, and duties.

This makes for a complex global system comprising a large number of stakeholders in and across the different layers whose many transactions consist of information flows that accompany physical flows, and in which both information and goods change hands many times between origin and destination.

To be able to assess the impacts of attacks to the supply chain, we now look for ways to describe the performance of the supply chain. Supply chain performance is composed of the levers available to improve system performance, which coincide with the qualities of supply chains valued by the stakeholders. Supply chain performance can be measured through five inherent capabilities:

- Efficiency, the system's core capability, is measured in terms of speed, cost, and volume of shipments.
- Shipment reliability ensures that goods arrive within a specified delivery window with a minimum of loss from theft and accident.
- Shipment transparency is the ability to know what is being moved through the system and is needed to ensure that cargo is legitimately represented to authorities and is legal for transport.
- Fault tolerance relates to the system's ability to respond to disruptions and failures of isolated components without bringing the entire system to a grinding halt.
- Resilience reflects the system's ability to return to normal operating conditions quickly after disruption of service.

Since each threat will differently affect supply chain performance, we need to assess the risk flowing from each individual type of threat by assessing vulnerability and consequence for each threat before we are able to monetize these risks.

3.1. THREAT ANALYSIS

Global commerce relies on the predictable performance of the container supply chain. A significant disruption at a critical facility, such as a channel, port, or exchange point, could disable for some time all business that relies on goods shipped through it resulting in a large economic impact. A recent study estimated that disruptions to the functioning of a port on the U.S. West Coast could result in economic damage ranging from $1-2 billion per day [9,12]. Other studies [1] estimate that the discovery of proposed attacks on U.S. ports, leading to a shutdown of U.S. ports, could result in economic damage to the tune of $58 billion. Terrorists could also use the container supply chain to deliver a chemical, biological, radiological, and nuclear weapon through a Trojan horse–style attack.

The scenarios listed in Table 1, below, represent a sample of maritime threat scenarios (not necessarily related to supply chains) that have been voiced in literature. The scenarios demonstrate the diversity of threats, vulnerabilities, and potential consequences of terrorist attack on maritime targets.

3.2. VULNERABILITY

Supply chains are open systems and thus inherently vulnerable to attacks. Theft and smuggling are indicative of this vulnerability. As can be seen in Figure 1, the

TABLE 1. Threat scenarios to the maritime sector [7,2,5]

Threat scenario	Description
Trojan horse	Use commercial cargo containers to smuggle terrorists, nuclear, chemical, or biological weapons, components thereof, or other dangerous materials across our borders.
Detonate fuel tanker	Attack a large ship carrying a volatile fuel (such as liquefied natural gas) and detonate the fuel to cause a massive in-port explosion.
Attack of oil tanker	Attack an oil tanker in a port or at an offshore discharge facility to disrupt the world oil trade and cause large-scale environmental damage.
Seize or attack ferries	Seize control of a ferry (which can carry hundreds of passengers) or a cruise ship (which can carry more than 3,000 passengers) and threaten the deaths of the passengers if a demand is not met.

(continued)

TABLE 1. (continued)

Threat scenario	Description
Sinking a ship in a port of channel	A ship is sunk or disabled in a narrow channel or port entrance in an effort to disrupt freight transportation. There are many scenarios for how this might be attempted including use of an onboard bomb, a USS Cole and M/V Limburg-style attack, or by having a diver place an explosive device on the hull of the ship.
Hijacking a ship	Terrorists board and commandeer a ship with the intent of ramming it into a facility, bridge, or other structure to cause potential freight disruptions, attract media attention, and kill and injure civilians.
Detonating a nuclear bomb	A container is used to transport a nuclear device into a port or near a city where it is detonated upon arrival. The container may arrive at its destination, where it is detonated on a ship or via truck or rail.
Detonating a dirty bomb	A container is used to transport a radiological device into a port or near a city where it is detonated upon arrival. The container may arrive at its destination, where it is detonated on an inbound ship, or via an outbound truck or train. The amount of high explosives in this scenario is relatively small, and the consequences are due to the radiological contamination and response.
Detonating a conventional bomb	A container is used to transport a high-explosive device into a port or near a city where it is detonated upon arrival. The container may arrive at its destination, where it is detonated on a ship or via truck or rail.
Internal threats	Employees with intimate knowledge of operations and facilities and access to transportation and port assets assist terrorists.
Maritime criminal cooperation	Terrorists mimic or collaborate with criminal enterprises (e.g., piracy, human trafficking, and drug smuggling).
Lack of visibility in noncommercial maritime activity	Terrorists capitalize on lack of monitoring of maritime activity: mines and other underwater attacks, smuggling by private craft with small payloads delivered outside ports, or attacks by small craft.
Environmental threats	Terrorists use seaborne traffic (including containers) to carry infectious diseases and other environmental threats.
Stand-off attacks from the sea	State and non-state groups develop capabilities of mounting short-range ballistic missiles and cruise missile attacks (possibly employing weapons of mass destruction).

threat is partly dependent on the vulnerability of a target; e.g., a target becomes more attractive if it is more vulnerable. Terrorist action is shifting to tend to focus on relatively easy to strike soft targets and inflicting economic damage in addition to performing attacks that may result in large numbers of casualties.

Hardening the defenses of supply chains requires the cooperation of all stakeholders (from Figure 2). Responsibilities of these stakeholders with regard to securing the supply chain have not been well defined, which creates ambiguity about who should pay for the investments that are required for protecting the supply chain. This ambiguity is exacerbated by the sector's low margins and a lack of belief that an attack might actually target supply chains.

Also, from a legislative perspective, it is not always clear who should take charge and establish security regulation. Conflicting, unclear, and overlapping jurisdictions of national and international regulatory and oversight authorities and lack of uniformity in the rules and their application for making transactions in different parts of the world hinder the establishment of such regulation.

A traditional instrument to prevent the transport of illicit cargo across our borders is Customs checks. In recent years technology to search containers has been improved (for instance through the introduction of x-ray scanners), which has also enabled the verification of cargos or scanning of a larger percentage of containers. In spite of these advances, still in the U.S. only 5 to 6% of containers are physically inspected and many other countries, especially in the developing world, are far from this figure. This causes the chances of intercepting or preventing an attack on or through supply chains to be relatively small.

If security measures are not well designed, they may well hamper supply chain performance as illustrated above. Additionally, operational and technological measures are often difficult to establish due to lack of standardization.

3.3. CONSEQUENCE

The global flow of containers represents the lifeblood of modern economies and enables our modern way of life. Over the past decades, container transport has seen explosive growth rates, with the world's largest container ports currently handling over 20 million containers (TEUs) annually. These growth rates have quadrupled the share of container transport in cargo transport to over a quarter[1] in the past 25 years. For the coming years, container transport is expected to continue to grow at an annual rate of about 6.6% [3], building to an estimated total port throughput of over 430 million TEUs in 2010.

[1] For non-bulk cargo 90% is transported per container [13].

These flows have rendered significant benefits to our economies in time and cost savings. With the proliferation of container transport, logistics costs dropped from 16.1% of GDP in 1980 to 10.1% in 2000 and the annual savings in the logistics costs of foreign trade amount to approximately $150 billion. These savings partly flow from the speed with which containers find their way in a port's hinterland. Due to the interdependency of supply chains and production processes, disruptions to these time-sensitive container flows can be potentially very costly.

A successful hit could result in significant damage. The fall 2004 congestion at the port of LA/Long Beach and the West Coast port shutdowns in 2002 have illustrated what kind of economic effects port congestion, although for reasons different from terrorist attacks, might have. Although comprehensive cost estimates for these types of events are typically not available, the potential damage can be very large due to ripple effects in sectors dependent on shipping[2]. In addition, an attack on the supply chain may inflict damage to the perception of supply chain reliability, negating the supply chain benefits presented above.

From the scarce available data on maritime terrorism, such as from the attack on the oil tanker Limburg and the USS Cole, we can see that attacks against ships mainly affect local insurance premiums and economies[3] and are relatively isolated.

3.4. RISK

Given the threat scenarios listed above, it appears that two types of supply chain attack scenarios are likely to pose the most risk: either using the relatively cloaked container supply chain to smuggle weapons facilitating attacks or attacking the supply chain itself as a way to damage the economy. Both scenarios meet the conditions for high-risk:

1. Terrorists can achieve their goals through attacking or misusing the supply chain.
2. An attack is not easily prevented as supply chains are difficult to protect.
3. The consequences of an attack are potentially significant (both in economic terms and in potential loss of life).

---

[2] The estimated direct cost of the 10-day shutdown of ports on the West Coast of the U.S. resulted in economic damage ranging from $0.5 – 19 billion.

[3] As a result of the attack on the Limburg, insurers bore a loss of $56 million and insurance premiums rose by 300%, resulting in additional costs of $150.000 per docking. Also, shipping bypassed Yemeni ports, leading to a 50% decrease in calls at Yemeni ports. The Yemeni economy suffered a severe impact, losing approximately $4 million a month.

TABLE 2. Security initiatives.

Focus of Measure	Security Measures
Port and Vessel Security	International Ship and Port Security (ISPS) Code Maritime Transportation Security Act (MTSA) 2002
Border Extension	Container Security Initiative (CSI) 24-hour Advanced Manifest Rule (AMR) 96-hour advanced notice of arrival
Supply Chain Management	Customs-Trade Partnership Against Terrorism (C-TPAT) Smart and Secure Trade Lanes (SST)
Technology Development/Deployment	Operation Safe Commerce (OSC)

4. Mitigation Measures

Policy makers have taken note of the risks described above and have enacted several measures to guard against them as can be seen in Table 2.

4.1. APPLYING RISK ASSESSMENT TO SECURITY MEASURES

In spite of these efforts, the question remains whether these measures have made us any safer than we were. To assess the effectiveness of these measures, we translate these initiatives into the performance indicators of the supply chain and typically find that the focus has been on threat reduction measures (detection and prevention) instead of consequence reduction (event response and recovery) (Figure 3).

The next step is to translate the measure's effect on supply chain performance in the elements composing risk (threat, vulnerability, and consequence) through a risk assessment. As mentioned before, verifying the content of containers has been the primary instrument used to intercept illicit trade including the interception of resources that may be used for terrorist activities. Improvements in scanning technology have enabled the scanning of a higher percentage of containers, although this percentage currently remains about 5% of total inbound containers. In addition, using current x-ray scan technologies and processes for inspections results in a need for many manual searches, which are typically false positives. For example, scanning does not in all cases provide a clear enough image to exclude the possibility that a container carries illicit goods, which necessitates time- and space-consuming manual searches[4]. The bulk of these searches, however, do not find any contraband.

[4] Approximately 5% of all containers are scanned; 5% of scanned containers must be hand-searched as a result of either an inconclusive scan image or a scan image that indicates the presence of illicit goods.

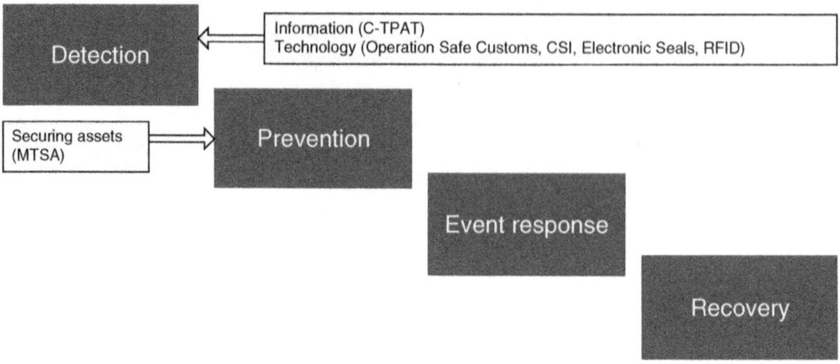

Figure 3. Emphasis in security initiatives to protect supply chains.

Not surprisingly, measures that have frequently been voiced as potential solutions to secure supply chains are the scanning of all containers entering ports and/or improving the technologies used in container scanning (both threat-reduction policies). Using the risk assessment methodology described previously, we can assess the effectiveness and feasibility of these measures and find the conditions under which a 100% container scanning would be viable, taking into account the cost and delays of screening 100% using different types of technologies to perform the scans and applying these policies to different port sizes.

4.2. CASE STUDY POLICIES

For this purpose, three policies were compared to the current practice of scanning approximately 5% of all arriving containers (the Base policy):

- Policy 1: Using current x-ray technologies to scan all containers.
- Policy 2: Using gamma-ray technologies to scan all containers.
- Policy 3: Using improved gamma-ray technologies to scan all containers.

In both the second and third policy, gamma-ray technology is deployed to scan all arriving containers. The improved scanning equipment used in Policies 2 and 3 is characterized by:

- A faster scanning rate.
- Lower equipment costs than the current technologies.
- A faster hand-inspection rate due to improved training for inspectors or improved signaling from the scanning equipment as to the region of the container requiring scrutiny in the Base Policy and Policy 1.

TABLE 3. Model parameters for cost-benefit analysis

Parameter	Base Policy	Policy 1	Policy 2	Policy 3
	5% inspection, same tech.	100% inspection, same tech.	100% inspection, slightly improved tech.	100% inspection, greatly improved tech.
Scanning Rate	20 TEU/hr.	20 TEU/hr.	30 TEU/hr.	30 TEU/hr.
False Positive Rate	5%	5%	5%	1%
Equipment Cost	$4.5 million	$4.5 million	$1 million	$1 million
Maintenance Costs	$200,000	$200,000	$90,000	$90,000
Operators per Scanner	4	4	4	4
Hand Searching Rate	1 cont./day	1 cont./day	4.8 cont./day	4.8 cont./day
Hand Inspectors per Team	5	5	5	5

Compared to Policy 2, Policy 3 assumes a lower false positive rate. Table 3, below, lists these and other relevant variables used in the model to assess the costs and benefits for the policies listed in that same table.

For each of these policies, the associated costs and benefits were compared; e.g., decreased costs associated with terrorist risk are offset against the cost of implementing the new policies and introducing new scanning schemes. In assessing the cost associated with terrorist risk, assumptions needed to be made to determine the cost of container delays and the direct cost of terrorist attacks. To generate robust answers for both variables, bandwidths were used with the costs of terrorist attacks varying between $1 billion and $1 trillion[5] and the costs of delays per container varying from $0 to $60 per hour.

As can be seen in Table 3, the main differences between x-ray (Base Policy and Policy 1) and gamma ray technologies (Policy 2 and Policy 3) are the scanning rate and equipment's cost. The difference in hand searching rate stems from the assumption that port officials will, with some policies, be better trained to perform these kinds of searches, speeding up the process.

4.3. CASE STUDY OUTCOMES

The derived model enables us to calculate with what frequencies or chances of attacks (within the predetermined bandwidths) the measures proposed in each policy are cost effective. The model therefore does not help in estimating the

[5] In determining the bandwidth for attack costs, attack sizes considered were: '93 WTC and '95 Oklahoma City at the low end of the scale—on a par with natural disasters—and a range of estimates for 9/11 or nuclear attack at the high end of the scale.

probability of attack, but provides bounds to help decision makers understand what factors might cause them to switch from a position of adopting a policy to one of not adopting a policy. Equation 1 explains how the breaking point of switching from one policy to another is determined in the model.

$$P(\text{Attack}) = \frac{\text{Change in Equipment, Labor, Delay Costs}}{\text{Attack Cost*(Change in Detection Rate)}} \tag{1}$$

Figure 4 gives an example of the outcomes of the model. It shows that the annual threshold of attack (the chance of attack) varies with:

- Attack cost
- Delay cost per TEU per hour
- Policy (technology) chosen

Using the listed parameters, it turns out that 100% inspection is only viable for large attacks (more than $10 billion in attack costs) or with greatly improved technology. If we consider a nationwide implementation under current technology, then depending on our assumed cost of attack and cost of delays, we require at least a 15–40% chance of attack if attack costs are high and even more if attack costs are low. Therefore, 100% scanning is not likely to be economically viable under current technology. Similar analysis shows that reducing the false positive rate makes 100% screening with new

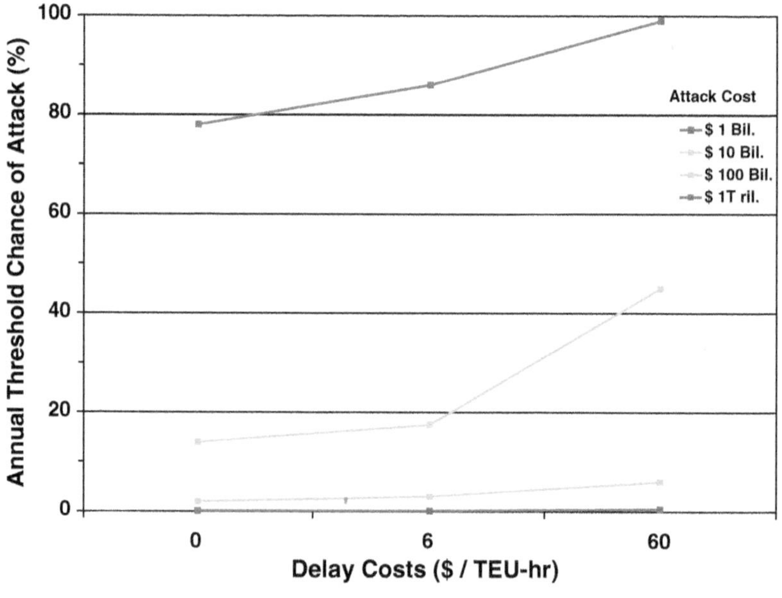

Figure 4. Attack cost as a function of chance of attack and delay cost per container per hour.

technology better on all accounts than 5% screening with current technology (e.g., equipment and delay costs decrease while security increases).

However, if restrictions on land and personnel are taken into account, adopting a policy of 100% scanning with current technology is not viable. For example, with current technology, a large port like Los Angeles would require over 400 hand search stations per shift to inspect containers to maintain a stable queue. Improved technology is not necessarily a panacea: improved technology still requires close to 100 hand search stations per shift at a port the size of Los Angeles. Port space is optimized for throughput and any regime of 100% scanning would require significant changes to port operations. Thus, the costs excluded from our model likely make 100% inspection unfeasible using the technology from the Base policy.

5. Findings

The proposed risk assessment methodology was successfully applied to assess risks to the security of supply chains to assist in prioritizing high-risk threats, and to assess security measures on their merits.

From assessing the risk associated with a multitude of threats that have been voiced in the wake of 9/11, two types of threat scenarios with regard to cargo and maritime terrorism are most relevant:

1. Maritime containers are misused to traffic illicit goods past our borders, especially when they involve potential nuclear detonation.
2. Extended disruption of operations at major ports.

Although many other potential maritime threat scenarios have been identified, including attacks on cargo vessels, the use of cargo vessels to block maritime corridors, and collaboration of terrorists and piracy syndicates, these are either unlikely to occur or unlikely to incur major damage (compared to the two scenarios mentioned above). This is corroborated by recent reports [7,15].

The economic consequences associated with any maritime assault could be severe. However, both the likelihood and the consequences in terms of loss of life of a dirty-bomb attack perpetrated using an illicit cargo container are relatively low[6]. Most container shipping scenarios present a low likelihood of inflicting human consequences (i.e., casualties), and the prospect is of relatively modest human consequences even where that likelihood is realized. Perhaps most notably, container shipping scenarios involving

---

[6] Reasons for a low risk assessment of dirty-bomb attacks include the difficulty of obtaining the materials and because for a small dirty bomb the consequences would be mostly economic and unlikely to be catastrophic.

nuclear detonations are less likely than the other scenarios we considered, but could entail far greater potential consequences in both human and economic terms, leading these types of threats to entail the greatest combination of likelihood and expected harm.

For scenarios involving nuclear detonation, effective risk management approaches must include securing nuclear materials at their points of origin. Current technologies and processes for checking cargo containers moving through the container shipping system are impractical and imperfect because of the large number of containers and the inherent errors (both false positives and false negatives) of inspection technologies. Applying the proposed methodology to assess the merits of scanning all inbound containers and/or improving the technology used to perform these scans shows that the introduction of new scanning technologies and innovative processes can overcome these problems if they result in less false positives. This would drive down the labor and land requirements associated with a large number of manual scans resulting from false positives.

The risks from extended disruption of ports are largely economic. These risks are most effectively reduced through planning to facilitate the restart of ports and container shipping systems in the wake of a terrorist attack or natural disaster. However, so far policy makers have not given event response and recovery measures enough detailed attention.

Security and efficiency are linked but distinct. In order to increase the system's fault tolerance and resilience, redundancies and slack need to be created. In order to make tradeoffs between the level of fault tolerance and resilience on the one hand and redundancies and slack on the other, security policies need to be assessed for their effects on supply chain performance (efficiency, reliability, transparency, fault tolerance, resilience).

Due to a focus on the maritime link in the supply chain, for example through programs such as ISPS, these specific links have been made more secure. However, to increase the security of the supply chain system, a more systematic approach to identify major terrorist threats and secure the system against these is required. The risk assessment methodology proposed in this paper could be a way to achieve this systematic approach.

6. References

1. BoozAllen Hamilton. Port security war game. 2003. For more information, see http://www.boozallen.com/publications/article/1440496.
2. Carafano JJ, Kochems A. Making the sea safer: a national agenda for maritime security and counterterrorism. Washington: The Heritage Foundation; 2005. Available at: http://www.heritage.org/Research/HomelandDefense/upload/74871_4.pdf

3. Economic and Social Commission for Asia and the Pacific. Regional shipping and port development strategies (container traffic forecast). New York: United Nations; 2005. Available at: http://www.unescap.org/ttdw/Publications/TFS_pubs/pub_2398/pub_2398_fulltext.pdf.
4. Flynn SE. America the vulnerable: how our government is failing to protect us from terrorism. New York: HarperCollins; 2004.
5. Frittelli JF. Port and maritime security: background and issues for Congress. 2005. Available from: http://www.stormingmedia.us/53/5373/A537354.html.
6. Gordon P, Moore JE III, Richardson HW, Pan O. The economic impact of a terrorist attack on the twin ports of Los Angeles Long Beach. In: Richardson HW, Gordon P, editors. The economic impacts of terrorist attacks. Cheltenham and Northampton: Edward Elgar; 2005.
7. Greenberg MD, Chalk P, Willis HH, Khilko I, Ortiz DS. Maritime terrorism: risk and liability. Santa Monica: RAND Corporation; 2006. Available at: http://www.rand.org/pubs/monographs/2006/RAND_MG520.pdf
8. Martonosi SE, Ortiz DS, Willis HH. Evaluating the viability of 100 per cent container inspection at America's ports. Santa Monica: RAND Corporation; 2005. Available at: http://www.rand.org/pubs/reprints/2006/RAND_RP1220.pdf.
9. Murray P. Remarks at the 10th Annual Washington Council on International Trade Conference, July 2005.
10. National Research Council. Risk assessment in the federal government: managing the process. Washington: National Academies Press; 1983.
11. National Research Council. Science and judgment in risk assessment. Washington: National Academies Press; 1994.
12. United Press International. Bush: port shut down hurts U.S. economy. 2002.
13. van de Voort MP. Maritime containers. Maritime Reporter and Engineering News. Dec 2003, 2004
14. van de Voort MP, Rahman A. Securing global supply chains. Port Technology International, 24th edition. 2005
15. van de Voort MP, O'Brien KA, Rahman A, Valeri L. "Seacurity": improving the security of the global sea-container shipping system. Santa Monica: RAND Corporation; 2003. Available at: http://www.rand.org/pubs/monograph_reports/MR1695/.
16. Willis HH. Analyzing terrorism risk. Santa Monica: RAND Corporation; 2005. Available at: http://www.rand.org/pubs/testimonies/2005/RAND_CT252.pdf.
17. Willis HH. Analyzing terrorism risk. Santa Monica: RAND Corporation; 2006. Available at: http://www.rand.org/pubs/testimonies/2006/RAND_CT265.pdf.
18. Willis HH, Ortiz DS. Evaluating the security of the global containerized supply chain. Santa Monica: RAND Corporation; 2004. Available at: http://www.rand.org/pubs/technical_reports/TR214/.
19. Willis HH, Morral AR, Kelly TK, Jamison JM. Estimating terrorism risk. Santa Monica: RAND Corporation; 2005. Available at http://www.rand.org/pubs/monographs/MG388/index.html.
20. Willis HH, Ortiz DS. Assessing container security: a framework for measuring performance of the global supply chain. Santa Monica: RAND Corporation; 2005.

CHAPTER 6

TRANSPORTATION OF DANGEROUS GOODS

A Challenge for Risk Analysts and Risk Managers

S. MENONI

Dipartimento di Architettura e Pianificazione
Politecnico di Milano
Via Bonardi 3
20133 Milano Italy

Abstract: The transportation of dangerous goods in Europe constitutes a serious problem. This paper addresses the complexity of road and train transportation networks and the challenges of multiple jurisdictions while applying the perspectives of goods logistics and industrial risk analysis to the analysis of safety issues. Legislative impacts are discussed, and examples from two provinces in Italy are presented. Recommendations include the development of protocols and agreements for transboundary cooperation and further research in this complex area of study.

1. Introduction

The transportation of dangerous goods (hazmat) constitutes a serious problem in Europe and in Italy in particular for a number of reasons.

First, the ubiquitous character of risk associated with hazmat transport must be considered: it reaches ports, railway stations, and customs and potentially touches virtually every railway and road network, not to mention airports and maritime or fluvial routes. This article, however, focuses on railways and road transport, which represent a major challenge for risk prevention policies.

Hazmat transport must be analyzed from two primary perspectives: goods logistics and industrial risk analysis.

In common with regular goods transport, hazmat transport is affected by just-in-time logistics practices. Such practices have significantly contributed to the creation of heavy traffic, by diverting goods from fixed-storage sites to moving systems. In the case of dangerous goods, as emphasized by several authors [9], risk may well be exacerbated by the introduction of increasingly strict regulations on hazardous chemical installations, since the introduction of Seveso Directive I in 1982. The complex legislative evolution of the Seveso

I. Linkov et al. (eds.), Managing Critical Infrastructure Risks, 97–110.
© 2007 *Springer*.

Directive—both at EU and national levels—has focused substantial attention on fixed installations at high risk. As a result, there has been a strong motivation to avoid controls and inspections by diminishing quantities of hazmat stored on site, thus creating a new demand for transportation.

Another important point providing strong similarities between common goods and hazmat logistics concerns the economic structure of a given region. Highly concentrated and large industrial areas will certainly produce huge traffic volumes, but, as a counterbalance, such traffic can be controlled and managed fairly easily.

However, countries like Italy present a rather dispersed productive structure. This dispersion leads to urban sprawl and to an inextricable mixture of various urban functions and industrial plants, including hazardous ones. In contrast to concentrated industrial areas, it is virtually impossible to distinguish industrial traffic from the heavy flow of commercial, residential, and other transport generated by urban sprawl.

Finally, hazmat transport is part of a more general market, which is characterized differently in EU countries. In countries like Italy, with a very large number of small and even micro firms involved in goods transportation (some of them owning just one truck), there are several obstacles to the rationalization of the whole sector. Rules and services are designed to maximize transport potential and minimize the eventuality of lorries or trains driving empty.

Hazmat transport, though, constitutes a sort of exception to the general rules aiming at maximizing benefits and resource exploitation on the one hand and minimizing costs on the other. In fact, what can be seen as a rational solution in terms of transport optimization, may translate into a risk when hazmat are involved. For example, there are fixed installations requiring a sort of exclusive service from the transport company. In the province of Como, Italy, for instance, a firm requires its contractors to carry the exact quantity of cyanide they need, in order to reduce the potential magnitude of an accident on site. Here, a risk avoidance measure for an individual installation translates into more trips by cyanide-carrying lorries—creating an increased risk of accidents along the road network.

This last point introduces another issue concerning the similarities between industrial risk analysis to be carried out for fixed facilities and risk analysis applied to the transportation of dangerous goods.

Industrial risk analysis is the essential activity at the core of any prevention policy regulating hazardous installations. The result of risk analysis is the basis for determining land use around those facilities, developing external emergency plans, and avoiding domino effects in a given environment.

Clearly an accident occurring to a truck or a train carrying dangerous goods can lead to exactly the outcomes described in the first article of the

Seveso Directive: fires, explosions, and hazmat release [12]. Yet, in many cases, accident scenarios analyzed for fixed installations contemplate the possibility that the triggering factor be a truck during loading and unloading operations.

Perhaps the intrinsic mechanism of failure that may occur in the transport system is far less complex in general with respect to the infinite variety of circumstances that may lead to an industrial top event. Nevertheless, this simplification is counterbalanced by the difficulties in envisioning the variety of external, environmental, and contextual factors characterizing an accident that may occur anywhere at any time.

2. Legislation, ADR, RID, UN Agreements

In Europe, hazmat transport is not regulated by a specific directive as fixed installation, and most countries lack specific legislation. There is however an international European agreement concerning the international carriage of dangerous goods by road (ADR) and by rail (RID).

These agreements require trains and trucks crossing borders to be equipped with specific labels with symbols and numbers (UN numbers) clearly codifying the type of danger and the class of transported material.

Other rules concern the features of trucks and trains; the composition of vessels, so as to avoid incompatible goods being carried close to each other; training and driving conditions that must be fulfilled by drivers; the type of documents on board; and finally the situations in which they can stop safely without creating a hazard for others.

In some countries, like Italy, those agreements have been extended to the entire territory, regardless of international borders, so as to enhance hazmat transportation safety in any area of the country. Nevertheless, neither agreement prescribes specific routes or hours for this type of transport. In other words, urban environments and sensitive areas are equally exposed to potential accidents, which may lead to life and health threats for communities or to severe contamination of natural resources. As noted by Oggero et al. [12] based on the Major Hazard Incidents Data Service [11], 78% of reported accidents involved dangerous materials release, which implies the largest areas of potential impact as far as the spatial dimension is concerned (Table 1).

In many countries, national laws give local and regional authorities the power to limit the transportation within given areas or within time intervals. France is a country that extensively uses such tools, setting restrictions in the crossing of urban centers as well as time limits for example along highways (e.g., lorries carrying ethylene oxide are prohibited from using the highway

TABLE 1. ADR classification of dangerous goods

ADR CLASS	HAZMAT
1	Explosive goods
2	Compressed or liquefied gas
3	Inflammable liquid
4.1	Inflammable solid
4.2	Self igniting substances
4.3	Substances becoming inflammable in contact with water
5.1	Comburent material
5.2	Organic peroxide
6.1	Toxic materials
6.2	Infectious materials
7	Radioactive
8	Corrosive materials
9	Other type of dangerous materials

connecting Italy to France along the Mediterranean Sea during daylight hours).

In Italy, such tools have been utilized only occasionally and do not constitute part of a coherent and comprehensive risk mitigation policy.

3. The Relationship between Hazmat Transportation and the Environment

Looking at the risk constituted by the transport of dangerous goods in deeper detail, three elements should be discussed:

1. The relationship between this risk and the urban and regional environment
2. The type of negative consequences for the population and the natural environment
3. The difficulties in managing an accident

3.1. RELATIONSHIP BETWEEN THE RISK AND THE URBAN AND REGIONAL ENVIRONMENT

It is clear that a very basic need is to have satisfactory information so as to map this type of traffic. In other words, what is needed is exact knowledge of routes preferred by heavy traffic carrying hazmat, quantities, and type of involved materials. Apparently this is not an insurmountable obstacle; however, as it will be shown, the situation is far more complicated than one may think at a first glance. The main problem derives from the fact that nobody is really responsible for tracking and keeping a register of this type of transport. In the absence of a national (though it would be preferable to have an EU or an international) inventory of such traffic, it will be extremely difficult, if not impossible, to develop satisfactory mitigation and risk reduction measures. In this respect, train and road transport should be kept separated.

In the first case, in fact, tracking what and how much is carried and how often is relatively easy, as train cargo authorities store information about the trains they manage. There are precise rules for combining train vessels: incompatible materials must be kept separated, and quantities and types of materials are registered at the origin station. Yet, information regarding the last year or the last five years cannot be taken as a prediction of what will occur in the future. This traffic is very dynamic, directly influenced by market constraints and suddenly changing demands. In order to perform precise analysis, the Cargo Division of Trenitalia was asked to perform specific surveys in two Italian provinces, the results of which often proved to be unpredictable on the basis of previous data records.

Of course the problem is greatly exacerbated when lorry traffic is considered. In this case, there is no authority in charge of keeping track. There are several projects investigating the technical feasibility of tracking trucks through satellite systems or electronic labeling. The critical issue, however, is not so much technical feasibility as the universal recognition of the need for such information and for the creation of an agency or authority in charge of maintaining the data.

In the absence of relevant information regarding the traffic of dangerous goods, it may be worthwhile to mention the recent experience of two northern Italian provinces: Milan and Como. These examples are relevant methodologically but also from a geographical point of view. The province of Como is at the border with Switzerland, constituting an important crossroad for both train and lorry traffic connecting Northern and Southern Europe. Milan province is at the heart of the first Italian region in terms of absolute number of Seveso facilities (38 at highest risk and 70 subject to notification. Note that those numbers are indicative—especially the latter—and subject to changes in legislation and classification rules).

The provincial level is responsible for producing a risk prevention program, in which the most relevant risks must be analyzed and countermeasures proposed.

In the case of Milan province, four sources of information have been used and laid over a map:

1. The Cargo division of Trenitalia provided information regarding railway stations where hazmat is stored for some time waiting either to be loaded

on trains or carried by road to its final destination. Railway stations often become very dangerous storage sites, without the inspections and controls that are compulsory for Seveso installations, despite the high risk they may present, especially when stations within urban centers are considered, as in the case of Milan province.

2. Regarding road traffic, a questionnaire has been developed in cooperation with Federchimica (the association of Italian chemical industries adhering to the Responsible Care program) [6], where the following questions were asked:

- What time does hazmat enter and exit the factory?
- What is the most frequently transported hazmat?
- What are the origin, destination, and probable routes of the hazmat?

The questionnaire was sent to the 120 industries under the Seveso Directive; 80 firms returned it completed to the public administration.

3. The third source of information derived from an inventory of roads and crossings where accidents are more frequent and can be therefore more dangerous for trucks carrying hazmat as well.

4. Finally, an important source of information was an inventory developed by the Firemen Regional Department with the contribution of the Politecnico di Milano in which all the accidents involving hazmat that occurred in the Lombardia Region in the years 1992–2002 were inserted, providing data on involved materials, location, and difficulties in intervention.

These four sources of information supported the elicitation of the following results:

- The most-transported ADR classes.
- The most-used hours of the day for charging and discharging lorries.
- The hours in which the majority of reported accidents occur.
- The routes most likely used by this type of traffic.
- Spatial clusters where most accidents occur.

In the case of the Como province, a supplementary crucial source of information was available, thanks to the involvement of the Customs Police in the project. Customs agencies keep track of everything that enters and exits the country as part of the fulfillment of fiscal requirements and assurance of the compliance of entering products with national laws and regulations.

Because of this rather different perspective, the codes used for dangerous goods do not correspond to UN numbers; safety issues are considered in the sense of determining the conformity of imported goods (e.g., toys) to the country's internal requirements, not to trace hazmat that may cause potentially high-impact accidents.

Despite Customs' lack of focus on general environmental safety issues, it was possible to analyze the transit of dangerous goods through the Chiasso-Como border since 2002. The analysis confirmed the high variability of hazmat transportation trends, in terms of both substances and quantities.

Furthermore, an interesting trend was observed: hazmat generally enters Italy by road or train and leaves the country only by road. Over the last years, the tendency is toward more extensive use of trucks in both directions.

3.2. NEGATIVE EFFECTS FOR THE POPULATION AND THE ENVIRONMENT

Another important field of concern relates to the consequences a given accident may have on the affected population. While since the Seveso Directive II the compatibility between hazardous plants and settlements must be considered and expressed in terms of safety distances, the same does not occur with roads and railway transportation.

If one is to apply the same concept that is adopted for fixed installations to transportation networks, instead of a circular area around the plant, a strip zone along roads and railways should be drawn and the most critical situations identified (Figure 1).

This has been done in a sub-area of Milan province, comprising the municipalities of Bollate, Senago, and Paderno Dugnano. First, a strip zone of 250 meters along the crossing roads (those along which hazmat transportation is more frequent) has been drawn. In the strip it is possible to recognize vulnerable objects, such as schools, nurseries, and hospitals. The underlying idea is to avoid the intersection of hazmat transportation with this type of vulnerable facility. This can be achieved in two ways.

An optimal route respecting predetermined constraints can be calculated using well-designed transport optimization software; whenever such a route exists, it can be prescribed as the only road to be used by such transport. In cases where such a path does not exist, time limitations can be prescribed; e.g., banning this type of traffic during rush hours, or when children start or finish school.

This application clearly considers safety as a primary concern for local and regional governments at the expense of other considerations (e.g., transportation costs). In fact, Erkut [4, 5] shows in his work that the problem of finding the best route for hazmat transport is multi-objective and

Figure 1. Detail of the map representing the strips along routes used by hazmat carriers. The various symbols refer to vulnerable objects in the area.

> ...the similarities between the routes found when using different risk criteria, such as population exposure minimization and accident probability minimization, are very low [and] the optimal route according to one criterion has very low tolerance for the other.

This type of prescription should be decided at large scales—at least regional, if not national and international—to avoid reducing the risk in one point of the network only to contribute to a dramatic increase in another. Furthermore, the continuity of the network must be guaranteed for transporters as well, therefore limiting the autonomy of decision of individual municipalities.

As for the consequences for the natural environment, very little can be found in the literature concerning the potential contamination effects and damage to the natural environment of accidents occurring in fixed installations [1]. The Seveso Directive has been focusing on human health only until the rather recent Seveso Directive III, which, despite initial premises, does not set a real dramatic change as far as the natural environment is concerned. After the Baia Mare accident in Romania, there was a large expectation for the introduction of a new field of concern in the legislation, looking at the potential contamination generated even by one individual accident. As this evolution has not been fully achieved by the Seveso Directive, certainly less attention is paid to the damage to natural systems provoked by transport accidents. Yet such cases can be found at least in the records of the Regional Firemen Department mentioned above as well as in the MHIDAS database [12]. For example, in managing a severe accident involving a hazmat lorry on a bridge, first interveners have neither time nor resources to avoid a major release of hazmat in river waters. Yet sometimes this type of release can be very dangerous for a given habitat and nearby communities.

3.3. MANAGING EMERGENCIES

The last point that will be mentioned here regards the difficulties in managing emergencies involving hazmat transportation. There are so many difficulties and challenges that all of them cannot be analyzed in this paper. Some of the most common will be mentioned.

The first is systemic in its nature. In general, hazmat transportation accidents will affect areas much larger than the exact zone where they occurred. Be it a quarter to be temporarily evacuated or an important highway that will be interrupted for hours, larger areas and systems will be involved in such occurrences.

In the analysis that has been carried out in the Como province, the police described the most common situation whenever the border with Switzerland must be closed for any reason: traffic must be diverted starting from provinces as southern as those of Bologna, Modena, and Piacenza in the Emilia Romagna region. Traffic originating from those provinces as from those situated farther north, must be either postponed to another day or directed to alternative routes, as far as the Varese province borders or even the Monte Bianco and Frejus tunnels.

Another difficulty that typically arises concerns police forces getting to the accident site: generally they are not adequately equipped or protected and very often the first responders are the least prepared to face a chemical accident (not to mention one involving radioactive substances, a case that has been reported in the Firemen database for the Como province).

Time is lost in the attempt to fully assess the exact substance involved, to take the right actions, and to call for specialized teams. In Italy, the latter are generally special units of fire brigades, fully equipped to respond to NBCR (nuclear, bacteriological, chemical and radioactive) threats.

In this regard, an interesting (and useful) tool is provided by the Set service of Federchimica which provides technical assistance to manage transportation chemical accidents [6].

As the study by Becker et al. [2] correctly puts it, psychological and social consequences of chemical spills and other types of accidents involving hazmat are rarely considered before or after an event:

> Victims of chemical or radiological accidents often live in a "permanent state of alarm and anxiety" [2].

This is due to:

> …considerable uncertainty about the consequences of exposure. Medical knowledge is frequently limited, and both contaminants and their resulting damage may be invisible. Further, potential long-term health consequences may take years or decades to develop. Thus it is not clear to people whether the worst is over or whether the worst is yet to come….[2]
> Response plans and protocols rarely devote adequate attention to the psychosocial effects of contamination incidents. When psychosocial content is included, it is usually limited to generic information about disasters, debriefing, and mental health. Plans rarely include specific information about contamination incidents and the complex psychosocial challenges—immediate and longer term—that they pose [2].

4. Adding Difficulties: Managing the Risk of Transportation of Dangerous Goods Across National Borders

Another relevant question that deserves to be tackled relates to the transregional and often transnational character of accidents involving dangerous goods. This may be the case when the accident occurs in a transboundary section, a tunnel, a custom station, a bridge connecting two countries, or when an international network is involved. In all those cases, the perturbation provoked by the accident, either minor or severe, is suffered by the two (or more) countries connected by the lifeline at stake. Secondly, emergency management will require some kind of cooperation and mutual agreement, if not help and mutual support. This is clearly true in the case of a tunnel like the Monte Bianco or the Frejus, where firemen of both sides are in charge of emergency activities in case of fires or accident. However, the scenario may involve many more authorities, depending on the severity and duration of the crisis, which will require information exchange at the very least, and mutual aid and resource exchange when the event goes far beyond the immediate response capabilities of forces at the site.

This type of transboundary activity can lead to strong cooperation and mutual solidarity, but, in other instances, can develop in contradictory actions and even conflicting views of the situation and its causes.

An interesting experience in this regard was gained during an Interreg project between Switzerland, Canton Ticino; and Italy, Como Province, in the years 2004–2005. The underlying idea of the project was to build shared protocols and agreements regarding emergency interventions requiring some kind of cooperation between the two countries and in particular between the two local governments.

During the project, it was decided to develop two scenarios involving dangerous materials, one occurring in the Chiasso railway station, the other in Monte Olimpino 2 tunnel. The latter will be described in further detail.

The Monte Olimpino 2 tunnel is an alternative for trains that do not have to use the Como San Giovanni station, thus connecting Italy directly to Switzerland along the mountain chain bordering the Como valley on the western side, avoiding the lake (Figure 2).

This route is regularly used for the transportation of goods, including hazmat, while a limited number of passenger trains use it.

The tunnel is 7 km long, with only one inspection well situated at a distance of 2 km from the southern entrance. The latter is located fairly close to a flat area that can be used as an emergency location for search and rescue teams and their equipment.

The northern entrance, a few hundreds meters from the Chiasso station, is rather difficult to access except by another train.

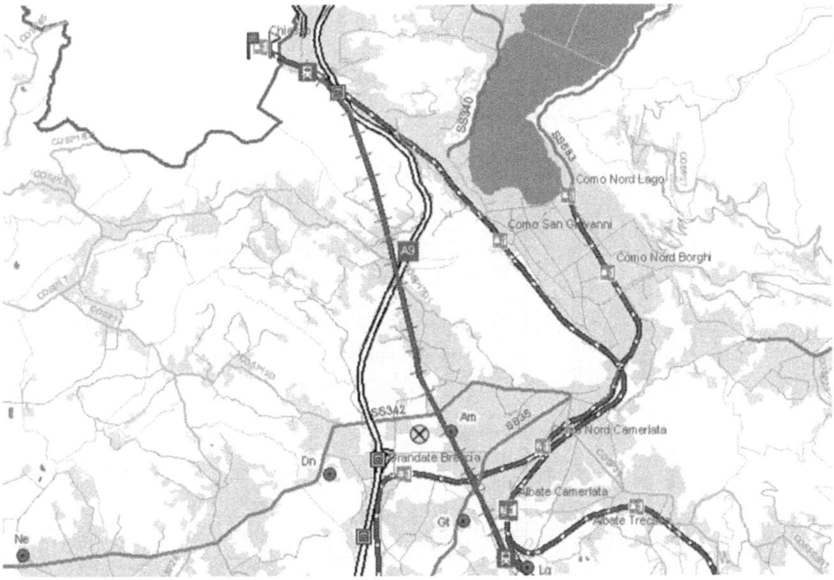

Figure 2. Representation of the Monte Olimpino 2 tunnel (in violet), in the Como Province.

Developing the scenario was not an easy task. A real accident involving toxic as well as inflammable goods would by very hard to handle because of the length of the tunnel and the lack of safety devices, such as water sprinklers and internal communication facilities (these are planned to be included in current tunnel repair works).

An interesting reference for building the scenario was provided by a real accident which occurred in a city tunnel in Baltimore in 2001 and extensively described in a U.S. Federal Emergency Management Agency (FEMA) report [10].

Although a city tunnel is rather different from a mountain one (mainly because of accessibility as well as other morphological aspects), the report provided important input to the team. The scenario was discussed in a couple of long meetings with the Swiss partners and representatives from a number of involved organizations expressing their view, until it reached its final version, schematized in Figure 3.

As can be seen, the scenario has been divided not only along the time axis, but also spatially, differentiating between a core area (the site of the disaster itself and its immediate surroundings) and an external context, from which responsible agencies must ensure the creation of positive conditions to manage the crisis. Information exchanges across the border are part of this context, to manage appropriately international train traffic variation, to provide adequate emergency response, and to carefully and honestly inform the population and the media.

Time development of the accident from the occurrence to the declaration of return to normalcy	**Description of the accident and involved communities**	**On site**	**External context**
		Direct intervention - Active organisations - First to be informed	***Coordination of resources coming from outside*** - Active organisations - Communication chains

Figure 3. Framework developed for the scenario along the Monte Olimpino 2 tunnel.

4.1. LESSONS LEARNED

A number of important lessons were learned from this experience. The first relates to the need for coordination among countries when engaged in the resolution of a "common" accident. At the very least, coordination means adequate information exchange between the affected countries. An interesting institutional example is provided in this case by the CCPD (Customs Police Cooperation Centre), an organism constituted by the EU to guarantee police cooperation along frontiers. The newly created organism between Switzerland (which previously adopted the EU model and applied it with France and Germany) and Italy permits, for example, the continuation of shadowing across borders.

But information is not enough; in the absence of a strong relationship between countries, it will be difficult to meet challenges and make difficult choices; e.g., suggest evacuation to the other country (because a toxic cloud is crossing the border). More obvious elements that need to be discussed in protocols and mutual agreements relate to establishing procedures for facilitating reciprocal aid at disaster scenes without imposing bureaucratic constraints that may result in serious losses. For example, the meetings determined that substantial hospital facilities exist on the Italian side, which could be used in case of a mass emergency, while Switzerland has more resources at hand to face chemical emergencies and large train fires.

A more general issue to be debated, closely linked to the discussed scenario, relates to whether to account for the most negative scenario, regardless of its likelihood. In a guideline provided by a joint commission of the National Italian Railways and the National Corp of Firemen, it is suggested that a collision between a train carrying hazmat and an oncoming passenger train is extremely unlikely and can therefore be dismissed. This assumption holds true only if hazmat and passenger trains never simultaneously pass through the tunnel. Whenever this occurs, possibility of an accident exists. A possible international and transboundary preventive measure is to change train schedules by adding as a guiding criteria the need to avoid the simultaneous presence of hazmat and passenger trains in a tunnel.

5. Some Conclusions

This article discusses some relevant safety issues related to hazmat transportation, with respect to long-term as well as short-term mitigation policies.

Long-term policies include satisfactory legislation and safety measures in all environments affected by hazmat transportation (i.e., tunnels, natural sensitive areas, crowded urban centers, and vulnerable elements); short-term strategies mainly refer to civil protection and emergency procedures, which are greatly challenged by the ubiquitous nature of this type of risk. Recent experiences in two Italian provinces have been briefly discussed to add new insight and propose mitigation tools.

Further research and practical experiences are certainly required to gain understanding of potential accidents and prepare appropriate responses once they have occurred.

Some studies have been carried out in this regard, but they still lack the necessary integration into a systemic view, which is required by a kind of risk that is strongly interlaced with communities and all types of natural and built environments.

6. References

1. Agenzia per la Protezione dell'Ambiente e per i servizi tecnici (APAT). Valutazione dell'impatto sull'ambiente degli incidenti rilevanti. Rapporto. 36/2004.
2. Becker S, Pitt R, Clark S. Environmental health, public safety, and social impacts associated with transportation accidents involving hazardous substances. Tuscaloosa: University Transportation Center for Alabama; 2000.
3. Carrara R, Fanelli R. Guida al trasporto di sostanze pericolose, come prevenire e gestire le emergenze nel tasporto su strada. Fondazione Lombardia per l'Ambiente. I manuali. 1999.
4. Erkut E. The road not taken. OR/MS. 1996; 23(6).
5. Erkut E, Vertyer V. Framework for hazardous materials transport risk assessment. Risk Analysis. 1995; 15(5):589-601.
6. Federchimica. SET Service. Available at: http://www2.federchimica.it/.
7. GESO. Bericht über: real-scale fire demonstration, Virgolo Tunnel. 2005.
8. Interreg Project Mis III-2a. Coordination of civil protection procedures between the Province of Como, Italy and the Canton Ticino, Switzerland. 2003-2005.
9. Marsili G, editor. La valutazione del rischio d'area. Il caso dell'area industriale di Mantova. Milan: Franco Angeli; 2000.
10. FEMA. US Fire Administration technical report series, CSX tunnel fire, Baltimore. 2001.
11. Major Hazard Incidents Data Service. Available at: http://www.hse.gov.uk/infoserv/mhidas.htm.
12. Oggero A, Darbra RM, Munoz M, Planas E, Casal J. A survey of accidents occurring during the transport of hazardous substances by road and rail. Journal of Hazardous Materials. 2006; 1(7).
13. National Italian Railways (FS) and the National Corp of Firemen (VVFF). Linee guida per il miglioramento della sicurezza delle gallerie ferroviarie - a cura del gruppo di lavoro misto FS-VVFF, internal document. 1997.
14. Universität für Bodenkultur Wien, University of Natural Resources and Applied Life Sciences, Vienna, Rapporto scientifico – Galleria del Virgolo, Nr. 875-05-004. Simulazione di incendio nella galleria del Virgolo, Bolzano / Alto Adige. October 2005.

CHAPTER 7

LIFELINE EARTHQUAKE VULNERABILITY ASSESSMENT

A Systemic Approach

S. MENONI, F. PERGALANI, M.P. BONI, V. PETRINI

Dipartimento di Architettura e Pianificazione
Politecnico di Milano
Via Bonardi 3
20133 Milano Italy

Abstract: This paper discusses the vulnerability of lifelines (major utilities and infrastructure systems such as those relating to water, energy, and communications) to earthquake. Direct, induced, and indirect impacts are considered. The inter- and intradependent character of lifelines is recognized and frameworks are introduced to facilitate the appraisal of lifeline vulnerability in the context of multiple interrelationships, stakeholders, and information sources as well as instances of incomplete or insufficient available data. The development of complete event scenarios is discussed, as a method of creating a holistic picture of potential impacts. The frameworks and methodology presented here are then demonstrated by application to the province of Brescia, Italy, an area subject to seismic activity. These vulnerability assessment techniques are proposed by the authors as an initial though highly integrated basis for future research.

1. Introduction

Lifeline earthquake engineering has been a recognized field of scientific research for more than 30 years. Studies concerning lifeline response to earthquakes noticeably increased in the 1990s, especially in Japan and in the U.S., which are exposed to strong ground motions and therefore seriously concerned about the disruption of the assets on which urban, social, and economic activities largely depend.

In Europe, and in Italy in particular, seismic studies have focused much more on seismological studies and the vulnerability of traditional housing, with fewer attempts to analyze the consequences of natural disasters on utility networks. In recent years something has changed, not only within utility management companies, but also in local and regional governments. Italian institutions have undergone a quite radical decentralizing process, resulting

I. Linkov et al. (eds.), Managing Critical Infrastructure Risks, 111–132.
© 2007 *Springer*.

in new responsibilities for local governments, many of which lack even the minimal technical and financial capabilities to manage the entire cycle of activities needed to prevent severe damage and mitigate disaster effects on the population. In the meantime, a much heavier burden has been placed on existing utility companies to provide reliability and continuity in maintaining service regardless of external disturbances. This is a requirement of modern businesses, many of which rely on communication lines and are constantly eager for energy and electrical power.

Towns and metropolitan areas have become increasingly dependent on lifelines; the latter have developed over time in a quite frantic and unplanned manner, thus creating concern for their vulnerability to extreme forces. Some lifelines have been substituted under the pressure of market innovation (e.g., the dramatic change in communication networks in the last 40 years) while others, old and obsolete, are still in operation after 60 to 80 years (e.g., water and sewage systems).

The time has come for European countries to start planning in the realm of urban and regional infrastructure, in order to reduce their vulnerability to natural hazards. This was done in New Zealand in the 1990s and has been suggested in forums and workshops dealing with urban response to calamities [13].

In this fast-evolving context, the Regione Lombardia administration supported a study of lifeline vulnerability to earthquakes, wishing to draw upon more than 10 years of continual exchange of research with scientific institutions in the field of seismic risk.

The study's accuracy has certainly benefited from the large amount of data that has been systematically gathered in recent disasters that affected densely populated metropolitan areas, such as Northridge in 1994, Kobe in 1995, and Taiwan and Izmit in Turkey in 1999.

The most interesting research has been produced in the last decade, with some attempts to link earthquake hazard analysis to urban and regional systems in order to address the many induced and indirect damages that are the inevitable consequence of public utility interruption. Aside from the more technical aspects of the physical features of nodes and networks, recently greater attention has been paid to mutual interconnections among lifelines, recognizing their inter- and intradependent character, as defined by Nojima [9].

Lifelines consist of multiple components, such as linear pipes, substations, control centers, and power generating systems, which adhere to highly hierarchical functions. Components at higher hierarchical levels are more vital than distribution systems for maintaining utility functions; however, nodes and linear parts are also highly integrated, making it very difficult to isolate them without considering their mutual connection.

Furthermore, different utility networks depend on each other: electrical power is necessary for pumping water stations, for remote control facilities, and for a number of communication and safety devices. Although electrical systems can be easily and quickly repaired—especially as compared to water or gas facilities—a light shaking is sometimes enough to put relays in electrical generators or transformation centers out of service, thus interrupting power for several hours, even when no severe damage has occurred in other parts of the system. Without electricity, several other utilities are at risk of paralysis.

Finally, lifelines are connected to the external environment. The latter can be the source of induced damage: landslides triggered by earthquakes or collapsed debris can obstruct and break utilities running beneath. On the other hand, several urban systems are the end users of lifeline services. In assessing the consequences of any natural or manmade hazard on lifelines, all these features must be taken into account.

In order to do this in the research presented here, a number of methodological premises have been determined as a ground for the common and integrated work of an interdisciplinary team, including seismologists, geologists, seismic engineers, computer scientists, and urban planners. First, it has been decided to follow a systemic approach, in order to capture the links existing among different lifelines, among the latter and the external environment, considered as both a potential source of damage and as the end user of lifeline facilities.

Second, particular stress was put on spatial factors, recognizing that lifelines cannot be studied as buildings—that is, as point-like objects, isolated from their geographical and spatial context. Because of the inter- and intradependence of lifelines, the effects of ruptures and interruptions may propagate hundreds miles away, depending on the hierarchical importance of the broken part and on the intrinsic pattern of the system. Lifelines are like veins in the human body, connecting seismic to non-seismic zones: the physical impact of earthquakes may be felt rather far away from the epicenter area.

This is the reason that—aside from results sought for each lifeline system—the spatial distribution of malfunctions in a wide enough geographical area has been carefully assessed and represented. This way it was possible to identify not only those utilities that may be mostly affected by earthquakes, but also in what areas the most severe consequences of lifeline interruption or malfunctioning can be expected.

The inter- and intradependence of lifelines led to another methodological decision: to go beyond estimating physical losses, including functional and organizational factors. This has been done to provide the largest range of outputs, to which alternative options of interventions may well correspond. It is believed that any risk assessment tool should respond to the need of finding good and affordable damage reduction and mitigation strategies. In the case of lifelines, it is fairly impossible to act only on physical vulnerabilities: entire networks cannot be substituted just to achieve better seismic performance, and it is unimaginable to carry out reliable surveys assessing fundamental parameters along thousands of miles of lifeline development, to detect where and at what stress degree each pipe or node is likely to break.

Inevitably, a more holistic approach has to be pursued, attempting first to identify priority areas or components for further and more detailed analysis, within a framework allowing for comprehensive scenarios and vulnerability appraisals.

This has been done considering physical, functional, and organizational factors as deeply interconnected to one another. Only in a limited number of cases, recognized as particularly crucial and vulnerable, at the same time using simulations and scenario techniques, physical retrofitting has to be proposed as a damage prevention program. In many other cases, softer solutions aimed for example at improving the organization and coordination among and within lifeline management companies, or reducing systemic vulnerability due to the lack of alternative paths or relative autonomy of vital systems, can be successfully promoted.

A much stronger effort then should be addressed to avoid those situations in which the inter- and intra- dependent character of lifeline systems lead to a genuine catastrophe. A variety of ways can be proposed: by adding safety devices to systems, by achieving stronger coordination between lifeline management companies and civil protection agencies, by finding specific solutions for those cases for which a failure in one area or in one component can produce much larger effects in other parts of the same systems or in other urban and regional systems.

The model combines two different perspectives: one internal to lifeline systems and the other considering the effect of lifeline damage on the external environment; that is, on these social, economic, and urban systems that depend on lifelines for their ordinary activities.

The final result that has been obtained from almost three years of research is a tool to generate what have been called "complete event scenarios," estimating not only the physical but also the functional and organizational failures originating in and from lifeline systems.

Scenarios here resemble much more a chain connecting physical ruptures suffered by pipes and plants to induced and secondary damage occurring in systems depending on lifelines than a sum of individual losses. In this sense, in complete event scenarios an attempt has been made to first estimate the physical ruptures in pipes and plants due to a given earthquake; then to forecast the effect of those breaks on the performance of each lifeline system; and finally to anticipate the consequences of utility interruption on urban and regional functions.

In a complete event scenario as intended here, the time dimension becomes as important as the spatial dimension of primary and secondary losses. The evolution of the event from the impact to the emergency phase and then to the reconstruction is followed, as proposed by Haas, Kates, and others [7]. Urban and regional systems depend on lifelines to different degrees, according to their changing needs as the disaster evolves from one phase to another. This is the reason distinct damage estimate procedures have been developed for the emergency and reconstruction stages. While in the first stage minimal lifeline performance is all that is needed to support victim assistance and rescue activities, return to normalcy at acceptable cost becomes the new priority in the reconstruction stage.

During the first 72 hours following an earthquake the following questions have been addressed: whether and under what conditions lifelines are still able to function—despite some degree of physical damage suffered by pipes and plants—and to what extent hospitals, civil protection centers, and fire departments can still provide service without or with limited amounts of water, gas, power, and communication.

In the reconstruction phase, not only the costs of lifeline repairs are evaluated, but also the economic and social costs borne by economic sectors and by the population as a consequence of service interruption or malfunction.

The paragraphs below are organized as follows: first the conceptual framework to construct complete event scenarios is presented and discussed; then an application in the area of the Garda Lake pertaining to the Brescia Province in Lombardia is shown. Recommendations deriving from the construction of the scenario in the area are then discussed as well as preliminary conclusions and suggestions for future research in the field.

As the vulnerability assessment tool is a particularly important part of the model, it will be explained in detail, as well as its application in the study area.

2. Looking Closer at the Scenario Approach

In the seismic field, as in other risk-related research, when scientists refer to scenarios they generally refer to a deterministic input which can be chosen either on a historical basis (taking for example the most severe event ever recorded or listed in the catalog in the area of interest) or as a synthetic event resulting from some model or code application. Then, the effects of this predetermined input are assessed in terms of wave propagation mechanisms, and the resulting damage to buildings or other artifacts is evaluated.

There are also other forms of scenarios, though less valued in the scientific arena: the more qualitative ones, consisting mainly of a sort of storytelling about the consequences of a strong ground motion in a given area at a given hour of a given day [11]. This kind of scenario, deriving from the economic and managerial ground, is of use for civil protection exercises but generally lacks a strong methodological basis. In fact, the stories tend to concentrate on elements that are more evidently vulnerable or those which proved to be a problem in similar areas or situations in the past.

Both types of scenarios provide valuable insight on the ability of a given community to cope; however both lack something. The first, though quantitative and therefore verifiable by others, does not take the systemic nature of our environment sufficiently into account, especially when large modern cities are considered. It would be virtually impossible to quantify completely and satisfactorily all the complex connections among different urban and regional systems, and between the latter and the socioeconomic systems.

On the other hand, the storytelling approach, though allowing for more flexible inclusions of aspects and elements deemed important in a real case, lacks the rigor of quantitative methods and relies too much both on back analysis and on the experience and variety of expertise included in the scenario-making team.

The authors are convinced that an integration of the two methods would prove much better, making it possible to adopt quantitative models when those are available, reliable, and provide meaningful output without sacrificing the richness of knowledge, information, and insight that has been gained from analyzing the impact of not only earthquakes but also other natural and manmade disasters on the built environment and social and economic systems. After all, certain situations, facts, and factors recur in any disaster, thus comprising a sort of invariable that can be treated as a sort of skeleton of scenarios, enabling the capture of particular and unexpected elements of a given specific circumstance or place.

In order to do this, it was necessary to strongly emphasize the structure of the methodology, looking for a framework that would be flexible enough to incorporate both kinds of scenarios and still have enough rigor to permit a certain degree of control and verification. It is a kind of modular method, based on the definition of risk which serves as a common groundwork for the interdisciplinary team collaborating in the research. The scenario is thus intended as the quantitative and qualitative description of the estimated direct and induced damage and the secondary organizational and systemic effects that may be provoked or triggered by an earthquake in a specific geographic context. The stress will be put on the damages and negative effects produced on, through, and by lifeline systems.

2.1. FRAMEWORK TO CONSTRUCT SCENARIOS OF THE EMERGENCY PHASE

Losses and damages that must be considered higher priority during emergencies include vital systems for search and rescue activities, as can be seen in the parameters selected in Figure 1. The arrows linking the various boxes represented in each column show the main relations among objects,

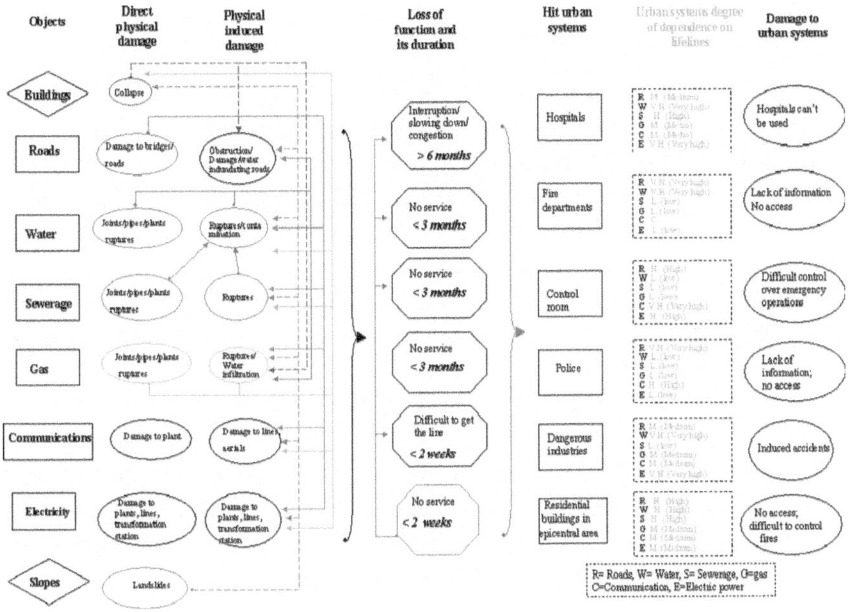

Figure 1. Framework to develop an event scenario during the emergency phase.

subsystems, and systems and the kind of damage and failure they may suffer in case of a strong tremor.

It can be seen that in the first column, lifeline systems are represented in rectangles and buildings and slopes are represented in rhomb-shaped boxes. This has been done to account for the induced damage they may provoke in lifelines while collapsing or moving, as is shown in the third column. In the second column, only the damage directly provoked by the earthquake on lifeline nodes, lines, and plants is considered.

Starting from the fourth column, systemic aspects are considered. First, the kind of malfunction and its maximum duration—derived from past earthquake reports—must be evaluated. Affected urban systems vital for managing emergencies, like hospitals and firemen, are identified in the fifth column. Each of those systems depends on lifelines to different degrees, as shown in the sixth column.

Finally, the damage to systems in terms of reduction of their performance is assessed in the last column. It is possible to follow several paths in the framework to assess different types of damage to lifelines and consequences on impacted systems, resulting in several forms of emergency management capacity reduction. In any case, the framework transforms direct physical damage suffered by lifelines and induced damage due to building failures and landslides into failures reverberating through the systems that are more crucial for managing emergencies.

2.2. FRAMEWORK TO CONSTRUCT SCENARIOS OF THE RECONSTRUCTION PHASE

Following the conceptual framework provided by Haas et al. [7], it is assumed that in the reconstruction phase emergency operations are already over and therefore quick recovery and return to normalcy are called for. Damage and failure are not measured any more in terms of time delay to emergency operations, but rather as social discomfort and economic costs associated with repairs and the lack of necessary utilities during several days as is shown in Figure 2. The first and second columns of this framework are very similar to those of Figure 1. Starting from the third column, things change as parameters expressing economic losses and social discomfort are analyzed. Reconstruction costs depend on the length and number of damaged conduits and pipes and on the availability of material and personnel for repairs. The type of fare and performance reduction is shown in the fourth column, while in the fifth column the impacted urban systems are reported. Economic activities, urban public transport, and other facilities depend on lifelines to different degrees and can persist without gas, water, electrical power, or communication only for a limited period of time before collapsing: this is considered in the sixth column. Social discomfort and economic costs to impacted urban systems are finally represented in the last column.

In order to actually run the two frameworks and build the complete event scenario following the time dimension through the emergency and

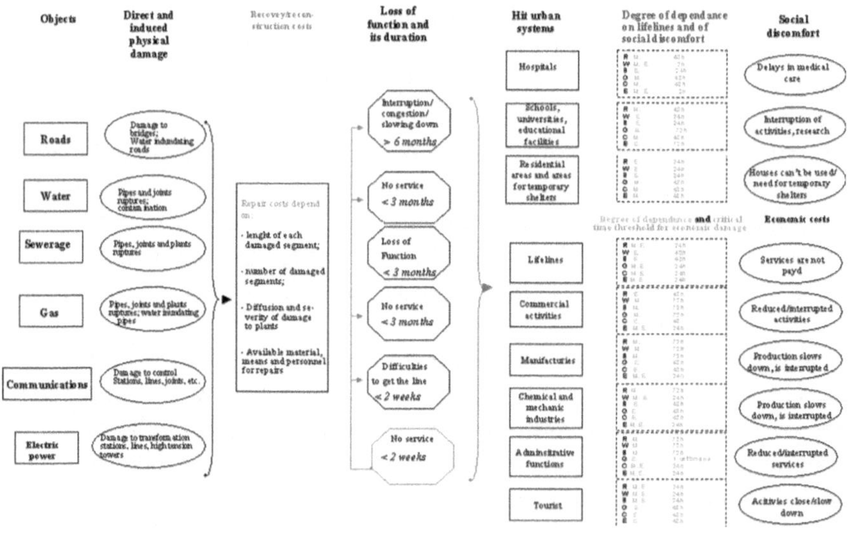

Figure 2. Framework to develop the continuation of the scenario in the recovery phase.

reconstruction phases, two main components must be defined in a given area: a seismic deterministic input on one side and the vulnerability of the exposed systems on the other (systems related to lifelines).

The combination of the seismic stress with vulnerabilities that are peculiar to a given set of systems and subsystems in a particular area provide the final forecast of damages.

As intended here, damages are not only those provoked by seismic waves and ground tremors, but also induced physical damage, such as that resulting from landslides, flooding, and fires triggered by the earthquake, as well as other losses and failures that are referred to here as systemic, due to the inter- and intradependence of various systems and particularly among urban and regional systems and lifeline utilities.

While a large body of literature exists regarding the choice or creation of a deterministic seismic input, the second ingredient, lifeline vulnerability must be discussed in further detail.

3. Vulnerability of Lifeline Systems

The notion of systemic vulnerability is the underlying concept of the proposed method, evaluating how a lifeline system is prone to damage or failure not only as a consequence of physical damage to one of its components, but also as the indirect effect of some physical, functional, or organizational failure suffered by other interconnected systems. Furthermore, the method takes into account the vulnerability of urban and regional systems to the loss of gas, water, power, and communication utilities.

The framework that has been proposed distinguishes lifeline vulnerability in the immediate post-event emergency and recovery/reconstruction phases. In the first phase, the resiliency of utilities to damage—which enables them to function in spite of some degree of physical loss and breakage—is appraised. In fact, to perform well during emergencies, lifelines must guarantee at the very least vital—though minimal—levels of service to facilities like fire departments and hospitals that are needed to cope with crises and take care of victims.

In the recovery/reconstruction phase, society seeks return to normalcy; therefore lifeline systems must be repaired as quickly and smoothly as possible to permit businesses and other services and urban functions to return to their pre-event level of activity.

Priorities, during both the emergency and the recovery phases, cannot be set only according to criteria provided by individual lifeline system companies. It is quite hard to define reliable and acceptable criteria without widening the scope of assessed needs and demands, which involve many urban systems other than lifelines, such as residential or industrial areas and public facilities. Without a deeper understanding of what happens during disasters, it is almost impossible to set priorities.

In order to account for the two different phases, distinct survey and assessment matrices have been developed. In the survey matrices, parameters to assess the vulnerability of lifelines in the emergency and reconstruction phases have been identified, while in the assessment matrices, thresholds of good and bad performance as well as weights expressing parameter importance have been set according to the different priorities identified for the two phases.

The survey and appraisal matrices are built according to the same structure; the first enables measurement of actual values in the area under examination, while the second produces a judgement of the situation in order to decide where to deepen the analysis and where the situation is most critical.

The structure to assess lifeline vulnerability during emergencies is shown in Figure 3: functional, organizational, and physical vulnerabilities are grouped respectively in the first, second, and third columns.

Functional factors explain lifeline vulnerability to malfunctions due to a variety of causes, many of which are not physical. Each lifeline is a hierarchical system as well: if crucial nodes fail, many customers will inevitably be cut off. Such nodes include, for instance, stations transforming electrical voltage and natural gas pressure, energy power stations, and water reservoirs.

Organizational factors considered in Figure 3 may consistently hamper search and rescue activities. Communication systems provide an enlightening example in this regard: the most frequent problem encountered in the first

Inter-dependent and intra-dependent systems	**Functional vulnerability factors** *	**Organizational vulnerability factors** +	**Physical vulnerability factors** #
Systemic links *Lifelines vulnerability induced by other systems*	* Dependance from other systems	+ Interagency cooperation	# Physical vulnerability of systems which lifelines are physically connected to
Inter-systemic lifelines dependance	* Dependance from other lifelines	+ Cooperation among lifelines companies	# Physical vulnerability of lifelines essential for the function of other lifelines
Vulnerability factors due to lifelines components	* Specific lifeline features	+ Organization within each company Number of companies managing each lifeline	# Vulnerability of each lifeline component
Siting factors *Accessibility*	* Direct access to broken parts of lifelines	+ Coordination between civil protection and agencies in charge of roads management Information regarding alternative accesses Availabe personnel, materials and means for urgent repair	# Physical vulnerability of roads
Vulnerability due to the physical contact among lifelines	* Function depending from physical contact with other vulnerable lifelines	+ Coordination among lifelines service suppliers	# Physical vulnerability of single components in contact points (joints)
Regional vulnerability *Urban and regional systems for emergency operations*	* How much do emergency systems depend on lifelines	+ Coordination among hospitals, civil protection, police and other agencies communication skills to the public	# Physical vulnerability of systems other than lifelines necessary during emergencies
Other urban systems	* How much do other urban systems depend on lifelines	+ Coordination among lifelines suppliers and managers of other systems	# Physical vulnerability of systems other than lifelines

Figure 3. Lifeline vulnerability assessment grid in the emergency phase.

hours following the impact of a disaster is overloading. Another important organizational parameter, critical for assessing coordination among those involved in emergency preparedness, is the number of companies managing lifelines or even the same lifeline in a given area. A few big firms are able to control interactions emerging among their utilities and others and are therefore more likely to sign protocols and mutual agreements for joining operations whenever needed. When the number of companies is large, this kind of spontaneous and centralized protocol is difficult to reach and therefore coordination tasks become demanding on local and governmental authorities.

The rows in Figure 3 are described below:

- In the first row, factors related to lifeline performance are considered; i.e., all the functional, organizational, and physical elements and conditions that may severely disrupt utilities ability to provide their service to the community.
- In the second set of rows, siting situations are examined. For example, sewers can pollute water when conduits break and leaks or infiltrations become more probable. Accessibility to each part of the network—especially those that may be easily broken—is another crucial issue: some components of lifeline systems lie in remote areas that are hard to reach even in ordinary times.
- The last group of rows indicates how vulnerable urban and regional systems other than lifelines are to the interruption of utilities such as gas, electricity, water, and communication.

The framework to assess lifeline vulnerability during reconstruction is shown in Figure 4: it contains those elements that proved to be more relevant to assess lifeline ability (or inability) to recover quickly and at reasonable cost. In this framework, physical vulnerability is no longer considered, as ruptures are supposed to have occurred during the emergency that is now over. Functional and organizational aspects remain important, although the actual parameters used have to be slightly modified. In organizational terms, for instance, coordination with public administrations and with institutions in charge of financial aid and technical support for reconstruction becomes as crucial as coordination among lifeline management companies.

The frameworks illustrated in Figures 3 and 4 constitute the skeleton upon which detailed survey forms and appraisal matrices have been designed: individual parameters have been identified for many of the elements of lifeline systems, but not for all of them. Lifelines in fact are a rather complex and diverse set of plants and technical devices: the frameworks are flexible enough to permit improvement of the analysis of individual elements and the addition of relevant information to particular types of plant components without having to redesign the entire structure.

Inter-dependent and intra-dependent systems	**Functional vulnerability factors** *	**Organizational vulnerability factors** +
Systemic links		
Lifelines vulnerability induced by other systems	Function dependance from other systems (**link** with physical vulnerability in the em.)	Coordination with PA, Health service and managers of dangerous plants
Inter-systemic lifelines dependance	Function dependance from other lifelines (**link** with physical vulnerability in the em.)	Coordination among lifelines manager
Vulnerability factors due to lifelines components	Individual features of each lifeline (**link** with physical vulnerability in the em.)	Organization of each company (personnel, means, material)
Siting factors		
vulnerability due to the physical contact among lifelines	Degree of function dependance from other vulnerable lifelines	Coordination among lifelines manager
Regional vulnerability	Degree of function dependance from lifelines of urban systems	Coordination among lifelines manager and with PA
Urban and regional systems for emergency operations	Degree of dependance of public services from vulnerable lifelines	Coordination among lifelines manager and public facilities managers
Other urban systems	Degree of dependance of industries from vulnerable lifelines	Coordination among lifelines manager and industries
	Degree of dependance of residential areas from vulnerable lifelines	Coordination among lifelines manager PA and Public health services
	Degree of dependance of areas for emergency shelters from vulnerable lifelines	

Figure 4. Lifeline vulnerability assessment grid in the recovery phase.

Especially with respect to physical vulnerability, many parameters have not been detailed in this research. Some will be accounted for in future improvement of the method, but others pertain to specific geographical contexts that entail variations in the kind of joints, pipes, and point-shaped elements used; it is virtually impossible (and perhaps not so useful) to include everything in one single framework.

As for physical vulnerability, the following systems have been analyzed in detail, based on previous or new research: slope-triggered movements and potential lifeline damage due to vulnerable buildings and bridges. While the assessment of potential lifeline damage from slope-triggered movements and vulnerable buildings has been described in previous articles [10,1], a fast method for assessing bridge vulnerability using parameters that can be assessed during a visual survey has been developed and is briefly discussed below.

3.1. A FAST METHOD FOR ASSESSING BRIDGE VULNERABILITY THROUGH A VISUAL SURVEY

In order to assess the vulnerability of bridges, a two-step method was developed. First a form was developed, identifying the most relevant parameters that could be evaluated in a visual survey, without any physical test on the bridge structure. Those parameters express the most important features that make a bridge more or less resistant to horizontal acceleration; e.g., material,

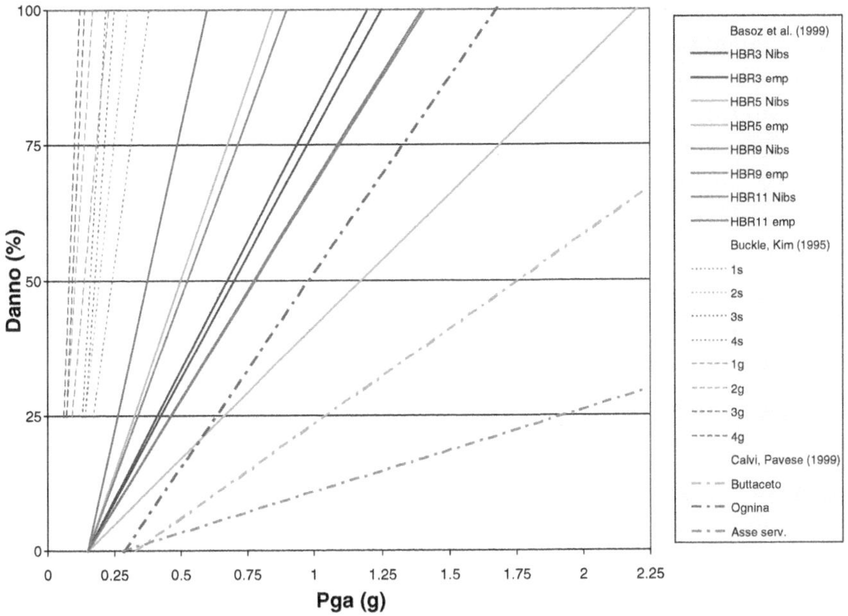

Figure 5. Fragility curves to assess bridges vulnerability to earthquakes.

dimension, length of bays, and shape. Those features may cover a wide range of options, with Italian structures generally more flexible and subtle than U.S. structures. This significantly limits the ability to correlate vulnerability indexes obtained for Italian bridges with damage that has been recorded in the U.S. (because of the different—more rigid—structural types, and because of higher seismic magnitudes).

Despite this limitation, as a second step of the procedure fragility curves for bridges were developed by referring to damage reports that are available in the scientific literature. As each fragility curve cannot be made to correspond to a specific bridge type, those curves were grouped in "fragility zones" (Figure 5).

Lastly, some Italian bridge types were assigned to a "fragility zone," to enable estimation of the expected level of damage once peak ground acceleration (PGA) was determined.

3.2. THE VULNERABILITY APPRAISAL PROCEDURE

In order to apply the two vulnerability assessment frameworks, measuring the indicators that have been selected, a rather complex set of forms has been prepared so as to allow consideration of the most important factors and issues that have been raised by literature and past experience until now. The resulting tool seems useful as it structures data and information that is already available and provides knowledge previously unavailable to interested agencies and companies. It constitutes an interesting cross-referencing framework for all managers in this delicate field, pointing at mutual dependency and the need for mutual agreements to make things work better in a disaster situation.

For the assessment to be accomplished, the model that has been developed requires various steps.

1. In the first step, separate forms for each lifeline must be completed with existing information. The forms also contain areas for declaring dependency on other lifelines as well as for expressing the degree of dependency of other urban systems on that particular utility.
2. The second step consists of four appraisal processes referring respectively to the following:

- Ability of a single lifeline to meet minimal required performance at the emergency.
- Time and expenditure needed in the recovery/reconstruction process.
- The capacity of urban systems necessary for providing emergency services to function despite some damage to lifelines.
- The capacity of all urban and regional systems to return to normalcy while restrictions in utility capacity are still in place.

For each parameter included in the matrices shown in Figures 3 and 4, at least two classes of roughly good or bad present conditions have been established by the research working group, on the basis of literature and post-earthquake reports.

Apart from classes of judgement, other considerations must enter into the assessment, related to the importance of each parameter with respect to all the others and to the hierarchical position of the assessed component. The first has been translated into a weight, while the second type, which has been indicated as "coefficient," takes into account the relative importance of the section of lifeline under examination. This coefficient distinguishes between high-voltage lines and local reticulation, between main roads and narrow secondary lanes, and between high-gas-pressure conduits and substations and distribution networks. It implies hierarchical distinctions that are very important in managing lifeline utilities.

The appraisal is then carried out by multiplying the score corresponding to each judgement class by the importance weight and by the hierarchical coefficient; partial results obtained for each parameter are summed up to get a final total score, representing the relative vulnerability of that particular lifeline in a virtual scale ranging between a minimal vulnerability level and a maximum, equivalent to the worst possible situation.

Different lifelines can be compared as far as their vulnerability status to help appraise the real situation, consisting—for example—of a perfect communication network depending on a highly vulnerable electrical system.

The method presented above provides a number of interesting features: the final vulnerability judgement can be always explained in the light of the parameters that most influenced it, thus clearly identifying problems. Secondly, it is possible to feed a GIS with the information generated through the assessment procedure, obtaining vulnerability maps for lifelines and their components in the region of interest.

Lifeline vulnerability analysis cannot be carried out without referring to geographical and spatial information; these are essential to preparing emergency plans and more generally any mitigation and damage reduction program. In urban contexts, geographical and spatial factors become even more important.

The main shortcomings of the proposed model are quite evident: it requires a large amount of precise and accurate data—a standard which can be hardly met by today's public administrations, at least in Italy. Furthermore private and public utilities are rather reluctant to provide data and plotted layers of their lines and stations, limiting to a great extent the potential for applying this kind of procedure. However, it is felt by the authors that growing concern about the stress of natural hazards on vital utilities—and the pressure on local authorities to make things function during crises—will change the present situation.

Public administrations at the regional level are considering—for example—the promotion of voluntary action to prevent risk at the municipal and provincial level, through mechanisms of quality assurance and shared responsibilities in the context of emergency aid and post-disaster funding.

4. Application in the Brescia Area

The method that has been described here was tested in the Brescia province, where most municipalities are exposed to medium/severe seismic tremors.

In order to test the method, a wider area has been considered, including the town of Brescia and several nonclassified settlements south and east of Garda Lake. The reason for this choice is that lifeline disruption and losses are not confined to the epicenter and core disaster area. Direct and indirect consequences have to be expected in a much larger context—the lifelines of which pertain to the same system as those that are likely to be damaged. This large area has been therefore subdivided into three zones, homogeneous as far as their geographical and urban features are concerned:

- The conurbation of Brescia and its immediate surroundings.
- The municipalities in the northern part of the lake coast, subject to the highest seismic intensities.
- The plains south of the lake and Brescia.

Several steps have been followed in order to test the method in all its parts. First, a vulnerability assessment was carried out considering a wide area in order to account for possible systemic effects. The result of this step is the evaluation of the vulnerability of lifelines and related systems in the area on a municipal basis; municipalities can be compared as well as different lifeline systems, to identify the most vulnerable systems and geographic zones.

As a second step, it was decided to restrict scenario building to the zone that has been recognized as most vulnerable among those assessed in the previous stage. In order to do this, a deterministic seismic input has been selected and the vulnerability of lifelines was analyzed and evaluated this time system-by-system and not just on a municipal basis. This required exact data and information regarding the location and technical features of those systems. Data was available only for the road and water systems and not for other lifeline utilities so the resulting scenario considers the extended effects of the damage virtually suffered by roads, water, and sewage systems only.

4.1. VULNERABILITY ASSESSMENT OF ALL LIFELINES ON A MUNICIPAL BASIS

In order to fulfill the first step of the procedure—that is, the survey of the parameters required in analytical form in order to evaluate the local situation—a workshop was organized with local authorities and representatives of utility companies in the Brescia province. Their contribution was fundamental to provide the necessary input for several parameters. Furthermore, in order to complete the analytical forms as thoroughly as possible, data and information kept by different administrations and agencies was gathered and structured according to the framework. This exercise proved to be useful in showing the need for a coordinating authority: individual companies, agencies, and administrations have very limited visions of the lifeline systems, strictly related to what they must control or provide to customers. However, in case of major events there is strong need for an agency in charge of managing lifeline coordination, not in detail but most importantly managing the many relations and links from which the most severe failures may derive.

The second step—i.e., the parameter evaluation—was carried out calculating the vulnerability index as explained in paragraph 3.2. Despite the fact that final vulnerability scores were assigned on a municipal basis and not to each individual lifeline, because of lacking or incomplete data, results can be considered significant already at this stage and provide important input to decision making.

First of all, in the whole province, water, sewage, and gas systems scored as the most vulnerable, confirming what has been reported in past earthquakes (Figure 6).

Salò was recognized as the most vulnerable zone, threatened by several landslides, which may represent a substantial source of induced damage for networks running in or beneath unstable slopes.

Furthermore, some systems—like the unique road linking all the municipalities along the lake to the main urban centers, and the water conduits which are locally fed and generally constitute a tree-like network—do not provide alternative routes in case of rupture.

Because this first general appraisal clearly indicated the Salò lifeline systems as the most vulnerable to earthquakes, it was decided to construct the complete event scenario in this area.

4.2. THE COMPLETE EVENT SCENARIO IN THE SALÒ AREA

In order to carry out a complete event scenario in the Salò area, the following ingredients were required:

1. A deterministic seismic input was selected.
2. The vulnerability of roads and water systems was evaluated in order to detect all critical elements and factors.
3. The vulnerability of urban systems which depend on lifelines for their normal function was appraised. With respect to the emergency stage, the vulnerability of hospitals and fire departments was considered; looking ahead at the reconstruction, the vulnerability of the local economic system was evaluated.
4. The scenario was constructed, combining the deterministic input, the possible induced hazards and the vulnerability, revealing a chain of subevents and failures in the area and in its surroundings.

4.2.1. Definition of the Seismic Input

To define the seismic input, the highest magnitude regional event (which occurred in the nearby Verona Province), magnitude 6.7 on the Richter scale, was moved to make the epicenter coincide with the city of Salò.

Applying attenuation laws, it was possible to obtain the maps of earthquakes in terms of PGA and other relevant parameters, such as the Arias intensity and the spectral intensity.

Those parameters were also used to evaluate the potential for induced landslides, using the Newmark method to assess at what critical coefficient

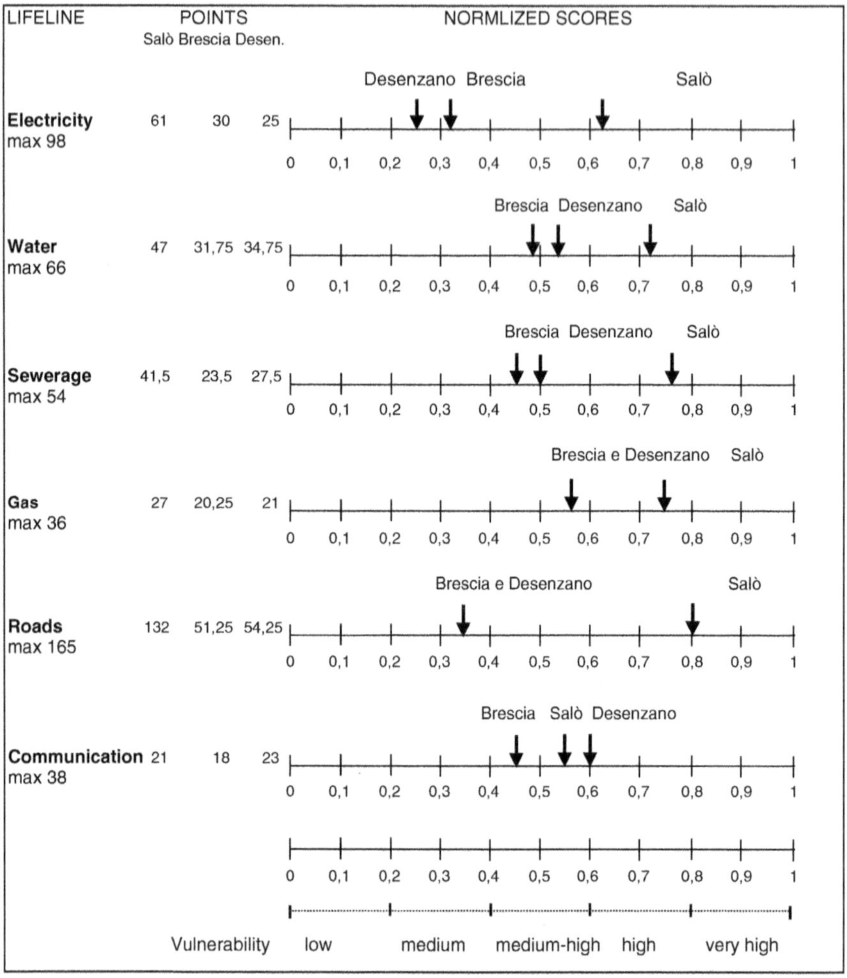

Figure 6. Results of the vulnerability assessment in the Brescia area.

slides would move and what the consequent displacement would be; secondly, a percentage method was developed to assess the quantity of material released by rock falls, covering zones at different distances and inclination with respect to the fractured rock slope.

4.2.2. The Vulnerability of Water and Road Systems; Vulnerability of Other Urban Systems to be Considered in the Emergency and in the Reconstruction

Factors of vulnerability related to road and water systems have already been discussed. In summary, the most critical elements derive from several sources of induced vulnerability, the uniqueness of accessways in the case of the single road running parallel to the lake and of conduits in the case of the local water network; and in the dependency of both on electric power because of the many traffic lights in urban centers crossed by the main road and the energy needed to pump water from wells.

As for the vulnerability of other urban systems crucial for emergencies, the hospitals in the area seem particularly dependent on external sources of water, gas, communication, and even electricity, as autonomous plants, where they exist, have a very limited capacity.

As for the economic sectors that have been analyzed to assess the vulnerability in the reconstruction and recovery phases, the industries and productive activities in the area, including tourism, have been considered and their degree of dependency on lifelines has been indicated in the sixth column of Figure 2. This degree of dependency clearly identifies a degree of vulnerability, as it measures for how long and how well those activities can carry on without normal utilities such as electrical power, gas, water, communication, and transportation. Those degrees of dependency have been set based on the available literature and a direct survey that was carried out in another Italian region (Emilia Romagna) to assess the capacity of medium to small family run businesses to respond to a similar extreme event.

4.3. DAMAGE SCENARIO

The main physical damage resulting from the deterministic input as defined in paragraph 4.2.1 to roads and water systems can be seen in the map shown in Figure 7.

They mostly derive from induced damage due to landslides and partial building collapse. Landslide displacements greater than 5–6 cm and building damage exceeding 60% have been considered thresholds for inducing significant damage to the road system. As seen in the map, only one bridge suffers a damage index greater than 50% as a consequence of the earthquake.

As a result of that damage, some municipalities north to Salò would be isolated and unreachable by fire brigades; victims could only be transported to hospitals by helicopter or ship.

The effects of road partial or complete interruption have been evaluated for the most important economic activities in two of the investigated municipalities: Toscolano Maderno and Salò. Because of its greater economic importance as well as its proximity to the epicenter, Salò suffers from larger economic disruption than Toscolano Maderno. Comparing different economic sectors, tourism and commerce are the most severely affected.

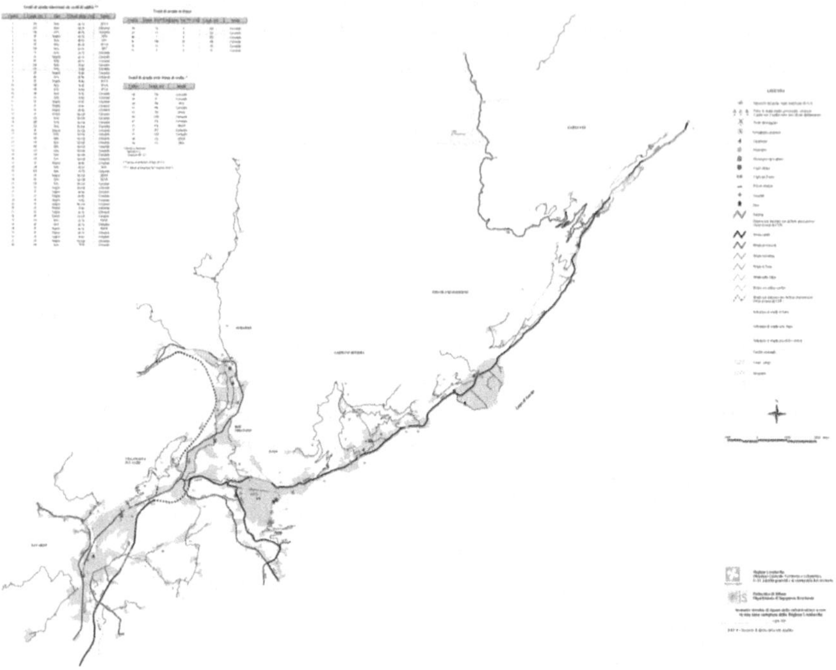

Figure 7. Map showing results of the damage scenario in the Salò area.

4.4. RECOMMENDATIONS FOR LOCAL AND PROVINCIAL AUTHORITIES RESULTING FROM THE VULNERABILITY ASSESSMENT AND THE COMPLETE EVENT SCENARIO

Recommendations can be grouped in two main streams depending on whether they are related to long- or short-term decisions. The first are more related to urban and land-use planning activities: the single road connecting the various municipalities along Garda Lake should be managed much more carefully than is currently the case. The risk of landslides in various sections should be recognized and adequate defense structures should be put in place. The priority in retrofitting vulnerable buildings should go to those standing along this single road, to assure as much as possible that an earthquake will not cut several settlements from relief and aid during the emergency.

In more general terms, a provincial bureau should take care of collecting and organizing the information related to lifelines and their mutual relations, in order to be really able to guarantee a sort of supervising and coordination activity in case of need. Such an authority should not enter in the detail or in the decisions pertaining to each utility company, but should make sure that protocols and agreements are signed to guarantee mutual support in case of emergencies and that vital facilities such as hospitals and fire departments are autonomous enough from lifelines to be able to carry out search and rescue activities at least in the first hours after an extreme event.

The vulnerability assessment and the construction of the complete event scenario also give some ideas concerning the economic sectors that could be most affected in case of earthquakes: public administrations could grant incentives to make them more resilient.

The second category of initiatives involves more short-term measures to update emergency procedures. For example, the notion of the particular systemic vulnerability of accessways to several municipalities and settlements should make civil protection authorities seriously consider alternative possibilities for reaching areas, in particular by lake. Other recommendations concern hospitals, which should prepare internal mass emergency plans and acquire equipment to be less dependent on external lifelines in case of serious disaster.

5. Conclusions

The paper has illustrated a methodology that has integrated different fields of study and research into a systemic framework. The final results are on the one hand a method to assess lifeline vulnerability, considering systemic and organizational factors as equally if not more important than physical factors, as the latter cannot be controlled over large surfaces and in the totality of lifeline networks and plants. On the other hand, a complete scenario was developed, showing the entire chain of losses and consequences that an earthquake may trigger in a complex environment, thus considering not only the physical damages but also the induced and secondary effects in larger areas and in different sectors of urban and community life.

As briefly discussed, the frameworks that were developed to assess vulnerabilities and to create scenarios are flexible and open enough to include new achievements and new knowledge whenever this becomes available.

It is the hope of the authors that such a method could constitute an initial though highly integrated basis for future research in this complex field of study.

6. References

1. Angeletti P, Pergalani F, Petrini V. A methodology for assessing seismic risk. Proc. Third National Earthquake Engineering Conference, March 27-31 1995, Istanbul, Turkey.
2. Basöz NI, Kiremidjian AS, King SA, Law KH. Statistical analysis of bridge damage data from the 1994 Northridge, CA, earthquake. Earthquake Spectra. 1999; 15(1).
3. Calvi GM, Pavese A. Seismic assessment of bridge piers. In: Faccioli F, Pessina V, editors. The Catania project: earthquake damage scenarios for a high risk area in the Mediterranean. Rome: GNDT; 2000.

4. Chang S, Shinozuka M, Moore II JE. Probabilistic earthquake scenarios: extending risk analysis methodologies to spatially distributed systems. Earthquake Spectra. August 2000; 16(3):557-572.
5. Centre for Advanced Engineering, New Zealand (CAENZ). Risks and realities: a multi-disciplinary approach to the vulnerability of lifelines to natural hazards. Christchurch: CAENZ; 1997.
6. Cochrane H, Howe C. Guidelines for uniform definition, identification, and measurement of economic damages from natural hazard events; special publication n. 28. Boulder: Natural Hazards Research and Applications Center; 1993.
7. Haas JE, Kates RW, Bowden MJ. Reconstruction following disasters. Cambridge: MIT Press; 1977.
8. Isoyama R, Ishida E, Yune K, Shirozu T. Seismic damage estimation procedure for water supply pipelines. Water Supply. 2000; 18(3):63-68.
9. Nojima N. Lifeline system malfunction and interaction. Proceedings of the World Urban-Earthquake Conference in Fukui. June 1998.
10. Petrini V. Overview report in vulnerability assessment. Proceedings of the Fifth International Conference on Seismic Zonation, Nice, France. October 1995.
11. Ringland G. Scenario planning: managing for the future. UK: Wiley; 1998.
12. Shinozouka M, Rose A, Eguchi R. Engineering and socio-economic impacts of earthquakes. Buffalo: MCEER; 1998.
13. The Institution of Civil Engineers. Megacities, reducing vulnerability to natural disasters. London: Thomas Telford; 1995.

CHAPTER 8

A NETWORK SECURITY ARCHITECTURE USING THE ZACHMAN FRAMEWORK

A.B. RAMADAN, M. HEFNAWI

National Center for Nuclear Safety and Radiation Control
Egyptian Atomic Energy Authority
Cairo, Egypt

Abstract: With the growth of information technology (IT) systems, network security is rapidly becoming a critical business concern. Due to the interconnectivity of IT systems, a comprehensive description of all of the key elements and relationships that make up an organization's network security is needed, which can be referred as network security architecture. The value of this architecture is often questioned by organizations in terms of its practical application. This paper has presents a new approach to the network security architecture in order to better describe and instantiate the Zachman Framework capabilities. The network security architecture of academic centers is discussed as a case study to show how a conceptual model can be applied to a real organization.

1. Introduction

The industry is moving toward more formal development and documentation of enterprise architectures based on Enterprise Architecture Frameworks. The term "architecture" has been used for many years within the information technology (IT) community to refer to various types of overviews that provide guidance to software systems and applications developers. The term is obviously a metaphor derived from the building trade [1,8]. Much like a homeowner designing a home, information technology managers work with an architect to provide an agreed upon architectural drawing of the enterprise's information systems and processes. This high-level architectural drawing does not change with tactical decisions to deploy improved technology since it is simply built around a framework of business processes and the information that they need [2].

Today, there is a growing movement among both business managers and IT managers to use the term "enterprise architecture" to refer to a comprehensive description of all of the key elements and relationships that make up an organization. Based on this, enterprise information architecture provides a

I. Linkov et al. (eds.), Managing Critical Infrastructure Risks, 133–143.
© 2007 *Springer*.

framework for reducing information system complexity and enabling enterprise information sharing. Since most enterprises have existing information systems, the architectural drawing provides the future state and facilitates the best possible strategy to remodel with the least amount of inconvenience to the business [1].

The rapidly growing interconnectivity of IT systems, and the convergence of their technology, renders these systems increasingly vulnerable to malicious attacks. Network attacks cause organizations several hours or days of downtime and serious breaches in data confidentiality and integrity. Depending on the level of the attack and the type of information that has been compromised, the consequences of network attacks vary in degree from mildly annoying to completely debilitating, and the cost of recovery from attacks can range from hundreds to millions of dollars [3].

This paper presents a network security architecture based on the Zachman Framework. The aim of this architecture is to organize the data, process, and technology around the points of view taken by various players instead of representing them as entirely separate entities. For this, we'll consider the Zachman Framework in more detail in Section 2. In Section 3, the relation between network security and the Zachman Framework is discussed. An example for designing security architecture of academic centers based on the Zachman Framework is presented in Section 4 and Section 5 is the conclusion of the paper.

2. The Zachman Framework

The Zachman Framework for Information Systems Architecture (ISA), defined in 1987, is a logical construct to define and control the interfaces and integration of all components of a system. The framework of the Zachman model enables systematic capture of system specific information from the various perspectives with respect to system architecture [4]. Figure 1 illustrates the Zachman model, tailored to support a network security system.

In this customization of the model, the system developers have an existing operational system in place.

The rows at the top are the most abstract and are oriented toward very broad goals and plans. If we were building a house, this layer would describe the diagrams, pictures, and plans the architect would discuss with the owner. The next level is more specific, but still abstract. These are the diagrams that the architect would discuss with the contractor. In a similar way, the top level of the Zachman Framework, labeled "Scope," is focused on the concerns of senior executives. The second is focused on the slightly more detailed concerns of business managers. Lower levels focus on concerns that business and IS managers work together on, and then, finally, on details that IS managers and developers work on [1]. The columns in the

	DATA	FUNCTION	NETWORK	PEOPLE	TIME	MOTIVATION
Planner	List of Things Important to the Enterprise	List of Processes	List of Locations	List of Organizational Units	List of Events	List of Business Goals
Owner	Semantic Model	Business Process Model	Network Logistics System	Work Flow Model	Master Schedule	Business Plan
Designer	Logical Data Model	Application Architecture	Distributed System Architecture	Human Interface Architecture	Processing Structure	Business Rule Model
Builder	Physical Data Model	System Design	Technology Architecture	Presentation Architecture	Control Structure	Rule Design
Sub-Contractor	Data Definition	Program	Network Architecture	Security Architecture	Timing Definition	Rule Specification
Functioning	Data	Function	Network	Organization	Schedule	Strategy

Figure 1. The Zachman Framework.

Zachman Framework represent different areas of interest for each perspective. The columns describe the dimensions of the systems development effort. The Zachman Framework has two very distinctive features that make it ideal for information modeling. The framework may be applied at any level of abstraction in the system development process, from a global enterprise, to a system, subsystem, or major module level. The framework also gives the modeler latitude in that any data representation technique can be used to model the inner workings of each cell. The system model becomes more implementation specific. However, the requirements traceability between layers can be maintained through backward references to upper layers of cells. This traceability is critical in security requirements engineering, where tracing a global access control requirement may translate into explicit setting of access controls on specific files or directories within an operating system.

The framework provides a taxonomy:

> …that helps us understand the perspectives of various players in the development of an information system and the descriptions of the system that can be produced during its creation [4].

The model is frequently used as a framework during information systems activities to support the solicitation, identification and mapping of the following information re-engineering associated with an information system's [4]:

- Goals, objectives, and environment.
- Customers served.
- Time constraints.
- Functional description.
- Information architecture.
- Supporting infrastructure.

In short, the Zachman ISA can provide a consolidated view of a system, to whatever level of detail a modeler chooses.

3. Network Security and the Zachman Framework

The objective of network security architecture is to provide the conceptual design of the network security infrastructure, related security mechanisms, and related security policies and procedures. The security architecture links the components of the security infrastructure as one cohesive unit. The goal of this cohesive unit is to protect corporate information [3]. The security architecture should be developed by both the network design and the IT security teams. It is typically integrated into the existing enterprise network and is dependent on the IT services that are offered through the network infrastructure. The access and security requirements of each IT service should be defined before the network is divided into modules with clearly identified trust levels. Each module can be treated separately and assigned a different security model. The goal is to have layers of security so that a "successful" intruder's access is constrained to a limited part of the network. Just as the bulkhead design in a ship can contain a leak so that the entire ship does not sink, the layered security design limits the damage a security breach has on the health of the entire network. In addition, the architecture should define common security services to be implemented across the network [7].

For security architecture modeling purposes, the first three levels of the perspective hierarchy (planner, owner, and designer) and the Network cell of the Builder's view are useful. They provide the consumer perspective of the system's end user, the perspective of the system "owner" or contracting entity, and the perspective of the designer, or systems engineer. In other words, the "as built" and used in daily operation perspective, the "as desired" operation perspective, and "as actually specified" perspective. The highest level, the Planner View, defines a clear and coordinated boundary (domain) of the system for the purposes of identifying the people, subsystems, and needs impacted by the system. The Owner's View captures the business and organizational relationships, and their external interfaces. It also documents sources of system requirements, including those derived from legacy systems. The Designer's View establishes and documents the security architectural design and provides a basis for system measurement. Finally, the Builder's View provides a detailed description of the design and methodology for monitoring and correcting system performance [2].

Similarly, the first three columns of the Zachman matrix (data, function, and network) provide the answers to what data assets the organization controls, how they are used and where they are located [5].

As shown in Figure 1, these are:

- *Data*: Each of the rows in this column address understanding of and dealing with an enterprise's data. This begins in Row One with a list of the security concerns of the enterprise and its directions and purposes. Row Two is a contiguous model of the security problems seen by the participants in the business. Also, relationships may be shown which themselves have attributes. Row Three provides more of an information-based perspective of the network security, resolving the rules and relationships, along with relationships containing their own attributes. Indeed, attributes are more exhaustively defined, and unique identifiers are specified.
- *Function*: The rows in the function column describe the process of translating the mission of the network security system of the enterprise into successively more detailed definitions of its operations. Where Row One is a list of the kinds of network security related activities the enterprise conducts, Row Two describes these activities in a contiguous model. Row Three portrays them in terms of data transforming processes, described exclusively in terms of the conversion of input data into output data.
- *Network*: This column is concerned with the geographical distribution of the enterprise's activities. At the strategic level (Row One), this is simply a listing of the places where the enterprise does business. At Row Two, this becomes a more detailed communications chart, describing how the various locations interact with each other. Row Three produces the network architecture for data distribution, itemizing the special security policy for the enterprise. In Row Four, this distribution is translated into the kinds of computer and network facilities that are required in each location to force the security policy.

4. The Network Security of Academic Centers: A Case Study

Academic centers, as major users of information and communication technology (especially Internet), also need security; however, because of their special structure and requirements, the traditional solutions and policies to limit access to the Internet are not effective for them. These institutions face concerns about the security of computing resources and information. The security problems in these environments are divided into two categories [3, 6]: problems with research information and problems with administrative information. Although the corporate and academic environments face common security problems they can't choose similar methods to solve them, because of their different structures. In a corporate environment, the natural place to draw a security perimeter is around the corporation itself. However, in an academic environment, it is very difficult to draw a perimeter surrounding all of the people who need to access information resources and only those people. This is mainly because of different types of information resources in these environments and also different users who want to access them. So if the security perimeter chosen is too big it includes untrusted people and if it is chosen too small it excludes some of the authorized people.

In addition, corporations can put serious limitations on the Internet connectivity in the name of security but research organizations simply cannot function under such limitations. First, trusted users need unrestricted and transparent access to Internet resources (including World Wide Web (WWW), FTP, Gopher, and electronic mail) located outside the security perimeter. Researchers rely on fingertip access to online library catalogs and bibliographies, preprints of papers, and other network resources supporting collaborative work. Second, trusted users need the unrestricted ability to publish and disseminate information to people outside the security perimeter via anonymous FTP, or WWW. This dissemination of research results and papers is critical to the research community. Third, the security perimeter must allow access to protected resources from trusted users located outside the security perimeter. An increasing number of users work at home or while traveling. Research collaborators may also need to enter the security perimeter from remote hosts.

If we consider these centers as an enterprise, the security architecture of their network can be designed based on the Zachman Framework. For the first four rows and first three columns of the framework the cells can be completed as described in the following sections:

4.1. PLANNER'S VIEW

An overall organizational policy would be implemented in the Planner's View. The first cell is the list of things important to the academic centers. Research groups often need to maintain the privacy of their work, ideas for future research, or results of research in progress. Administrative organizations need to prevent leakage of student grades, personal contact information, and faculty and staff personnel records. Moreover, the cost of security compromises is high. A research group could lose its competitive edge, and administrative organizations could face legal proceedings for unauthorized information release. On the other hand, academic and research institutions are ideal environments for hackers and intruders and many of them are physically located in these places and they are highly motivated to access and modify grades and other information. There are several reports of break-ins and deletion of data from educational institutions [3, 6].

The second cell in this row is the list of the processes important to the enterprise. This can also be divided into two categories: academic processes, such as examinations; and research processes, such as conducting projects and disseminating information.

The next cell (the network cell) is the location of the academic center. It is much easier to develop a network security architecture for universities with a central campus than universities with several branches.

4.2. OWNER'S VIEW

The next level down, the Owner's View, considers the groupings of data and means of access available to both internal and external users. For the first cell (data), we can see three categories of information in a university:

- The information officially disseminated by the university (such as news and events articles).
- The information gathered and used by network users.
- The information not allowed to be publicly disseminated.

Based on the above categories, three types of function servers (second cell) may be proposed in the university:

- Public servers, which are used to support information dissemination.
- Experimental servers, which are used for researchers and students to develop and test their own software packages and protocols.
- Trusted servers, which are used for administrative purposes or keeping confidential information.

These servers are the places where functions occur with respect to the data [9].

The other requirement of an academic environment is to let its trusted members access the resources of the network from outside of the security perimeter (e.g., from home or on trips).

Another problem that causes serious trouble for the university is network viruses. These viruses are distributed through the network after users access special sites. Proxy servers can be used to control this problem. Of course these proxy servers should be transparent.

The network cell of the framework in this layer is shown in Figure 2.

4.3. DESIGNER'S VIEW

At the next level, the Designer's View, we introduce mechanisms to protect the network. To achieve the goals described in owner's view, the logical data model (first cell) of the proposed network security policy was designed based on seven basic rules:

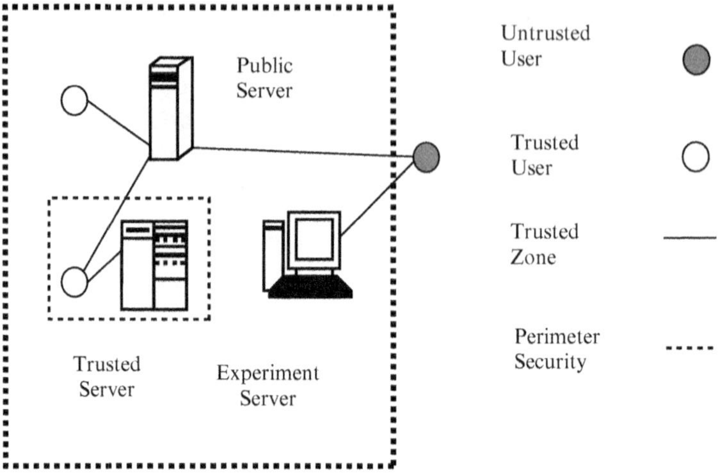

Figure 2. Network layer in Owner's View.

1. Packets to or from the public servers are unrestricted if they are from authorized ports. The authorized port is the port that the special service is on. Of course, each public server should be protected itself. The server-level security means to enforce stronger access controls on that level.
2. Packets to or from the experimental servers are unrestricted. These servers can be located outside of the security perimeter.
3. Packets to or from the authorized ports of trusted servers are allowed only from or to the authorized clients inside the security perimeter.
4. All of the outgoing packets are allowed to travel outside after port address translation. The incoming packets are allowed if they can be determined to be responses to outbound requests.
5. The packets to or from trusted users of hosts outside the security perimeter are allowed.
6. All of the requests from particular applications such as http should be passed through a proxy server.
7. All the packets to or from outside the security perimeter should be passed through the Intrusion Detection System (IDS).

Rule 1 is based on our need to support information dissemination in a research environment. We have to separate the public servers from our trusted hosts and protect them at the server level and accept the fact that they may be compromised, so we should have a plan to recover them from information kept securely behind the security perimeter.

Rule 2 follows from our recognition that researchers and students sometimes need to develop and test insecure software packages and protocols on the Internet. Of course they should be alerted that their server is not secure and their information may be corrupted.

Rule 3 is based on the fact that we want to protect confidential information. These servers are our most important resources to be protected and we put them in a special secure zone.

Rule 4 follows from our recognition that open network access is a necessary component of a research environment. On the other hand we don't want to allow the users to set up Internet servers without permission. The address translation prevents outside systems from accessing internal resources except those listed as public servers.

Rule 5 grants access to protected resources to users as they work from home or while traveling, as well as to collaborators located outside the research group.

Rule 6 is based on the need to block some Internet sites that contain viruses.

Rule 7 follows from our recognition that the above rules should be monitored somehow. An IDS can be a proper tool to monitor the network and check if there is any violation of proposed rules. The network cell is shown in Figure 3.

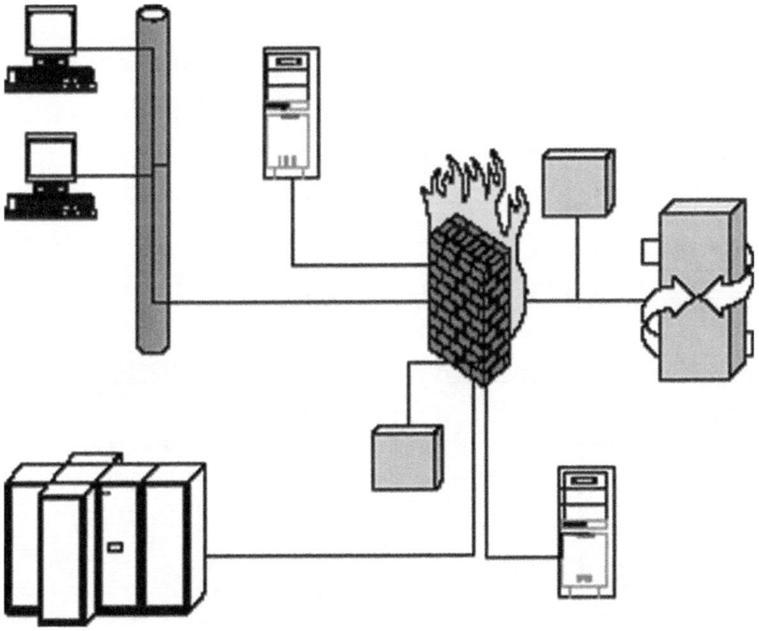

Figure 3. Network layer in Designer's View.

4.4. BUILDER'S VIEW

Finally, the Builder's View describes how technology may be used to address the information processing needs identified in the previous rows. For network security purposes, mainly the network cell is needed. Generally, two ways can be proposed to implement the designed network: first, to use hardware firewalls (such as Cisco PIX or Watchguard) and caches; and second, to use general purpose servers with proper software packages as cache, proxy, and firewall. In our case study in the Egyptian Atomic Energy Authority we used a server with Linux operating system (Redhat 7.3 upgraded to Redhat 8.0) with a normal hardware specification (800 MHz CPU, 1 GB RAM). We used SQUID as the transparent proxy and cache server, and IPTABLES as the firewall for packet filtering, in which the different zones of the network were defined. We also used Network Address Translation of the IPTABLES for implementing the rules in our design. Of course each server in the network had also its own security rules and guards. For restricting access to special websites (mainly to avoid viruses) SQUIDGUARD software was utilized. We used SNORT as our IDS. The network cell is shown in Figure 4.

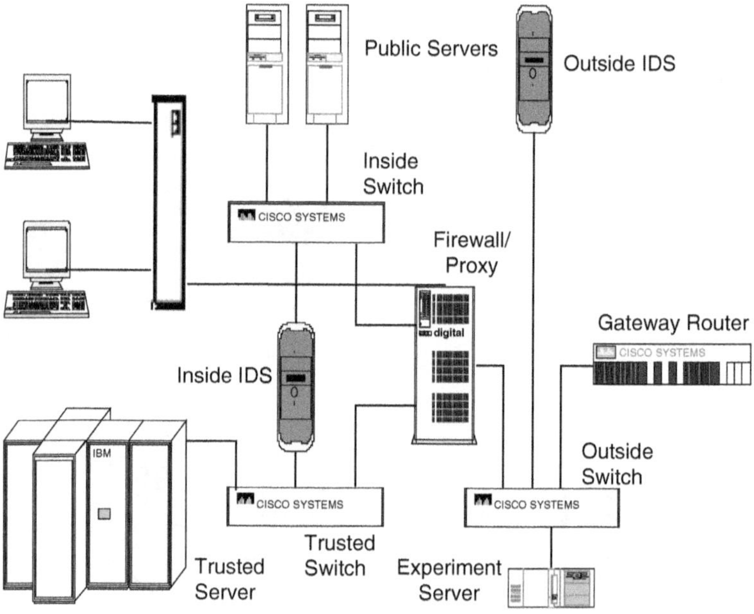

Figure 4. Network layer in Builder's View.

5. Conclusion

The Zachman ISA framework for systems modeling provides a commonly used technique that can be applied to network security architecture modeling early in the system requirements definition process. By applying the top three levels of the Zachman hierarchy, it is possible to develop descriptive security architecture. They provide the "as built" and used in daily operation perspective, the "as desired" operation perspective, and "as actually specified" perspective. Similarly, the first three columns of the Zachman matrix (data, function, and network) provide the answers to what data assets the organization controls, how they are used, and where they are located. Academic centers, as one of the major users of information and communication technology, present a good case study for applying our proposed architecture. The key point of the research is to design the network security architecture of these centers based on a framework so it provides the consumer perspective of the system's end user, the perspective of the system "owner" or contracting entity, and the perspective of the designer or systems engineer simultaneously.

6. References

1. Harmon P. Developing an enterprise architecture. Business Process Trends. Available at: http://database.ittoolbox.com/documents/document.asp?i=2385. 2002.
2. DeLooze LL. Applying security to an enterprise using the Zachman Framework. SANS Publications. Available at: http://www.sans.org/rr/paper.php?id=367. 2001.
3. Mohajerani MR, Moeini A. An approach to a new network security architecture for academic environments. Proc. of the 21st International SAFECOMP Conference, Italy. 2002.
4. Henning R, Corporation H. Use of the Zachman architecture for security engineering. Proc. of the 19th National Information Systems Security Conference, Baltimore, MD. 1996.
5. Hey DC. A different kind of life cycle: the Zachman Framework. Essential Strategies Inc. Available at: http://www.essentialstrategies.com/documents/zachman2000.pdf. 2000.
6. Greenwald M et al. Designing an academic firewall: policy, practice and experience with SURF. IEEE Proceedings of 1966 Symposium of Network and Distributed Systems Security. 1996.
7. Ramachandran J. Designing security architecture solutions. Hoboken: John Wiley and Sons; 2002.
8. Heaney J et al. Information assurance for enterprise engineering. Proc. of the 9th Conference on Pattern Language of Programs, Monticello, Illinois. 2002.
9. Rosenthal M, Coopers P. Three-zone model to depict enterprise security & technology architectures. 28th Annual Computer Security Conference, Washington D.C. 2001.

PART 3

ECOLOGICAL RISKS IN HARBORS AND COASTAL AREAS

CHAPTER 9

CONCEPTUAL FRAMEWORKS TO BALANCE ECOSYSTEM AND SECURITY GOALS

S.E. APITZ

SEA Environmental Decisions, Ltd.
1 South Cottages, The Ford, Little Hadham
Hertfordshire SG11 2AT UK

Abstract: Emerging European legislation is changing the scope of water management from the local scale to basin scale. The focus is shifting from sectoral, issue-by-issue management to the protection of aquatic ecosystems, as well as the terrestrial ecosystems and wetlands linked to them. There has also been a movement from addressing problems in isolation on land, in freshwaters, in estuaries or the coastal zone, to integrating these zones, and extending the ecosystem approach to whole shelf areas. Ecosystem protection will thus affect how many human activities are regulated and managed in coastal and port areas, but legislation is also designed to balance these ecosystem objectives with socioeconomic needs and goals. Sustainable protection of ecosystems requires an expansion of traditional ecological risk assessment methods, in order to address multiple risk drivers on multiple spatial and temporal scales. If one accepts the Belluck et al. [8] definition of environmental security, which:

> involves actions that guard against environmental degradation in order to preserve or protect human, material, and natural resources at scales ranging from global to local…

then the goals of this legislation can be defined as environmental security. However, the current climate of anxiety about terrorism and extreme events often results in a situation where rare but dramatic events (such as terrorist attacks and extreme storms) are not addressed in the same frameworks as the more mundane issues such as contaminant control and habitat degradation. There is a need to develop decision frameworks in which these seemingly disparate issues are addressed together in support of regional budgeting, decision making, and management. To that end, vulnerabilities must be identified and ranked, and decisions must be developed based upon a number of issues including scenario probability, preventability, causality (human-caused or natural), time scale (gradual or sudden), and potential costs and

I. Linkov et al. (eds.), Managing Critical Infrastructure Risks, 147–173.
© 2007 *Springer*.

risks. Depending on these assessments, prevention strategies and response strategies (whether a scenario is unpreventable or if prevention fails) must be developed.

1. Environmental Security and Sustainability

The term "environmental security" means different things to different people, depending upon their background, objectives, and focus. A search for this term in the published literature and the Internet brings up a myriad of issues, but the bulk seems to cluster in two categories, depending upon the timescale of discussion. If one looks at the short- to mid-term, references to environmental security generally reflect concerns about the potential negative consequences of extreme natural or anthropogenic events such as storms, floods, wars, and terrorism, and if one looks at the longer term, references generally reflect concerns about negative effects of more gradual events such as global warming, sea level rise, ecosystem degradation, and the consequent losses of ecosystem services, equity, and thus socioeconomic and political stability. As discussed in Wenning et al. [42], a slight revision of the definition of environmental security provided by Belluck et al. [8] seemed the most appropriate for the range of issues addressed at the workshop:

> Environmental security involves actions that guard against environmental degradation in order to preserve or protect human, material, and natural resources at scales ranging from global to local <u>in a sustainable manner</u> [new text and emphasis added].

This modified definition suggests that environmental security requires a balance between human, environmental, and socioeconomic needs. Thus, it is clear that environmental security as a concept is closely linked to the objectives of sustainable development. These links have been explored at length by the Millennium Ecosystem Assessment (MES), which concludes that human security is dependent upon a healthy and functioning ecosystem. According to MES [27]:

> Human well-being and progress toward sustainable development are vitally dependent upon improving the management of Earth's ecosystems to ensure their conservation and sustainable use. But while demands for ecosystem services such as food and clean water are growing, human actions are at the same time diminishing the capability of many ecosystems to meet these demands. Sound policy and management interventions can often reverse ecosystem degradation and enhance the contributions of ecosystems to human well-being, but knowing when and how to intervene requires substantial understanding of both the ecological and the social systems involved. Better information cannot guarantee improved decisions, but it is a prerequisite for sound decision-making.

Under such a framework, ecosystems and human well-being are inextricably linked. In fact, MES [27] contrasts the dimensions of human well -being (freedom of choice and action, security, good health, sufficient material wealth for a good life, and good social relationships) with their obverse, the dimensions of ill-being (powerlessness, vulnerability, ill health, material lack, and bad social relations), and defines the linkages between well-being and ecosystem services (defined here as the benefits people obtain from ecosystems). Figure 1 illustrates the complex linkages as defined in their conceptual approach.

The perception by some that there is a conflict between environmental protection and economic development is, in this framework, largely a consequence of two factors—a disconnect in timescales between immediate benefits of some economic choices and the longer-term risks to sustainable delivery of ecosystem services and a disparity between those who reap benefits and those who bear the risks [27]. However, over time, continued unsustainable exploitation will result in a loss of services for all sectors of society. Thus, the provisioning of sustaining goods and services that we obtain from natural ecosystems is a strong economic justification for the conservation of biological diversity [15]. While one can argue that there is a moral obligation to protect biodiversity and ecosystem health, these frameworks make clear that it is also essential for the maintenance of human well-being and security.

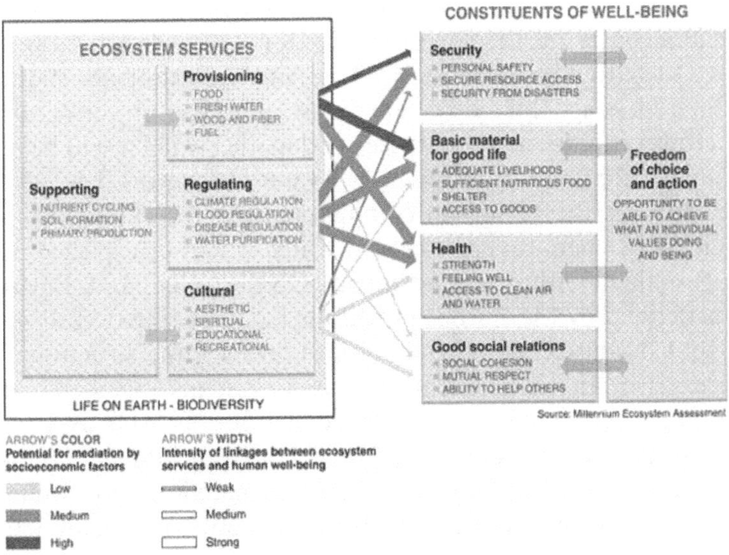

Figure 1. Ecosystem services and their links to human well-being [27].

2. Ecosystem Health

Ecosystem health has been defined many ways, including:

- The maintenance of biodiversity and ecosystem integrity [1].
- a) Lack of a distress syndrome, b) stability over time, and c) resilience to change [32].

Costanza et al. [12] proposed that ecosystem health can be defined using three components (V-O-R):

- a) **V**igor (throughput or productivity of the ecosystem; an ecosystem should be actively producing).
- b) **O**rganization (species diversity and complexity of trophic and other interactions should be maintained over time).
- c) **R**esilience (an ecosystem's ability to maintain structure and patterns of behavior in the face of stress).

These approaches are appropriate for describing the ecosystem health concept, but do not define ecosystem health in terms of easily measurable quantities. While Costanza proposed a set of indices to define the V-O-R terms, they are difficult to quantify even in relatively well understood systems. Furthermore, it is difficult to address these issues in a complete way if one is only assessing or managing risk in specific sectors, rather than whole ecosystems of estuaries, lakes, regions, and watersheds.

Current and emerging EU ecosystem-based policy will result in ecosystem-based management of the environment from land to the open sea, but estuaries and coasts are a region of overlap for several directives and thus will be subject to particular attention [2, 3, 5]. Various European statutes define their ecosystem goals differently, depending upon their objectives, drivers and criteria. However, since many of these overlap, they will have to be reconciled. Good Ecological Status is defined in the Water Framework Directive in terms of the quality of the biological community, and the characteristics of the hydrological and chemical environment which would be expected under conditions of minimal anthropogenic impact. On the other hand, Environmental Status, as defined by the European Marine Strategy (the overall state of the environment in marine waters, taking into account the structure, functions, and processes of the constituent marine ecosystems together with natural physiographic, geographic, and climatic factors, as well as physical and chemical conditions including those resulting from human activities in the area concerned) allows for human use and sustainability, and requires that the ecosystem and socioeconomic impacts and benefits of human activities are balanced. Although these directives do not specifically mandate "ecosystem health," this term is used extensively and interchangeably with terms such as "ecosystem status" in discussions and supporting documents on most of these statutes, and must be addressed and defined. For instance, in an evaluation of how one might assess eutrophication in support of the above statutes, "undesirable disturbance" is defined as a "perturbation of an ecosystem that appreciably degrades the health or sustainable human use of that ecosystem" [14]. On the other hand, others have objected vehemently to the use of the term "ecosystem health"—Suter [35] called it a "bandwagon," applying a "value-laden metaphor," rather than a observable property, the misuse of which leads to inappropriate tools and erroneous management actions, such as the development and use of indices which combine several ecosystem variables into a single index value. He suggests that the terms "environmental sustainability" or "quality" are preferable. Kapustka and Landis [24] warn against an "ecotheocracy," redolent with the concepts of grand design, the goodness of nature and the evil of humans, in which the base principles of ecosystem health, integrity, stability, balance of nature, recovery, and restoration are unacceptable as science-based assessment endpoints; inappropriate measurement endpoints, and ignore the fact that ecosystems are nonlinear, have "memory," and that change is inherent. Even if measures are carried out to "remediate" systems, structures may never return to "original" conditions. These authors suggest that since ecosystems are too complex to fully protect or understand, we must choose a set of real, measurable, and valued properties of each ecosystem that one wishes to protect, rather than compiling an index of generic and heterogeneous ecosystem properties. While these are compelling points, it is clear that the ecosystem health concept is embedded in policy and in the public consciousness, and may be here to stay. Most likely, risk assessors and managers will have to translate science and conceptual frameworks into ecosystem health terms in a clear way, remaining mindful of their weaknesses, but remembering that they have resonance for the public and decision makers.

However, meaningful measures of these terms remain elusive, especially in dynamic estuaries, the site of many ports and harbors [26, 9]. If we are to successfully address the role of the ecological and socioeconomic structures in ports and harbors in the evaluation and maintenance of GEcQ and GEnQ, and in terms of ecosystem health, it is important that we expand our current tools for evaluating ecological risk (currently primarily focused on contaminants) to broader risk factors. Thus, there is clearly a need for the development of ecological measures and technology that can evaluate the environmental status and health of sediments and benthic and pelagic ecosystems, and the potential impacts of both proposed developments and measures carried out to mitigate the impacts of past and projected activities. Furthermore, there is a need to assess how the maintenance of sustainable functioning of these ecosystems plays a role in the services expected of ports and harbors. Under the umbrella of ecosystem health or status, the V-O-R concept, even if not fully quantifiable, may provide a useful narrative framework around which we can organize various assessments of sediment-related ecosystem structure, function and status.

3. Definition and Evolution of Risk Terminology and Concepts

As is addressed throughout this book, the task of ensuring environmental security, however it is defined, involves the identification, ranking, and management of risks, hazards, and vulnerabilities. This process is not just a scientific one, but involves the evaluation and balancing of scientific, societal, sociological, economic, and political issues. As such, the balancing of complex, often competing issues should be the result of open and clear decision processes that involve as many affected sectors of society as are possible. Clear communication requires a clear definition of terms. However, as with most terms, risk terms are defined differently in different fields, so some relevant definitions are presented here.

The International Organization of Standardization (ISO) [20, 21, 22] has rather generically defined a number of risk assessment terms, summarized by the Organization for Economic Cooperation and Development (OECD) [29]:

- **Hazard:** Potential source of harm.
- **Risk:** Combination of the probability of an event and its consequence. The term "risk" is generally used only when there is at least the possibility of negative consequences.
- **Risk assessment:** Overall process of risk analysis and risk evaluation.
- **Risk analysis:** Systematic use of information to identify sources and to estimate the risk. Information can include historical data, theoretical analysis, informed opinions, and the concerns of stakeholders.
- **Risk identification:** Process to find, list, and characterize elements of risk. Elements can include source or hazard, event, consequence, and probability.
- **Risk estimation:** Process used to assign values to the probability and consequences of a risk.
- **Risk evaluation:** Process of comparing the estimated risk against given risk criteria to determine the significance of the risk. Risk evaluation may be used to assist in the decision to accept or to treat a risk.
- **Risk criteria:** Terms of reference by which the significance of risk is assessed.

ISO [22] has also defined vulnerability and the related concept of threat in the context of information systems and information security, but the definitions probably have use for environmental security as well:

- **Vulnerability:** A weakness of an asset or a group of assets than can be exploited by one or more threats.
- **Threat:** A potential cause of an incident that may result in harm to system or organization.

Chambers [11] defined the concept of vulnerability in social science terms, in a manner that is relevant for environmental security discussions:

- **Vulnerability:** The exposure to contingencies and stress, and difficulty in coping with them. Vulnerability thus has two sides: an external side of risks, shocks, and stress to which an individual or household is subject; and an internal side which is defenselessness, meaning a lack of means to cope without damaging loss.

Thus, while risk refers to a defined hazard and a number of endpoints, vulnerability refers to a single endpoint and various hazards [29]. This issue is particularly of interest in the sphere of environmental security, as vulnerabilities are unevenly distributed. For instance, depending upon the hazard, old or infirm people, specific ethnic communities, farmers on low-lying land, specific types of structures, and specific species or ecosystems may be more vulnerable. These must be properly identified if proper hazard prevention and response strategies are to be designed.

Ensuring environmental security or sustainable development requires an understanding of the complex interrelationships between various aspects of human and ecological systems. These complex issues have resulted in an emerging body of literature examining systemic risks (e.g., OECD [30]). Renn and Klinke [33] state that the term "systemic risk" denotes the fact that:

> ...risk to human health and the environment is embedded in a larger context of social, financial and economic risks and opportunities. Systemic risk combines natural events—partially altered and amplified by human activity, such as the emission of greenhouse gases—economic, social and technological developments with policy-driven actions both at the national and at the international level. This interdisciplinary field requires a new form of risk analysis, which geographically or functionally integrates data from various sources into one analytical approach. Consequently, systemic risk requires a holistic perspective to combine the identification of hazards, risk assessment and risk management. Investigating systemic risks therefore goes beyond the usual analysis of causes and consequences, and focuses instead on the interdependencies and relationships between various risk clusters.

Often, risk assessment (understanding) is considered a scientific process that should be independent of, but should inform decisions on risk management (action). In order to avoid bias, "value-free" risk assessment, the scientific analysis and characterization of adverse effects of environmental hazards, often excludes the analysis of perceived risk, risk comparisons, and analysis of the social and economic effects of regulatory decisions (e.g., National Research Council (NRC) [28]). Risk management then includes the activities of identifying and evaluating alternative regulatory options, balancing the scientific inputs of the risk assessment and the broad social, economic, ethical, and political issues to selecting among options. The conceptual distinction between risk assessment and risk management can insulate scientific activity from political pressure and maintaining the analytic distinction between the magnitude of a risk and the cost of coping with it, but for complex problems, there are many problems with this approach. This problem was already recognized by NRC [28] when they stated:

> Separation of the risk assessment function from an agency's regulatory activities is likely to inhibit the interaction between assessors and regulators that is necessary for the proper interpretation of risk estimates and the evaluation of risk management options. Separation can lead to disjunction between assessment and regulatory agendas and cause delays in regulatory proceedings.

NRC [34] examined the issue of how risk assessment should inform decisions and presented an alternative view, in which, for complex (or systemic) problems, risk analysis should be a deliberative, collaborative, and iterative process (Figure 2). For complex problems, this approach is of much more use, as it will always be impossible to address all aspects of risk, and thus assessments must be designed to inform the decision process.

To support this process, Stern and Feinberg [34] proposed expanded risk decision definitions, with a special emphasis on the participatory process:

- **Affected parties.** People, groups, or organizations that may experience benefit or harm as a result of a hazard, or of the process leading to *risk characterization*, or of a decision about *risk*. They need not be aware of the possible harm to be considered affected.
- **Analysis.** The systematic application of specific theories and methods, including those from natural science, social science, engineering, decision science, logic, mathematics, and law, for the purpose of collecting and interpreting data and drawing conclusions about phenomena. It may be qualitative or quantitative. Its competence is typically judged by criteria developed within the fields of expertise from which the theories and methods come.

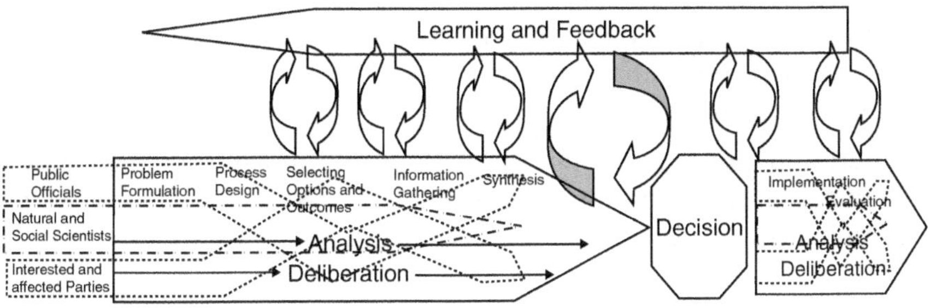

Figure 2. A schematic representation of the risk decision process, emphasizing the continuous involvement of all parties, and the iterative and deliberative nature of the process (adapted from Stern and Fineberg [34]).

- **Broadly based deliberation.** *Deliberation* in which participation from across the spectrum of *interested and affected parties,* by policy makers, and by specialists in *risk analysis* is sufficiently diverse to ensure that the important, decision-relevant knowledge enters the process, that the important perspectives are considered, and that the parties' legitimate concerns about the inclusiveness and openness of the process are addressed. Such deliberation involves the participation or at least the representation of the relevant range of interests and values as well as of scientific and technical expertise.
- **Deliberation.** Any process for communication and for raising and collectively considering issues. In the process leading to *risk characterization,* deliberation may involve various combinations of scientific and technical specialists, public officials, and interested and affected parties, and may be formalized (as in mediation) or occur in informal settings. It may be used both to increase understanding and to arrive at substantive decisions. In deliberation, people discuss, ponder, exchange observations and views, reflect upon information and judgments concerning matters of mutual interest, and attempt to persuade each other. Deliberations about risk often include discussions of the role, subjects, methods, and results of *analysis*. Bargaining and mediation are specific deliberative processes, as are debating, consulting, and commenting.
- **Hazard.** An act or phenomenon that has the potential to produce harm or other undesirable consequences to humans or what they value. Hazards may come from physical phenomena (such as radioactivity, sound waves, magnetic fields, fire, floods, explosions), chemicals (ozone, mercury, dioxins, carbon dioxide, drugs, food additives), organisms (viruses, bacteria), commercial products (toys, tools, automobiles), or

human behavior (drunk driving, firing guns). Hazards can also come from information (e.g., information that a person carries a gene that increases susceptibility to cancer may expose the person to job discrimination or increased insurance costs).

- **Interested parties.** People, groups, or organizations that decide to become informed about and involved in a *risk characterization* or decision making process. Interested parties may or may not also be *affected parties*.
- **Problem formulation.** An activity in which public officials, scientists, and *interested and affected parties* clarify the nature of the choices to be considered, the attendant hazards and risks, and the knowledge needed to inform the choices. Problem formulation sets the agenda for the other steps leading to a *risk characterization:* process design, selection of options and outcomes to consider, gathering and interpreting information, and synthesis.
- **Risk.** A concept used to give meaning to things, forces, or circumstances that pose danger to people or to what they value. Descriptions of risk are typically stated in terms of the likelihood of harm or loss from a *hazard* and usually include: an identification of what is "at risk" and may be harmed or lost (e.g., health of human beings or an ecosystem, personal property, quality of life, ability to carry on an economic activity); the hazard that may occasion this loss; and a judgment about the likelihood that harm will occur.
- **Risk analysis.** The application of methods of *analysis* to matters of risk. Its aim is to increase understanding of the substantive qualities, seriousness, likelihood, and conditions of a hazard or risk and of the options for managing it. Although risk analysis is sometimes conceived to be relevant only to gathering, interpreting, and summarizing information about certain possible consequences of a hazard, analysis has other uses in risk characterization.
- **Risk characterization.** A synthesis and summary of information about a *hazard* that addresses the needs and interests of decision makers and of interested and affected parties. Risk characterization is a prelude to decision making and depends on an iterative, analytic-deliberative process.

It is clear that simple views of risk assessment cannot alone be used to inform the complex decisions society will have to make to preserve security and sustainability. Rather, methods of deliberation and communication will have to be developed and refined to allow society to address emerging, systemic, and complex risks to their well-being and security.

3.1. TYPES OF RISK

When examining potential risks to security, ecosystem health, or sustainability, a number of features help characterize the risk. A few of these are discussed below.

- **Undesirable disturbances and relevance of impacts.** It is important to determine whether an impact is relevant before a management action is considered. Aquatic ecosystems are dynamic by nature, and this in fact is often considered an important feature of ecosystem health. Thus, even a change may not be damaging to the ecosystem (or its services), and care must be taken to separate anthropogenic impacts from natural variability (especially in tidally influenced areas). Defra [14] provides an extensive discussion on the differences between measurable changes and undesirable disturbances in the context of eutrophication, but many lessons can be brought to other applications. The parameters below must be considered carefully if predicted or measured impacts are to be considered to be of relevance, and if they might trigger management actions.
- **Anthropogenic vs. natural.** There are, for most risk pathways, both natural and anthropogenic drivers. For example, sediments would move into, out of, and through a system, whether humans were there or not (in fact, dredged areas that are not natural will silt up again). It is important to understand both types of drivers for processes.
- **Controllable vs. uncontrollable.** Whether natural or anthropogenic, some processes will be controllable, some will not. Natural inputs can be stopped or diverted (although there will be impacts elsewhere). Some anthropogenic processes can be stopped or controlled; others, for political, territorial, financial or technological reasons, cannot.
- **Pulse vs. press.** An impact or driver may be sporadic (pulse) or chronic (press). In general, pulse processes don't cause undue impact (with notable exceptions); press processes have longer cumulative effects. The timescales of processes must be clarified.
- **Spatial scale.** An impact on a small site may be less important than one on a larger area. Of course, not only ecological impacts are of relevance. For instance, while severe shallowing on a small section of a waterway may be ecologically insignificant, it may, if in an ecologically or navigationally important area, be considered of great significance. Furthermore, ecological damage in a sensitive area or to a sensitive or endangered species (e.g., the depressed mussel) may be considered of importance regardless of the scale of the impact. On the other hand, smothering of a small area of a large biome may be deemed of little relevance.

Although there are endless distinctions that one can make, in terms of environmental security three broad categories of risk can be identified [41] by using the above criteria:

1. Natural or manmade gradual changes that slowly erode economic and environmental sustainability, and, in some cases, may even be irreversible.
2. Natural catastrophic events that, to some extent, may be predictable and thereby possible to plan response and protection measures.
3. Manmade catastrophic events, which are typically sudden and unpredictable.

These categories are broadly similar to those identified and addressed by OECD [29].

The first category, natural or manmade gradual changes, threaten the maintenance of resources or ecosystem services (clean water, land, property, trees, crop viability) in the face of environmental changes such as global warming, sea level rise, build up of contaminants, etc. They have a number of distinguishing features:

- Erode both environmental and economic sustainability over time.
- Are often manmade (but may be perceived as natural due to the time delay between actions and effects), but are gradual and inexorable.
- Impacts are often reasonably predictable (in broad strokes) but causes are difficult to address.
- Decisions to address them may involve trying to prevent or remediate the problems, or trying to protect resources in the face of change.
- Vulnerability to these hazards is often unevenly distributed, both locally and globally.
- Governments need to plan for both prevention and response, and maybe prepare to protect limited resources from invasion/threat.
- These are extensive, press, pressures.

The second type of risk is natural, catastrophic impacts as a result of natural disasters. These are addressed and managed differently than the first category, and have different distinguishing features:

- To address the impacts of these events on environmental security, one must determine the likelihood of a number of relatively predictable events such as major storms and earthquakes and protect against impacts such as dam break, chemical spills, and explosions.
- Vulnerabilities may be unevenly distributed and must be identified.
- Scenario development is straightforward, probabilities of the natural disasters are matched to impacts of potential scenarios, and then prevention technologies and their costs are considered.
- For the most unlikely scenarios, prevention may not be a choice, but response contingencies should be considered.

- Chemical plants might be moved away from storm tracks or faults, but still be built.
- These are pulse pressures, and can be localized or extensive.

The third category, manmade catastrophic events, are the results of intentional or unintentional human activities such as terrorism and industrial or shipping accidents. While the consequences can be similar to the former category, they have some distinguishing features that require different environmental security approaches.

- Especially for intentional impacts, these events are more difficult to predict than natural catastrophes as people can contrive to achieve improbable things.
- A security focus will be more on identifying vulnerabilities and then figuring out how to prevent them or how respond to events.
- A major concern is either the detection of bombs/people or the creation of barriers, but simply moving something out of a storm track—for example—won't protect it.
- This category is not too different from the second category, but may involve an entirely different set of detection/prevention tools.
- These events produce pulse pressures, and can be localized or extensive.

For each of these risk categories, decision makers must identify risks, hazards, and vulnerabilities; examine possible scenarios, with their associated ecological, social, and economic threats; and then rank the threats using various models. For relatively simple problems, models to define risk based upon severity (hazard) and probability (exposure) can be used. However, for more complex systemic risks, and for other categories of risk, a number of other tools are used, including hazard analysis and identification of critical control points for food safety; catastrophe models and community vulnerability assessments for natural disasters; failure analysis and preliminary hazard assessment for complex systems; and system risk and vulnerability analysis, terrorism risk analysis, and information security risk assessment for security issues. Each of these approaches is designed for a different type of question, and they each have specific applications, strengths, and weaknesses, which are summarized in OECD [29].

However, for systemic risk and environmental security, new tools will need to be developed to address these disparate issues in a manner that allows society to make informed choices about how they will preserve various aspects of human well-being. To achieve this, the German Advisory Council on Global Change (WBGU) [18] sought to identify what aspects of various systemic risks could be used to categorize risk classes. After an extensive consultation and deliberation, the Council selected several risk criteria [33]:

- **Extent of damage.** Potential adverse effects measurable in natural units, such as deaths, injuries, or production losses.
- **Probability of occurrence.** Estimates of the relative frequency of a discrete or continuous loss.
- **Incertitude.** An overall indicator of uncertainty components.
- **Ubiquity.** The geographical extent and distribution of potential damage.
- **Persistency.** The temporal extent of potential damage.
- **Reversibility.** The possibility of restoring the situation to the state before the damage occurred; for instance, reforestation or water treatment.
- **Delay effects.** The latency between an initial triggering event and the actual occurrence of damage, the measure of which could be of a physical, chemical, or biological nature.
- **Violation of equity.** The discrepancy between those who benefit and those who bear the risks.
- **Potential of mobilization.** Potential violation of individual, social, or cultural interests and values that generate social conflicts and psychological reactions by individuals or groups who feel afflicted by the consequences. These could also result from perceived inequities in the distribution of risks and benefits.

Given various combinations of the above criteria, it is clear that countless subcategories can be identified. However, the objective of WBGU [18] was to develop a holistic framework which would allow decision makers and managers of complex and systemic risks to class risks and select tools for their communication and management. To that end, six risk subcategories were identified, with specific characteristics that suggested various management types and strategies. These classes were named after illustrative characters or stories in Greek mythology, which are summarized in Renn and Klinke [33]. Figure 3 lists these classes, their characteristics, examples, management types, and strategies.

For example, Cassandra risks are high-probability, high-damage events with a high delay effect, such as the impacts of an increase in sea level rise due to climate change. Because people do not immediately perceive the effects of their actions, and because the populations most vulnerable to the hazards are not necessarily the same ones causing them, these issues must be managed using a discursive, rather than only scientific, strategy. Thus, management strategies include consciousness building and contingency planning. On the other hand, uncertain probability but potentially high-damage Cyclops risks such as extreme events require science-based management and strategies including ascertaining probability, reducing disaster potential, and planning

Risk Class	Characteristics	Examples	Management type	Strategies
Type 1 Sword of Damocles	probability low (towards 0); damage high (towards infinite); confidence intervals of p and d low	nuclear energy, chemical plants, dams, meteorite impacts	Science-based	• Reducing disaster potential • Ascertaining probability • Increasing resilience • Preventing surprises • Emergency management
Type 2 Cyclops	probability uncertain damage high; confidence interval of p high; confidence interval of d rather low	floods, earthquakes, volcanic, eruptions, AIDS, El Nino		
Type 3 Pythia	probability uncertain damage uncertain (potentially high); confidence intervals of p and d high;	increasing greenhouse effect, endocrine disrupting substances, release and spread of transgenic plants	Precautionary	• Implementing precautionary principle • Developing substitutes • Improving knowledge • Reduction and containment • Emergency management
Type 4 Pandora's box	probability uncertain; damage uncertain (only presumptions); confidence intervals of p and d uncertain (unclear); persistency high (several generations)	ozone destroying substances		
Type 5 Cassandra	probability rather high damage rather high; confidence interval of p rather high; confidence interval of d rather low; delay effect high	anthropogenic climate change for vulnerable areas	Discursive	• Consciousness building • Confidence building • Public participation • Risk communication • Contingency management
Type 6 Medusa	probability rather low damage rather low (exposition high); confidence interval of p rather high; confidence interval of d rather low; potential of mobilization high	electromagnetic fields; MMR immunizations		

Figure 3. WBGU risk classes, their characteristics, examples, and management type and strategies. Adapted from Renn and Klinke [33].

emergency management. Given these different characteristics and risk classes, scientists and managers from different fields can begin to understand why there is no one-size-fits-all approach to risk management, and may be able to better select, design and communicate across fields and to the public in support of informed and focused decision making.

WBGU [18] then mapped the above risk classes in terms of normal, intermediate, or intolerable risk. Furthermore, specific risks were characterized based upon the various risk criteria listed above, and then implications for risk communication and management for the specific systemic risk were discussed. This detailed study provides a conceptual approach that should help focus and inform environmental security decision approaches.

4. Environmental Security and Decisions

Ecosystem-based policy will affect how human activities are regulated and managed in coastal and port areas. Legislation seeks to balance ecosystem objectives with socioeconomic needs and goals. On the longer time scale, it is clear that environmental security, sustainability, and human well-being in all sectors, including the protection and maintenance of ports and harbors, will require a balance between ecological and human needs, which are inextricable. These are complex issues requiring the development of interdisciplinary and cross-sectoral assessment, communication, and decision tools. However, much of the work that addresses what is currently considered "environmental security" does not explicitly address the issues of concern in this chapter. Can we expand both our risk and our security paradigms to resolve this artificial division?

Currently, much of the focus of environmental security is on rare but dramatic events (such as terrorist attacks and extreme storms), and these are not addressed in the same decision, policy, or funding allocation frameworks as more "mundane" issues such as contaminant control and habitat degradation. On can also argue that an overemphasis on human-induced rare events can cause a misallocation of resources, as many natural impacts, whether sudden or gradual, are historically more damaging. There is a need to develop decision frameworks in which these seemingly disparate issues are addressed together in support of regional budgeting, decision making, communication and management. Sustainable protection of infrastructure and ecosystems requires an expansion of traditional ecological risk assessment, to address multiple risk drivers on multiple spatial and temporal scales. Many of the documents addressed above provide insights and tools into addressing these complex, systemic problems.

A common problem in decision making is that decision makers are unclear as to what their objectives are, which makes the assessor's work especially difficult. This is a strong argument for the discursive approach to risk characterization recommended by Stern and Feinberg [34]. For the many-headed hydra of environmental security, the process should begin with the definition of some objectives, and, most likely, the development of potential scenarios. To develop environmental security decision frameworks, a few questions should be asked:

- Are we protecting against everything?
- At what spatial and temporal scale?

- What is controllable; what is not?
- Are we developing preventions, tracking changes, or selecting responses?

All of the above issues must be addressed to provide environmental security, but unless they are clearly separated in a decision hierarchy, they will get muddled. For a specific hazard or risk, there are a number of decisions that must be addressed. The simplest categories of these are illustrated in Figure 4. For various types of hazards, scenarios should be developed, and vulnerabilities and risks must be identified, characterized, and ranked. Once this has been done, for each potential risk, decisions should be based upon scenario probability, preventability, causality (human-caused or natural), time scale (gradual or sudden), and potential costs and risks. Prevention strategies and preparedness and response strategies (whether a scenario is unpreventable or if prevention fails) must be developed. Depending on the probability, vulnerabilities, and costs of a given risk, emphases will be put on different measures.

While Figure 4 is a simple diagram that seems to state the obvious, a review of many papers and presentations on environmental security, at a number of meetings and workshops, made clear that a large number of authors, addressing a specific aspects of environmental security and decision making, did not state explicitly what sort of decision they were informing, or what the role of the decision they were addressing was in an overall vision of environmental security. However, the science, communication, and decision tools being used will

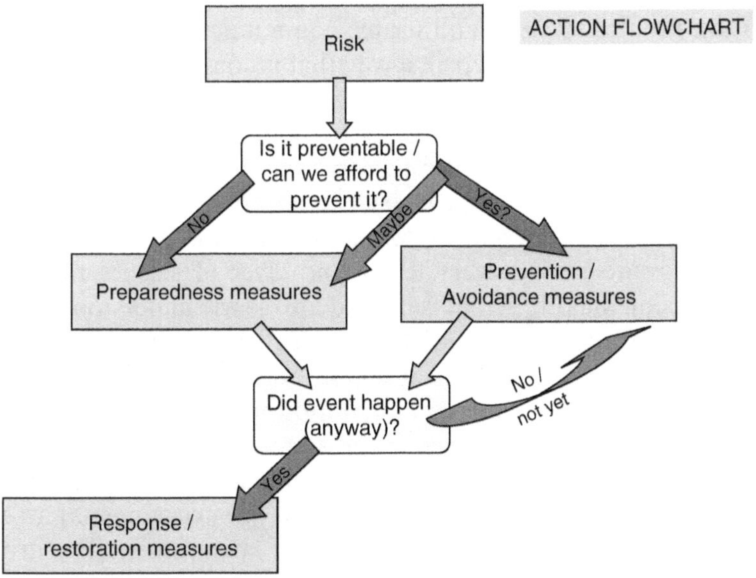

Figure 4. Action flowchart identifying decisions that must be made for a given risk.

be applied differently to the different decisions, and thus which decision is being addressed (and its relationship with other decisions) should be made clear.

5. Emerging Tools in Support of Ecosystem-Based Management, Which Can Be Adapted for Security Issues

Successful implementation of environmental security, in its many guises described above, will require integration to an unprecedented degree: integration of environmental, political, and socioeconomic objectives from the catchment basin to the coast and, ultimately, to European seas; of the various water and land uses, including their functions and values; different skills and disciplines; previous and emerging legislation and policy into common and coherent frameworks; technical, socioeconomic, and legislative instruments; stakeholders in decision making; and the different decision-making levels, affecting ecosystem status and management among many nations. This integration will require extensive collaboration and research to adapt current systems of environmental assessment and management to the basin and ecosystem level. Such integration in support of complex problems is a nontrivial problem, which pushes the boundaries traditionally addressed by many research fields, agencies, and policies. However, this chapter has reviewed a number of efforts that have begun to address how we might better achieve this task.

Still, many significant challenges remain, not the least of which is the development of better tools to link the complex science that should underlie the decision process to clear, and usable communication and decision tools for the public and policy makers in a way that is compelling and understandable to both groups. Traditionally, many research scientists have scorned the perceived oversimplification necessary to communicate across disciplines and with potential users of their research, preferring the communication and company of their own disciplinary peer group. Similarly, the public and policy makers cannot generally, in terms of time, resources, or background, be expected to understand every detail and layer of science that underlies informed decision making. What is needed are clearly understood conceptual approaches and frameworks to build a bridge between these end members, with a clear, logical, and unbroken chain linking and translating the fundamental science and its underlying assumptions to the applied issues and decisions they inform. Many decision tools are addressed in chapters of this and other books. This section will briefly summarize some frameworks and approaches currently being used or developed in environmental assessment and management that may be of use in task of environmental security.

An emerging issue in ecosystem management is the assessment and management of multiple stressors, rather than a single stressor at a time.

Assessing and managing such issues is a complex task. In the United States, the EPA favors a stressor identification (SI) process [38] as a means of identifying and differentiating causes of observed biological impairments in watersheds where there are multiple candidate stressors (e.g., sedimentation, low dissolved oxygen, and toxics). The SI framework highlights the conceptual links between diverse natural and anthropogenic stressors and drivers on the ecological functions of a watershed. The approach begins by characterizing the nature of the impairment and proceeds by identifying candidate causes, eliminating improbable causes, and identifying the most probable causes based on the strength of evidence. A logical framework is used to establish strength of evidence based on spatial and temporal patterns, available knowledge of the stressors, and plausibility arguments. This sharpened focus enables industry and regulators to consider appropriate mitigation measures and to assess the efficacy of such measures in reducing potential short- and long-term impacts [39,40]. Furthermore, the SI approach can be used to support communication with environmental managers and other stakeholders as part of a public outreach process.

The SI process is a logical framework which helps organize lines of evidence to establish and manage the specific drivers of given biological impacts. The DPSIR Framework, endorsed by the European Environment Agency and OECD provides a mechanism for analyzing and communicating the conceptual links between many human activities and environmental problems. Particularly useful for policymakers, DPSIR offers a basis for analyzing the interrelated factors that impact the environment. The DPSIR approach defines the interactions between various parameters, including:

- *Driving forces*, such as industry and transport, which produce
- *Pressures* on the environment, such as polluting emissions, which then degrade the
- *State* of the environment, and which then
- *Impact* human health and ecosystems, causing society to
- *Respond* with various policy measures, such as regulations, information, and taxes, which can be directed at any other part of the system [16].

Drivers may be defined as human activities, and the pressures, states, and impacts are reflected in the damage done to the environment by those activities; responses focus on reducing those impacts by controlling or eliminating drivers and pressures, improving states, or reducing or preventing impacts. Sustainable management decisions require that responses seek to balance environmental, regulatory, and socioeconomic goals [5,23].

For instance, White et al. [43] produced a DPSIR framework to address the issues necessary to manage sediments within the Norfolk and Suffolk

Broads, a unique area of water, grazing marshes, fen, and woodland, and home to some of the rarest plants and creatures in the UK. This area is Britain's largest protected wetland, having similar status to a national park. The Broads Authority (BA) was set up in 1989 to conserve, and enhance the natural beauty of the Broads, promote the enjoyment of the Broads and protect the interests of navigation. Sustainably achieving this complex and sometimes conflicting set of objectives requires that the BA develop a long-term sediment management strategy. A critical component of such a strategy is the understanding and management of sediment dynamics in the catchment, and this DPSIR framework addresses various issues impacting this component (see Figure 5):

- **Drivers**, both human and natural, that affect sediment sources, sinks and transport.
- **Pressures** on sediment dynamics as a result of those drivers, and their characteristics, including extent, intensity, frequency and length of these pressures.

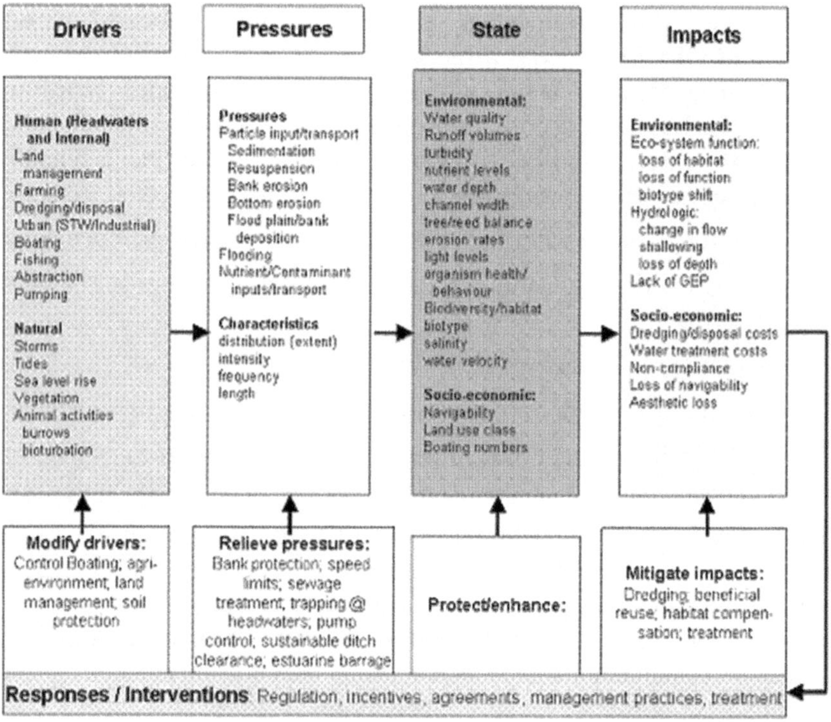

Figure 5. Sediment management DPSIR developed for Broads Authority Sediment Management Strategy [43].

- **Measurable states**, both environmental and socioeconomic, which are affected by sediment dynamics.
- **Impacts**, both environmental and socioeconomic, of changes in these states.
- **Responses/**Interventions that can modify drivers, relieve pressures, protect or enhance states, or mitigate impacts.

While this DPSIR framework can help focus data gathering, further research, or management efforts, it is difficult for scientists to see how the specific processes they study are reflected in this policy-level overview. Another approach to the DPSIR framework is to overlay a "horrendogram" [17] identifying the mechanistic links between various processes (something that is a critical component of conceptual site models and various other environmental assessment frameworks) on a DPSIR framework, to begin to identify the links between the science and the higher-level policy issues. Figure 6 shows an example of this approach, in which a conceptual framework for processes affecting Venice Lagoon ecological quality is projected on a DPSIR framework.

In this approach, the arrows between layers represent various processes that affect states and impacts that may have differing importance. Some links may, after further investigation (perhaps using something like the SI framework) prove to be unimportant relative to other processes,

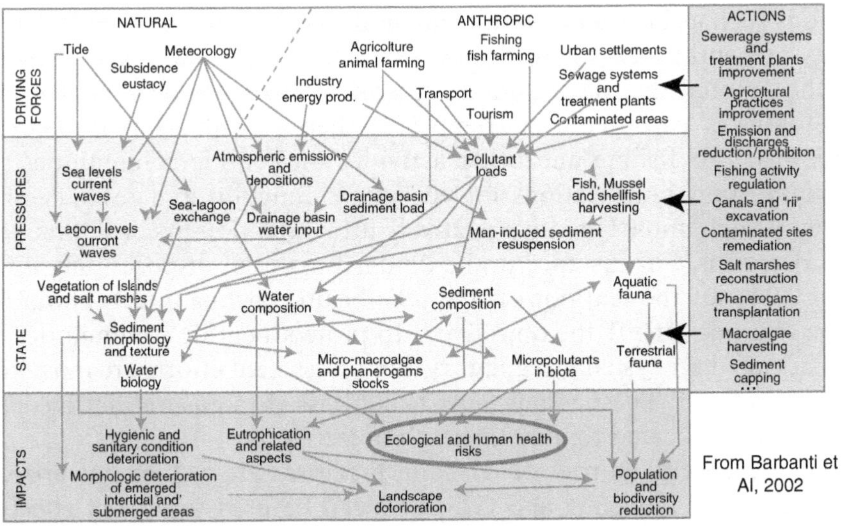

Figure 6. For Venice Lagoon, this conceptual diagram lays out potential impacts to a complex web of receptors—note that both natural and anthropogenic drivers are considered [36]. Reprinted with permission.

while other processes may dominate. Thus, it is important to identify both natural and anthropogenic processes that may impact various states, and to determine their relative importance (which may vary in space and time). Furthermore, some processes, while important, may be more or less well understood scientifically. An examination of the potential importance of various processes, and the degree to which potentially important processes are understood, can help focus both research and management efforts.

For instance, due to the need to both dredge and dispose of sediments to maintain waterways and to manage the impacts of extensive human activities, sediment management is an important function of port and harbor authorities. Until recently, most regional sediment risk prioritizations and site-specific sediment Ecological Risk Assessments have focused primarily on the risk of contaminants in sediments on associated organisms. Occasionally, some food chain effects are explicitly or implicitly addressed, but rarely are contaminated sediments assessed within an overall ecosystem model. Sediments figure extensively in the Marine MES [37], but contaminated sediments are not the dominant issue. Rather, habitat change and loss, due to changes in sediment inputs, whether reductions (resulting in the loss of beaches, storm protection, nutrient inputs, etc.) or increases (resulting in lake, reservoir and wetland infilling, coral reef smothering, etc); eutrophication and reductions in nutrient inputs, and disturbance due to development and fishing practices were major drivers, with significant consequences for biodiversity and ecosystem function. For example, hard-bottom and soft-sediment seafloor habitats are severely impacted by fishing methods such as bottom trawling and dredging. Because of the high biodiversity of benthic ecosystems in freshwater, estuarine and marine systems, and their global significance in the critical ecosystem processes of cycling and storage of carbon and nutrients [13], there is evidence that the extensive impacts of fishing and other activities on benthic communities may have important ramifications for ecosystem function and resilience [37]. However, such impacts receive relatively little attention in coastal management decisions. For instance, while dredged material disposal only affects about 0.04% of the seafloor off the UK, beam trawling affects about 67% of the seafloor [31]. If the objective is to protect benthic communities and the associated ecosystems, regulatory and assessment efforts are most likely misappropriated. These issues must be addressed to ensure that resources are properly allocated to meet society's objectives.

If, for a specific impact of concern, processes are well characterized in space and time, modeling and GIS tools can be used to predict effects of various impacts or responses. For instance, Parker et al. [31] have developed a GIS-based DPSIR tool for predicting and evaluating the effects of various natural and human disturbances on benthic community function. For instance,

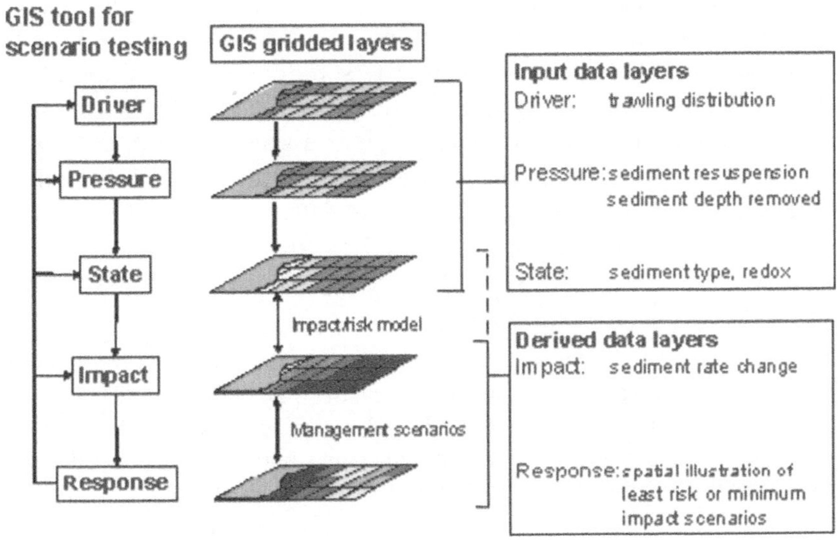

Figure 7. Measured disturbances are combined with modeled impacts to predict responses to various scenarios. From Parker et al., 2004. Reprinted with permission.

Figure 7 illustrates a scenario in which GIS-gridded layers are used to examine the effects of various fishery management approaches on benthic community function (in this case, nutrient cycling) using satellite data on trawling distribution, information on sediment state, and models predicting how benthic function responds to disturbance. Similarly, composite disturbances from several pressures can be modeled and mapped together to predict the aggregate impact of various scenarios. In this way, fundamental research on benthic function and process is linked with regional data to help inform management decisions.

Clearly, when management decisions are being made at a large scale, assessing multiple stressors on multiple receptors in various habitats, it becomes increasingly difficult to determine what the dominant issues might be throughout a region. Although there have to date been very few environmental management strategies based upon catchment or watershed-scale considerations, it is impossible to realistically assess the relative impacts of various activities and potential mitigation actions unless they are examined at this scale. Unlike management carried out at a specific site, regional-scale management must address a complex mix of risk scales, sources, drivers, and receptors. Given the realities of limited resources and mitigation options, it is clear that any expectation of complete risk removal will not be met [2, 5,10]. Rather, risk mitigation at the regional scale (and, realistically, even at all but the most simple of sites) seeks to provide maximum net risk reduction within the region with the resources available [5]. Successfully achieving such a goal will require good conceptual models, but will also require a clear understanding and definition of net risk within a catchment or region. While there are well-established methods for assessing ecological or human health risk at a specific site or region, if only one type of pressure is considered at a time, it remains difficult to effectively inform decisions which have to address cumulative or net risk to various receptors at various scales. One approach to addressing such complexity is to apply the relative risk model (RRM) to develop regional-scale risk assessments at a landscape scale, defined by Landis [25] as:

> A risk assessment (which) deals at a spatial scale that contains multiple habitats with multiple sources of multiple stressors affecting multiple endpoints, and the characteristics of the landscape affect the risk estimate. Although there may only be one stressor of concern to the decision maker, at a regional scale the other stressors acting upon the assessment endpoints are to be considered.

Using DPSIR-based models, and a regional-scale risk assessment tool such as the RRM, which allow for the combination and ranking of disparate risk sources and drivers at many sites within a region in a transparent way, it should be possible to rank and prioritize mitigation and management actions such that they most effectively reduce net risk in a region. A proper selection of mitigation approaches will be driven by a regional prioritization of sites, a rigorous evaluation of uncertainty, a matching of available technologies to site characteristics, and an evaluation of various socioeconomic and regulatory issues [5, 7].

6. Conclusions

Sustainable management decisions must balance complex issues. Clearly, ecological objectives must be balanced against socioeconomic and regulatory goals. However, we must also project potential risks into the future and assess how we hope to prevent and/or respond to either natural or human-induced impacts upon our safety and environment. These risks can be gradual, (e.g., climate change or resource depletion), or sudden (e.g., a hurricane or terrorism), and can be preventable or uncontrollable. All prevention has a cost which must be balanced against the cost of consequences. Risk assessment and management tools, as well as decision and communication tools, can be adapted to allow society and decision makers to allocate resources in support of their environmental security goals, and many approaches that begin to examine and integrate the complex aspects of systemic and multiple-stressor risk have been reviewed here.

Whether it is to inform a contaminated sediment site assessment, a dredged material disposal application, or an Environmental Impact Assessment for an offshore wind farm, there is a need for scientists to consider the relationship between our sectoral task and larger environmental security, sustainable development, and ecosystem management goals. Depending upon the processes being addressed, different spatial and temporal scales must be understood. Whether we apply models which focus on the particulate organic and nutrient flows between the catchment, pelagic, and benthic system (e.g., Harris [19]), or whether these involve complex "horrendograms" showing the links between natural and anthropogenic pressures and various parts of the ecosystem [17], there is a need to explicitly link the science, management, and policy of security and ecosystem management. To this end, we should be aware that we are not carrying out port and harbor management, but the port-specific aspects of ecosystem management. We need to develop regional conceptual frameworks that identify, and provisionally quantify, these processes, to inform the prioritization of research, assessment, management, and regulatory efforts. Even if all issues cannot be addressed in every task, research, assessment, and management must be couched in larger conceptual models that address the interactions between benthic and pelagic ecosystems, as well as the processes that might be affecting them (and the services they provide) in the larger catchment. These challenges are not trivial, but otherwise there is a risk that we will continue to spend huge amounts of time and money without making any difference to overall ecosystem health, security, or sustainability [4].

7. Acknowledgements

SEA acknowledges partial support by the 6th Framework Programme for Global Change and Ecosystems (1.1.6.3) project "Integrating new technologies for the study of benthic ecosystem response to human activity: towards a Coastal Ocean Benthic Observatory" (project #505564).

8. References

1. Australian and New Zealand Environment and Conservation Council: Agriculture, Resource Management Council of Australia and New Zealand. Australian and New Zealand guidelines for fresh and marine water quality – volume 1: the guidelines—national water quality management strategy paper #4. Canberra. 2000.
2. Apitz SE. Editorial: new and recent developments in soil and sediment management, policy and science: do we need a journal of soils AND sediments? Journal of Soils and Sediments. 2005; 5(3):129-133.

3. Apitz SE. Editorial: Managing European sediments: can we expand our ecological risk assessment paradigms? Journal of Soils and Sediments. 2006; 6(1):1.
4. Apitz SE. Sediments and ecosystem health. SETAC Globe - Sediment Corner. 2006; September-October:21-23.
5. Apitz SE, Carlon C, Oen A, White S. Strategic frameworks for managing sediment risk at the basin and site-specific scale. In: Heise S, editor. Sediment risk management and communication. Amsterdam: Elsevier; 2007.
6. Apitz SE, Davis JW, Finkelstein K, Hohreiter DL, Hoke R, Jensen RH, Kirtay VJ, Jersac J, Mack EE, Magar V, Moore D, Reible D, Stahl R. Assessing and managing contaminated sediments: part I, developing an effective investigation and risk evaluation strategy. Integrated Environmental Assessment and Management. 2005; 1:2-8.
7. Babut M, Oen A, Hollert H, Apitz SE, Heise S, White S. Prioritisation at catchment scale, risk ranking at local scale: suggested approaches. In: Heise S, editor. Sediment risk management and communication. Amsterdam: Elsevier; 2007.
8. Belluck DA, Hull R, Benjamin SL, Alcorn J, Linkov I. Environmental security, critical infrastructure and risk assessment: definitions and current trends. In: Kiker GA, Linkov I, editors. Environmental security and environmental decision-making. Amsterdam: Springer-Verlag Press. Forthcoming 2006.
9. Bortone SA, editor. Estuarine indicators. Boca Raton: CRC Press; 2005.
10. Bridges TS, Apitz SE, Evison L, Keckler K, Logan M, Nadeau S, Wenning RJ. Risk-based decision-making to manage contaminated sediments. Integrated Environmental Assessment and Management. 2006; 2:51-58.
11. Chambers R. Vulnerability, coping and policy. IDS Bulletin. 1989; 20(2).
12. Costanza R, Norton B, Haskell B, editors. Ecosystem health: new goals for environmental management. Washington, DC: Island Press; 1992.
13. Covich AP, Austen M, Bärlocher F, Chauvet E, Biles CL, Inchausti P, Dangles O, Statzner B, Solan M, Moss BR, and others. The role of biodiversity in the functioning of freshwater and marine benthic ecosystems. Bioscience. 2004; 54(8):767-775.
14. Defra. Understanding undesirable disturbance in the context of eutrophication, and development of UK assessment methodology for coastal and marine waters: Stage 1 - what is undesirable disturbance? Unpublished report. Edinburgh: Napier University; 2004.
15. Dobson A, Lodge D, Alder J, Cumming GS, Keymer J, McGlade J, Mooney H, Rusak JA, Sala O, Wolters V, and others. Habitat loss, trophic collapse, and the decline of ecosystem services. Ecology. 2006; 87(8):1915-1924.
16. European Environmental Agency. Conceptual framework: how we reason. Copenhagen. 2005.
17. Elliott M. The role of the DPSIR approach and conceptual models in marine environmental management: an example for offshore wind power. Marine Pollution Bulletin. 2002; 44:iii-vii.
18. German Scientific Advisory Council for Global Environmental Change (WBGU). World in transition: strategies for managing global environmental risks. WBGU) annual report 1998. Berlin: Springer; 2000.
19. Harris GP. Algal biomass and biogeochemistry in catchments and aquatic ecosystems: scaling of processes, models and empirical tests. Hydrobioligia. 1997; 349:19-26.
20. International Organization for Standardization (ISO). Safety aspects: guidelines for their inclusion in standards, ISO/IEC guide 51. Geneva. 1999.
21. ISO. Risk management- vocabulary-guide for use in standards, ISO guide 73. Geneva. 2002.
22. ISO. Management of information and communications technology security. Part 1: concepts and models for information and communications technology security management. ISO/IEC 13555-1, Geneva. 2004.
23. Joziasse J, Heise S, Oen A, Ellen GJ, Gerritts L. Sediment management objectives and risk indicators. In: Heise S, editor. Sediment risk management and communication. Amsterdam: Elsevier; 2007.

24. Kapustka LA, Landis WG. Ecology: the science versus the myth. Human and Ecological Risk Assessment. 1998; 4(4):829-838.
25. Landis W. Regional scale ecological risk assessment using the relative risk model. Boca Raton: CRC Press; 2005.
26. Livingston RJ. Trophic organization in coastal systems. Boca Raton: CRC Press; 2003.
27. Millennium Ecosystem Assessment. Ecosystems and human well-being: general synthesis. Washington, DC: Island Press; 2005.
28. National Research Council. Risk assessment in the federal government: managing the process. Washington, DC: National Academy Press; 1983.
29. Organization for Economic Cooperation and Development (OECD). Denmark: assessing societal risks and vulnerabilities. Paris. 2006.
30. OECD. Emerging systemic risks: final report to the OECD Futures Project. Paris; 2003.
31. Parker R, Aldridge J, Eastwood P, Houghton C, Mills C, Kershaw P. The ecosystem effects of sediment disturbance: development and application of a GIS based disturbance impact assessment tool. Lowestoft: The Centre for Environment, Fisheries and Aquaculture Science; 2004.
32. Rapport DJ, Gaudet C, Calow P, editors. Evaluating and monitoring the health of large-scale ecosystems. Heidelberg: Springer-Verlag; 1995.
33. Renn O, Klinke A. Systemic risks: a new challenge for risk management. European Molecular Biology Organization Reports 5. Special Issue:S41-S46. 2004.
34. Stern PC, Fineberg HV, editors. Understanding risk: informing decisions in a democratic society. Washington, DC: National Academy Press; 1996.
35. Suter GWII. A critique of ecosystem health concepts and indexes. Environmental Toxicology and Chemistry. 1993; 12:1533-1539.
36. Thetis. Rapporto sullo stato attuale dell'ecosistema lagunare veneziano. MAV-CVN Technical Report. 98.T339-REL-T022.1. (in Italian). 2000.
37. United Nations Environment Programme. Marine and coastal ecosystems and human wellbeing: a synthesis report based on the findings of the Millennium Ecosystem Assessment. Nairobi. 2006.
38. US Environmental Protection Agency (USEPA). Stressor identification guidance document, EPA-822-B-00–025. Washington, DC. 2000
39. USEPA. Handbook for characterizing causes, 6th Ed. Available at: http://www.epa.gov/caddisp1/worksheets/IllustratedGlossary.pdf. 2004.
40. USEPA. CADDIS step-by-step guide. Available at: http://cfpub.epa.gov/caddis/guide.cfm. 2005.
41. Wenning RJ, Apitz SE, Belluck DA, Chiesa S, Figueria J, Filip Z, Ignatavieius G, Kiker MK, Lahdelma R, MacDonell M, Mannini A, Raissouni B, Sinka D, Xenidis Y. Environmental security: approaches and tools. In: Kiker GA, Linkov I, editors. Environmental security and environmental decision-making. Amsterdam: Springer-Verlag Press. Forthcoming 2006.
42. Wenning RJ, Apitz SE, Baba A, Citron M, Elliott K, Al-Halasah N, Kaputska L, Lazarescu M, Mohamed M, Rutjes R. Understanding environmental security in ports and harbors. In: Linkov I, Wenning RJ, editors. Environmental security in ports and harbors. Amsterdam: Springer-Verlag Press. Forthcoming 2007.
43. White SM, Deeks L, Apitz SE, Freeman M. Desk based study of the sediment inputs to the Broads catchment, with the identification of key inputs and recommendations for further targeted research and management to minimise inputs: final report to the Broads Authority, phases I and II. Cranfield University, Cranfield, March 2006.

CHAPTER 10

INVASIVE SPECIES

A Real, but Largely Ignored Threat to Environmental Security

L. KAPUSTKA

Golder Associates Ltd
1000 940-6th Avenue SW
Calgary T2P 3T1 Canada

G. LINDER

USGS/BRD/CERC
HeronWorks Field Office
Brooks, OR 97305 USA

Abstract: Environmental security tends to focus on the dramatic, partly because of the immediately obvious effects. Meanwhile, decidedly nondramatic actions pose potentially greater long-range threats. One such subtle or nondramatic action would be the intentional introduction of invasive species that would be capable of markedly altering functional characteristics of a critical ecological system (e.g., coastal wetlands, interior croplands, and interior waterways). The effects from an insertion of an invasive species might require several years before being realized; and then the costs of containing or rehabilitating the affected ecosystems may be prohibitive. Regulatory procedures to thwart the spread of invasive species vary considerably among jurisdictions. Both the patchwork regulatory status and the relatively poor underlying scientific/technical understanding of invasive potential provide major challenges to managing this problem. Observations from a recent ASTM symposium on invasive species and activities within the ASTM Standards Development process to address this issue are presented.

1. Introduction

Ecological systems provide the underpinning for cultural and economic prosperity of all humans. Agrarian societies and subsistence groups understand the linkages, but many if not most citizens of industrialized societies would be surprised to learn how closely their prosperity is dependent on the functioning of dynamic ecological systems. Invasive species are one of the greatest threats

I. Linkov et al.(eds.), Managing Critical Infrastructure Risks, 175–188.
© 2007 *Springer*.

to biodiversity and cost countries billions of dollars in lost revenue, property damage, and eradication expenses [12]. Only recently have inroads been made into documenting the economic importance of wildlands [1, 5]. Efforts are underway to expand our thinking about the values of ecological systems and the various resources so that environmental decisions can concurrently take into account both monetized and nonmonetized values [14].

When considering the relationships between economic and cultural prosperity and ecological systems, it becomes clear that great changes in ecological functions are disruptive to economic systems. Catastrophic events such as droughts, floods, and unseasonal cold snaps, disrupt food production with great economic consequences. Similarly, the spread of disease (e.g., West Nile virus, avian flu) or the threat of a disease (chronic wasting disease) often triggers draconian responses at great economic cost. Likewise, there are many prominent examples of introductions of exotic species that have had major economic consequences (e.g., kudzu, Dutch elm disease, zebra mussel, various rodents), whereas others have touched off great media frenzy (e.g., recent occurrence of bighead carp and snakeheads in North American rivers and streams)—the lists are seemingly endless. The purpose of this paper is to draw attention to the insidious threat that invasive species pose to economic and social systems. We do so by examining the types of problems that invasive species may pose, pointing to examples where the magnitude of impacts have been estimated, discussing the ecological dynamics involved in the establishment and spread of invasive species, highlighting ways of controlling the spread of these species, and touching upon the research and regulatory challenges they pose.

2. Types of Invasive Species Problems

Impacts from invasive species can be classified into two categories: those that compete with valued resources and those that facilitate the spread of diseases. Within the valued resources category, the most obvious locations for effects of invasive species are coastal waters and estuaries. Here, the affected species include finfish, shellfish, and the foods and structures that marine resources depend upon. However, connections to the interior lands via waterways and overland transportation systems provide conduits to inject invasive species into the forests, rangelands, and agricultural lands. As such, the whole of terrestrial and aquatic resources potentially are exposed to invasive species.

Many invasive species carry disease agents that attack domestic crops, forests, or fish and wildlife. In some cases, the disease agents require multiple hosts (such as some parasitic diseases of fishes or certain rust diseases in plants) that can result from the introduction of all of the intermediate hosts or the occurrence of a closely related species that serves as the alternate host.

There is growing concern regarding the adaptation of invasive diseases to new hosts, such as appears to be the case with the avian flu; historically there have been similar issues with brucellosis spreading from wildlife into domestic herds, hoof-and-mouth disease, and chronic wasting disease.

3. Magnitude of Economic Impacts

Efforts to document the monetary impact of invasive species have for the most part been incomplete in that tools for assessing the impact have been restricted to estimates of reduced "commodity value;" e.g., loss of value in fisheries or loss of crop value. Alternatively, containment cost has been used to estimate economic consequences linked to a species invasion. Regardless of focus, the estimated costs are often staggering. One study by Cornell University estimated the annual damage from displacement of local animals and plants by invasive species is more than US$138 billion [18]. Other monetized costs include the estimates that the Asian gypsy moth required $20 million for control in the States of Oregon and Washington [19] and that the Asian longhorned beetle damage to wood products, maple syrup, commercial fruit, tree nurseries, and tourist industries reached $41 billion.

Often the costs have not been estimated, but rather the effects of invasives have been catalogued. For example, at the ASTM Symposium on Invasive Species in 2005, Naomi Mdzeke summarized the magnitude of problems resulting from multiple introductions of exotic species through ports around the Cape of Good Hope, which was a major replenishment stop for European ships sailing to and from the East. The list of invasive alien plants that arrived from Australia and Central and South America includes various trees, scramblers, and brambles that have spread into the semi-arid savannas, alluvial plains, and seasonal watercourses; 117 are well established and an additional 84 species are gaining in abundance. A number of aquatic invasive species occur over prolifically over large areas, including hyacinth (*Eichhornia crassipes*), water lettuce (*Pistia stratiotes*), Kariba weed (*Salvinia molesta*), parrot's feather (*Myriophyllum aquaticum*), and red water fern (*Azolla filiculoides*).

Though the situation in South Africa appears to have developed gradually over decades—even centuries—there are documented cases of rapid establishment of invasive species. One example is on Palmyra, an equatorial atoll ~1,500 km south of Honolulu that is owned by The Nature Conservancy. The timeline for realization of effects from an exotic ant species spanned only two years. Alien ants were noticed in 2002 in an area that supported Pisonia trees, a favorite nesting and roosting location for seabirds. Ten months after the ants were observed, the trees were heavily infested with two species of scale insects tended by the ants. Within a year there were large gaps in the forest where trees had died due to defoliation by the scale [20].

Port security and the role that port activities play in mediating unintended or maliciously intended species invasions was considered by ASTM symposium participants affiliated with the U.S. Coast Guard (Coast Guard) (R. Everett, Environmental Standards Division (G-MSO-4), Coast Guard, Washington, DC). The Coast Guard has been directed by Congress to develop a national regulatory program to reduce, and eventually prevent, introductions of non-indigenous aquatic species into U.S. waters via vessel operations. Under the authorizations provided by the National Aquatic Nuisance Prevention and Control Act of 1990 and the National Invasive Species Act of 1996, mandatory and voluntary ballast water management (BWM) programs have been established for the Great Lakes and the rest of the U.S., respectively. The Coast Guard is working to develop discharge standards for treated ballast water, given shipping operations cannot avoid using ballast water; e.g., when cargo is loaded onto a ship, the weight of the ship increases, and water from ballast tanks is discharged into the surrounding waters to offset the increased weight of the cargo. Water from other international ports previously pumped into a ship's ballast tanks may contain numerous aquatic organisms, including viruses, bacteria, algae, jellyfish, crabs, mollusks, and fish. If the organisms within a ship's ballast tanks survive the trip to the next destination, they may be released with the ballast water into waters in which they do not naturally occur. If these nonnative organisms survive and spread throughout their new environment, they may become invasive species. In this way, ballast water can accidentally introduce harmful microalgae and other organisms into the environment.

Invasive species may be harmful to a new environment for several reasons. Depending on the species life history attributes, they may negatively affect human health and economic activities; e.g., by competitively reducing the abundance and diversity of native species and changing the ecological dynamic of native ecosystems. Invasive species are transported through ballast water and create environmental problems throughout the world, including a range of species presenting challenges for BWM programs such as those the Coast Guard is charged to develop and administer. For example, European zebra mussel (*Dreissena polymorpha*) was introduced to the Great Lakes in the mid-1980s through ballast water collected at a freshwater European port. Zebra mussels clog water pipes and foul underwater structures, resulting in billions of dollars in control and cleanup costs. Harmful algae such as dinoflagellates (e.g., *Pyrodinium* spp. and *Alexandrium* spp.) were historically transported from Southeast Asia to Australian waters, where the species cause "red tides." Some species can cause paralytic shellfish poisoning and harm local shellfish industries. Asian green mussels (*Perna viridis*) have recently become established in ports throughout North America, where green mussels clog the pipes of power plants and desalinization plants and can compete with native species. These are but a few examples of the more than 3,000 species of plants and animals that are unintentionally transported daily around the world in ballast water [10, 21]. Incorporate malicious intent associated with "invasion by design" and the potential for adverse effects and unbounded consequences clearly suggests that prevention is paramount to managing species invasions, given control and mitigation measures for invasive species in their various guises undoubtedly lead to long-term programs that pale in comparison to the success of species capable of breaching security countermeasures once intended or unintended challenges are initiated.

In the continental U.S., there have been some 4,500 species of nonnative plants, animals, and microbes that have become established. Of these about 15 percent are considered to be invasive (i.e., recognized as reaching the level of being a pest) [19]. For the most part the economic costs have not been computed, but one indication of the impact would be the quantity of pesticides used to contain weedy plants, pest insects, rodents, fungal disease, and to a lesser extent pest birds. World pesticide expenditures totaled more than $32.5 billion in 2000 and nearly $32.0 billion in 2001 [22].

4. Ecological Dynamics

There are a number of imperatives from ecology that are important to consider when evaluating threats from invasive species. In this section, we discuss stochasticity, scale (both spatial and temporal), characteristics or traits of invasive species, and metapopulation dynamics.

4.1. STOCHASTICITY

Stochasticity, the quality of lacking any predictable order or plan, is a central characteristic of ecological systems. Ecological systems are also chaotic; that is, they are dynamic systems that are sensitive to initial conditions. Because of this, we can only hope to identify patterns of response and assign probabilities of recurrence of particular patterns.

In the realm of invasive species, that means that we cannot predict with certainty that the introduction of an exotic species into a new area will be detrimental. Moreover, "initial conditions" (i.e., those existing at the time of the introduction) influence the outcome, such that repeated introductions may fail, but with a future event, if conditions are favorable for establishment, the exotic species may become a pest.

As a consequence of stochasticity, description of a scenario leading to establishment of an invasive species requires that events be expressed as probabilities; and for the most part, the probabilities are multiplicative. For example:

1. Species A has a 0.001% chance of establishment per introduction event.
2. If established, Species A has a 5% chance of expanding its range.
3. Species A has a 10% chance of impacting a valued resource.
4. Resulting in a probability of an adverse effect occurring of $0.00001 \times 0.05 \times 0.1$ or 0.0000005%.

Complexities of this nature are easily dismissed in some circles as irrelevant, especially when more dramatic threats of terrorism are daily headlines. Nevertheless, the examples of economic damage resulting from invasive species discussed above began for the most part as innocuous events. An introduction made with the intent to inflict economic injury—by a party with considerable knowledge of the species niche requirements—would greatly increase the likelihood of establishment, the most limiting factor in the chain of events in the invasive species life history.

4.2. SCALE—TIME AND SPACE

Ecology, the distribution and abundance of organisms and the way they interact with the environment, is all about scale, both in terms of space and time. The ecological systems we observe around us, and upon which our economies are built, are the products of ecological processes that have played out over decades, centuries, and even millennia. As these systems are undergoing change constantly, the short-term trajectories that we observe in most studies may give false indication of long-term trend. A fortuitous change that coincides with one hypothesis can be misleading. The assessment of potential impacts from invasive species must be done in a manner that provides an appropriate landscape perspective and must be modeled over sufficient time to get a proper analysis of likely consequences. Thus forecasting risks from the introduction of exotic species at ports requires being cognizant of the connectivity of transportation corridors into continental interiors. Modern transportation from port facilities allows packages to reach mid-continent locales within two to three days [17]. However, manifestation of the events leading to the stage where an invasive species might have an adverse impact (i.e., establishment, spread, and impact) may occur over years or decades. This lag time between introduction and manifestation of effects is a critical feature that must be considered in characterizing the risks of invasive species, but it also contributes to the difficulty of elevating such risks into the consciousness of decision makers in systems where dealing with more immediate needs is rewarded.

Additionally, interrelated functions linked to occurrence of stochastic events, and the temporal and spatial scales serving as a stage for these events to potentially occur, assure that even low-probability events such as a species invasion are likely to happen, yielding high-consequence outcomes in the absence of directed countermeasures. Simply stated, when time and space are practically unbound, even low-probability events occur. While probabilities of transport, release, and invasion may be small—e.g., less than 10^{-6} or 10^{-9} for any single occurrence—given life history attributes of invasive species related to their fecundity, and the number of trials realized in the challenge process (e.g., the number of reproductive members within an incursion event such as an exchange of ballast water, the number of exchanges of a given volume of ballast water, and the repeated trials through many port visits and revisits), outcomes leading to a successful invasion are inevitable unless countermeasures are in place as risk management practices intended to offset these scale-related attributes that potentially lead to long-term success in the invasion process. Again, malicious actions intended to thwart a system's inherent capacity to fend off invasive species incursions increase the likelihood of successful invasions.

4.3. PROPERTIES OF INVASIVE SPECIES

As a prelude to managing efforts against the introduction of exotic species that may become invasives, it is important to profile the characteristics of species that enable them to become invasive. In general, invasive species share several of the following traits that enable them to gain a foothold in a new location and then to expand their range and increase in numbers to the point where they can compete against or otherwise inflict harm to valued resources. These traits include:

- **High fecundity.** Organisms that produce large numbers of offspring can "flood" a new environment not only with numbers but also greater genetic diversity, thereby increasing the chance of establishing a foothold. This can occur from a single gravid female having been fertilized (or carrying sperm) prior to being introduced into the new area. Or in the case of plants, a single seed pod or an inflorescence having many seeds can reach areas without having had any individuals residing in the area.
- **Reproductive flexibility.** Organisms that have different reproductive styles enhance the ability to become established in new areas. Many species of plants and insects toggle back and forth between asexual and sexual forms of reproduction. Many are hermaphroditic and others exhibit remarkable traits of gender plasticity, a situation in which female phenotypes dominate the populations until certain (often unknown) environmental triggers transform some individuals into male phenotypes that effect

sexual reproduction bringing about increases in genetic recombinations. Collectively, these reproductive traits increase the likelihood of establishment when only a single or very few individuals are introduced into the new area.

- **Cryptic patterning, pleomorphism, and mimicry.** Species that have greater capacity to hide have a better chance of surviving when introduced into a new locality. The ability to blend into the background reduces the likelihood of being consumed by higher trophic animals, and therefore they have greater likelihood of reproduction (either sexually or asexually). The ability to take on different somatic shapes (pleomorphism) enables plants and animals to be blend into their surroundings or be disguised and somewhat unrecognizable among unrelated taxa. Mimicry can be especially effective in being disguised among the local flora or fauna. Each of these traits can confound efforts to eradicate these types due to the difficulties of locating and identifying the exotic individuals.
- **Genetic diversity.** Though this may be more speculative than actually a proven trait, it makes good sense theoretically that individuals with broad inherent tolerance ranges can survive under a broader range of environmental conditions. Similarly, individuals with greater acclimation potentials can adjust to new environments more easily than individuals with limited acclimation potential.
- **Facile "hitchhiker."** Many invasive species arrive via hitchhiking. Insects, spiders, fungi, and some vertebrates arrive in shipments of raw timber or produce. The restrictions on foods transported across borders are a clear example of the seriousness of this avenue for introductions. According to Mark Hay, an ecologist at Georgia Tech:

> The notion has always been that these organisms are spreading because they've escaped their natural enemies, but the problem is that they're following their natural enemies [29].

4.4. METAPOPULATIONS AND SPREAD ACROSS THE LANDSCAPE

In recent years, theoretical ecology has led to great improvements in the general understanding of population dynamics [3]. A very important aspect of assessing risk at the population level is the interactive nature of metapopulations [9]. Theoretical underpinnings [13, 7] established the counterintuitive situation that occurs when a stressor is applied to one subpopulation in a metapopulation network, namely that the effect is observed first in the most distantly connected subpopulation. That is, the subpopulation that has the stressor applied continues to maintain fairly constant population densities whereas the most distant subpopulation (distance measured as the ability to immigrate and emigrate) declines and eventually is extirpated. Thus, if the only observations are those taken at the place where the stressor is applied, one can easily be misled into believing that a stressor has no effect on the population. The implications encompass both determining effects on populations impacted by invasive species and determining the efficacy of measures to eradicate the invasive species.

5. Control Mechanisms

Control measures available to minimize the risks from invasive species begin with efforts to exclude entry through eradication actions once an invasive species has been established. These include regulations that ban outright the entry of species and any materials that the target species might be associated with (as a hitchhiker). Enforcement of bans typically occurs at border crossings; thus, the effectiveness of such bans is directly linked to the level of inspections and confiscations administered. Clearly, where crossings occur that do not pass through secured checkpoints (e.g., along a waterway or a secondary road), opportunities for illegal or incidental transport can be significant. Another level of control occurs on those items having restricted entry regulations. In such cases, quarantines or analogous containment practices can be effective in lowering the incidence of introductions.

Interrelationships between control measures as a risk management tool, and research and development targeting control of invasive species is illustrated by ongoing efforts focused on issues related to ballast water. For over a decade, the International Maritime Organization of the United Nations (IMO) has been working on the development of international regulations for managing ballast water in vessels to minimize the risk of nonindigenous aquatic nuisance species invading coastal waters around the world as a result of ballast water discharges. In 2003, IMO agreed to the International Convention for the Control and Management of Ships' Ballast Water and Sediments (the Convention) intended to minimize the spread of nonindigenous, aquatic organisms by requiring BWM. Under the Convention, all new and existing vessels with ballast tanks will be required to implement BWM procedures and meet specific standards when on voyages entering a nation's waters from beyond its Exclusive Economic Zone, more than 200 nautical miles from the nearest land. The Convention specifies both (i) an interim ballast water exchange standard, and (ii) ballast water performance standards. The ballast water exchange standard would require at least 95 percent volumetric exchange of ballast water in a vessel's ballast tanks. This volumetric exchange must attain compliance with performance standards aimed at reducing concentrations of viable organisms (expressed in terms of the number of organisms per unit volume of discharged ballast water) discharged from their ballast tanks to the levels specified in the Convention's regulations. At present, no technologies have been demonstrated aboard vessels that would meet the standards established by the Convention, but Coast Guard and the National Oceanic and Atmospheric Administration have been involved in testing treatment technologies. Many hope the Convention's standards will spur scientific research to develop sufficient technologies and the data needed to effectively manage ballast water to the prescribed standards before the Convention's deadlines.

Eradication of invasives comes into play once an exotic species has become established and realization of adverse effects has occurred. We know of no incidence where eradication has eliminated an exotic species; the best one can hope for realistically is to lower the abundance of the exotic species to the point where the economic and social consequences are tolerable. In that eradication practices often involve use of a pesticide or other chemical control agent, there are other environmental risks that must be weighed in the overall planning for controlling the invasive species. Again, research and development focused on eradication, or more practically speaking, control of invasive species is illustrated by recent developments in the control of zebra mussel. For example, the recent ASTM symposium focused on invasive species included reports on development of a microencapsulated "BioBullet" capable of controlling zebra mussels that now infest many waterways [4]. The BioBullet uses the encapsulation of an active ingredient in microscopic particles of edible material and relies on the mussels' natural filtering ability to remove and concentrate the particles from the water during feeding without stimulating the valve-closing response. In addition, the particles may be engineered to break up and dissolve within a few hours, reducing risks of polluting the wider ecosystem with the encapsulated toxicant. BioBullets have demonstrated the effectiveness of a toxicant in the control of biofouling filter-feeders. Future work may yield other control agents such as the BioBullet with targets on other aquatic nuisance species.

6. Research and Regulatory Challenges

One important precursor to understanding and managing the risks from invasive species is to gain better understanding of historical cases. To this end, there has been an excellent compilation of reports that assess the status of invasive species in various countries [23]. This site includes papers on forest health and biosecurity with special emphasis on the status of invasive forest tree species [28].

In another compilation of invasive species [24] there were 95 case studies, including:

- Ecoregion conservation in Cambodia, Lao PDR, and Viet Nam.
- Inter-institutional initiative for biodiversity conservation, Pampas Region, Argentina.
- Accidental and intentional introduction of exotic species in the Black Sea.
- Alien invasive species and agriculture in Turkey.
- Alien invasive species in Israel.
- Alien organisms in Germany.
- Alien species that threaten ecosystems, habitats, or species (alien bird species in Lebanon, alien insect species in Lebanon).
- *Ambrosia artemisiifolia* L. in China.
- Case studies from the marine environment; introduction to the Baltic Sea and the Swedish West Coast.
- Case study on the Black Striped Mussel in Australia.
- Consequences of the invasion of a predatory cladoceran. Report for International Council for the Exploration of the Sea in Estonia.

These excellent compilations provide a good introduction, much as good natural history descriptions of species are essential starting points for other types of risk characterization. These data can be used to extrapolate critical features onto GIS based platforms to identify likely areas that might be prone to establishment of invasive species as has been done to predict future distribution of selected invasive species from China into the U.S. and from the U.S. into China [25].

Along with the continued development of basic and applied ecology, there are efforts underway to develop standard guides such as the ASTM Draft Standard Guide for Conducting Hazard Assessment-Critical Control Point (HACCP) Evaluations [26]. HACCP evaluation has become increasingly applied to natural resource management and environmental engineering problems, particularly as hazards may be managed; e.g., with respect to the safety of processes or release of materials or products to the environment. HACCP should be an integral part of management practices focused on engineering or resource management practices to develop aquatic, wetland, and terrestrial habitats for human use (e.g., agriculture or construction activities) or to enhance habitats for fish and wildlife. HACCP is a systematic and preventive approach that addresses biological, chemical, and physical hazards through anticipation and prevention, rather than through end-product inspection and testing or retrospective engineering solutions necessitated because of previous undertakings. The HACCP system is intended for managing risks and safety concerns associated with a wide range of materials, products, and management practices with an emphasis on a total systems approach to improve environmental quality.

HACCP emphasizes control of a process as far upstream in the processing system as possible by utilizing operator control or continuous monitoring techniques, or a combination of both, at critical control points. The HACCP system uses the approach of controlling critical points in any process to reduce or eliminate risks and prevent safety problems from developing. The identification of specific hazards and measures for their control to ensure the safety of a process, material, or product through prevention reduces the reliance on end-product inspection and testing (e.g., for agrichemicals), remedial measures (e.g., related to construction practices), or mitigation measures as part of a control program (e.g., quarantine or disinfection for control of invasive species). Any HACCP system should be capable of accommodating change, such as advances in equipment design or developing alternative resource management practices, changes in processing procedures, or technological developments.

Beyond these scientific and technological efforts, there are also many policy and political developments that are needed, such as the establishment of multilateral agreements that can be put in place to improve cooperation on preventing the spread of species. Such agreements could be linked to and enforced under agreements by signatories to the Convention on Biodiversity [27].

Addressing the issues surrounding invasive species as a legitimate environmental security threat poses many challenges. As with any feature of ecological systems, the issue of invasive species threats is characteristically complex. Complexity comes from the inherent ecological relationships embodied in population dynamics—though complex, the state-of-the-science enables identification of patterns and assignment of probabilities to various outcomes. The larger problems to overcome are the inherent limitations of the political realities that tend to set higher priority on dramatic and near-term consequences over those that are less dramatic and potentially entail long lag times (several years) before the impacts are realized. Fortunately, multilateral agreements under the Convention on Biological Diversity and bilateral activities among many signatories of the Convention are making progress on both the technical and the political issues pertaining to the spread of invasive species.

7. References

1. Balmford A, Bruner A, Cooper P, Costanza R, Farber S, Green RE, Jenkins M, Jefferiss P, Jessamy V, Madde J, Munro K, Myers N, Naeem S, Paavola J, Rayment M, Rosendo S, Roughgarde J, Trumper K, Turner RK. Economic reasons for conserving wild nature. Science. 2002; 297:950-953.
2. Campbell DE. Proposal for including what is valuable to ecosystems in environmental assessments. Environ. Sci. Technol. 2001; 35:2867-2873.

3. Caswell H. Matrix Population Models, 2nd edition. Sunderland: Sinauer Associates, Inc.; 2000.
4. Aldridge DC, Elliott P, Moggridge GD. Microencapsulated biobullets for the control of biofouling zebra mussels. Environ. Sci. Technol. 2006; 40:975-979, 10.1021/es050614+S0013-936X(05)00614-0, Web Release Date: December 22, 2005.
5. Daily GC. Nature's services: societal dependence on natural ecosystems. Washington, DC: Island Press; 1997.
6. Dietz D, Ranganathan J, Banks RD, editors. Green ledgers: case studies in corporate environmental accounting. Baltimore: World Resources Institute; 1995.
7. Macovsky LA. Test of the action at a distance hypothesis using insect metapopulations. Master's Thesis, Huxley College. Bellingham: Western Washington University; 1999.
8. McCormick RJ, Zellmer AJ, Allen TFH. Type, scale, and adaptive narrative: keeping models of salmon, toxicology and risk alive to the world. In Kapustka LA, Galbraith H, Luxon M, Biddinger GR, editors. Landscape ecology and wildlife habitat evaluation: critical information for ecological risk assessment, land-use management activities, and biodiversity enhancement practices. West Conshohocken: ASTM International; 2004.
9. Menzie C, Bettinger N, Fritz A, Kapustka L, Regan H, Møller V, Noel H. Populations protection goals. In Barnthouse L, Sorensen M, editors. Population-level ecological risk assessment. Pensacola: SETAC Press; in press.
10. National Research Council Commission on Engineering and Technical Systems. Stemming the tide: controlling introductions of nonindigenous species by ships' ballast water. 1996; Washington, DC: National Academies Press.
11. National Research Council. Valuing ecosystem services: toward better environmental decision-making. 2004; Washington, DC: National Academies Press.
12. Pimental D, Lach L, Zuniga R, Morrison D. Environmental and economic costs of nonindigenous species in the U.S. BioScience. 2000; 50:53-65.
13. Spromberg JA, Johns BM, Landis WG. Metapopulation dynamics: indirect effects and multiple discrete outcomes in ecological risk assessment. Environ. Toxicol. Chem. 1998; 17:1640-1649.
14. Stahl Jr RG, Kapustka L, Munns Jr WR, Bruins RJF. Valuation of ecological resources: integration of ecological risk assessment and socio-economics to inform environmental decisions. New York: Elsevier; in press.
15. USEPA. Improved science-based environmental stakeholder processes. EPA-SAB-EC-COM-01-006. August, 2001.
16. USEPA. Understanding public values and attitudes related to ecological risk management: an SAB workshop report of an EPA/SAB workshop, EPA-SAB-EC-WKSP-01-001. September 2001.
17. van de Voort M, Willis H, Ortiz D, Martonosi S, Rahman A. Policy considerations in securing the global containerized supply chain. In: Linkov et al., editors. Risk management tools for port security, critical infrastructure, and sustainability. New York: Springer Publishing; 2006.
18. Computational technologies project. ESDCD News. 2003. Available at: http://esdcd-news.gsfc.nasa.gov/2003.Winter/02.invasive-species.html.
19. National Invasive Species Information Center. Available at: http://www.invasivespeciesinfo.gov/.
20. Invasive species. Malamalama: the magazine of the University of Hawaii system. Available at: http://www.hawaii.edu/malamalama/2004/09/f2_invasive.html.
21. National Invasive Species Information Center, USDA Agricultural Library. Ballast water. Available at: http://www.invasivespeciesinfo.gov/aquatics/ballast.shtml.
22. USEPA. World and U.S. pesticide expenditures. 2000–2001 pesticide market estimates. 2001. Available at: http://www.epa.gov/oppbead1/pestsales/01pestsales/sales2001.html#2_1.
23. Food and Agriculture Organization of the United Nations. Alien invasive species website. Available at: http://www.fao.org/forestry/site/27082/en/.
24. Convention on Biological Diversity. Invasive alien species case studies. Available at: http://www.biodiv.org/programmes/cross-cutting/alien/cs.aspx.

25. USGS. International program website. Available at: http://edcintl.cr.usgs.gov/invasiveinfosheet.html.
26. ASTM. WK7844 new hazard analysis and critical control point (HACCP) protocol for preventing the spread of aquatic nuisance species. Available at: http://www.astm.org/cgi-bin/SoftCart.exe/DATABASE.CART/WORKITEMS/WK7844.htm?L+mystore+ytdh9331.
27. Convention on Biological Diversity. Available at: http://www.biodiv.org/convention/articles.asp. 2002.
28. Haysom KA, Murphy ST. The status of invasiveness of forest tree species outside their natural habitat: a global review and discussion paper-forest health and biosecurity working paper FBS/3E. Rome: Forestry Department, FAO (unpublished); 2003.
29. Dennis O'Brien. SUN REPORTER. March 10, 2006.

CHAPTER 11

EVALUATING RISKS FROM CONTAMINATED SEDIMENTS AT INDUSTRIAL PORTS AND HARBORS

R.J. WENNING

ENVIRON International Corporation, 6001 Shellmound Street, Ste. 700, Emeryville, CA 94608 USA

M.T. SORENSEN

ENVIRON International Corporation, 1600 Parkwood Circle, Ste. 310, Atlanta, GA 30039 USA

V.S. MAGAR

ENVIRON International Corporation, 123 N. Wacker Street, Ste. 250, Chicago, IL 60606 USA

Abstract The management of surface waters and sediments is one of several activities at commercial and industrial shipping ports and harbors critical to maintaining environmental quality that safeguards surrounding communities and the environment. This chapter discusses a possible framework for assessing and managing risks to the aquatic environment, focusing primarily on sediments. Risk assessment provides a useful foundation for understanding the environmental benefits, residual hazards, and engineering limitations of different management strategies, as well as identifying and ranking management options. Understanding the important pathways for contaminant exposure, the human and wildlife populations potentially at risk, and the possible hazards associated with the implementation of different engineering options contributes to informed risk management decision making with regard to short- and long-term effectiveness and implementability of different sediment management strategies.

1. Introduction

For commercial and industrial shipping ports, which are typically located in heavily populated urban areas and sensitive coastal environments, the protection of critical infrastructure, industrial tenants, surrounding communities,

I. Linkov et al. (eds.), Managing Critical Infrastructure Risks, 189–205.
© 2007 *Springer*.

and the environment are of paramount concern. Port facilities handle a variety of potentially sensitive or dangerous activities and materials that can have serious environmental consequences if not managed carefully [31]. In the context of providing safe and secure operations, risk assessment is increasingly used as part of vulnerability assessments to prioritize the hazards posed by different port activities and to classify potentially dangerous materials and cargo [32]. This provides environmental managers and emergency services providers with critical information on the potential threats posed by environmental accidents or uncontrolled releases of chemical substances to land, water and sediment quality, port workers, and public health, as well as help to identify appropriate security and prevention measures [31].

Wenning et al. [32] describes three challenges when applying risk analysis to environmental security at ports and harbors:

- Inter.pret available site-specific data and derive, with a high degree of confidence, quantitative predictions of the range of possible damages and potential threats to human health and the environment.
- Predict the potential risks using models and assumptions that reflect meaningful and plausible scenarios.
- Inform decision makers about the nature of the risks and help environmental managers and emergency services providers develop disaster prevention/response scenarios that effectively prevent or mitigate unacceptable risks.

If these challenges can be overcome, the risk assessment process provides port authorities and both local and national government authorities with much needed tools to prioritize their investments in security system upgrades or maintenance.

A recent review of environmental practices at commercial and industrial shipping ports in the U.S. [22, 3] indicated security failures and accidental releases by industrial tenants as among the most common causes of soil, water, and sediment contamination to coastal waterways and bays. Improvements in environmental practices, in particular storm water collection and treatment, diesel particulate emissions to air, and ballast water management, were identified as among the most urgent needs to reduce sources of environmental contamination at port facilities. The nature and extent of pollution varies widely at different ports depending on the specific circumstances responsible for the contamination.

The full extent of the environmental challenges facing ports in developed and undeveloped countries is poorly documented. It is well known, however, that rivers, harbors, and estuaries supporting commercial shipping ports frequently contain contaminated sediments. It is also well known that pollutants typically found in sediments can directly harm aquatic life, degrade essential habitat, bioaccumulate in fish and shellfish tissues, and migrate through food webs to birds and mammals, including humans [25,18, 10, 9]. The potential for adverse effects on human health and the environment are compelling reasons prompting many coastal communities and commercial shipping ports worldwide to invest in sediment management programs and upgrades of security systems and critical infrastructure.

Whether or not sediment management is warranted depends on the magnitude of direct and indirect human health risks, ecological threats to aquatic biota, and the extent of risk reduction that can be achieved by removal or containment of the contamination [29, 10]. Understanding the risk posed by contaminated sediments can be difficult and uncertain, often relying heavily on surrogate measures and predictive models of environmental effects.

To date, common practice typically consists of using simplistic and often overly conservative baseline risk assessments to support the selection of a remediation strategy to reduce or eliminate the perceived unacceptable health risks or ecological threats. Typically, the efficacy of different remediation options is given only cursory evaluation. Until recently, the removal of pollutant mass by dredging was presumed by most regulatory agencies and stakeholders to accelerate ecological recovery and prevent the possibility for higher risks in the future due to unforeseen extreme events that mobilize the contamination.

More recently, port authorities, governing authorities, and stakeholders recognize that a range of remedy alternatives can and should be considered and compared as part of the decision-making process. The analysis of remedy alternatives should balance several factors such as the forecasted environmental benefits and impacts, costs, and implementability [29, 30].

This chapter reviews the importance of risk assessment and management at commercial shipping ports and discusses approaches that can lead to cost-effective and environmentally beneficial environmental management strategies, particularly for sediments and surface waters. While past practices at ports often resulted in considerable alteration and damage to the natural environment, ports have become more conscious of and responsive to the need to minimize impacts on coastal resources and surrounding communities. Although the dynamic nature of coastal environments and port operations complicates evaluation of environmental risks and identification of appropriate management strategies, these hurdles can be overcome or taken into consideration with adequate site-specific information. This chapter addresses topics that can be used by port authorities, governing authorities, and stakeholders to achieve these goals.

2. Distinguishing between Risk Assessment and Risk Management

Risk assessment in the context of sediment remediation is the process that evaluates the likelihood that adverse effects may occur, either short-term or long-term. Risk may take several forms (e.g., the likelihood that the population of a fish species will be affected because of the introduction of petroleum or metal contamination to surface waters or the likelihood that sediment contamination will mobilize in response to disturbance and migrate elsewhere or become more bioavailable to aquatic life). A broad range of physical, chemical, and biological studies inform the assessment process. Figure 1 provides examples of in situ, laboratory, and literature-based assessment tools that can be useful to quantify the direct risks or the geochemical conditions that govern those risks [20, 15].

The level of effort associated with the use of these tools should be commensurate with the magnitude of the real or perceived environmental challenge, the anticipated costs (financial, safety, security, or otherwise) of managing the challenge, and the potential impacts of the remedy activity itself [19].

In the context of contaminated sediments, risk management is the activity that makes use of the information gathered from site-specific investigations, risk assessment, and the lessons learned from other waterways facing similar challenges. Kiker et al. [13] describes the transition to risk management as the second phase of risk-based decision making, integrating the results of risk assessment with technical, political, legal, social, and economic issues to develop and implement risk reduction and prevention strategies.

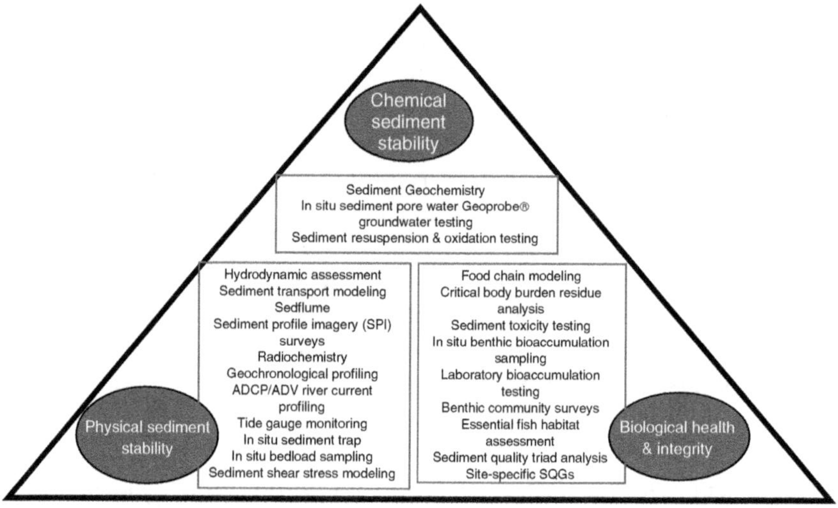

Figure 1. Physical, chemical, and biological studies that inform risk assessment and risk management.

3. Defining Contaminated Sediment Risks

Assessing or managing risk implies that the sources of risk itself are understood. Several sources of risk can be identified either directly or as a consequence of contaminated sediments [18,1], including:

- Human health impacts from eating fish/shellfish, wading, and swimming.
- Ecological impacts on wildlife and aquatic species.
- Loss of recreational and subsistence fishing opportunities.
- Loss of recreational swimming and boating opportunities.
- Loss of traditional cultural practices by Indian tribes and others.
- Economic effects of loss of fisheries.
- Economic effects on development, reduction in property values, or property transferability.
- Economic effects on tourism.
- Increased costs of drinking water treatment, other effects on drinking water, and other water uses.
- Loss or increased cost of commercial navigation.

As discussed by USEPA [24, 29], risk management decision making should consider the net risk reductions provided by implementation of different sediment management options, which include source control and technologies such as monitored natural recovery (MNR), capping, and environmental dredging. Different options (or combinations of options) may also contribute to risk by changing contaminant exposure processes in the sediment or surface water; generating new sources of exposures related to post-remedy transport, processing, treatment, and confinement; habitat modification/destruction; and activities or conditions related to human safety [30,17].

As indicated in Table 1, contaminant risks from sediment remediation projects should be considered on two time scales:

- Adverse risks may be evident during the active remedy phase (i.e., remedy implementation) primarily due to resuspension and other short-term increases in chemical exposure to aquatic organisms.
- After remediation is complete, there also may be longer-term contaminant risks resulting from residual contamination (i.e., residual risk), transport of releases beyond the project area, or a new sediment physical-chemical equilibrium that alters ecological exposure conditions [30].

Remedy implementation is generally expected to have short-term environmental impacts primarily from increases in the potential for exposures to chemicals resuspended in the water column or transported off site. In most situations with sediment contamination from historical chemical releases, disequilibrium between the water column and sediments results in chemical concentrations in the water column that are far lower than those in the sediment interstitial water. Resuspension can bring this more highly contaminated interstitial water into the water column, as well as allow desorption of contaminants from sediment particles into the water column. The resulting increase in water column exposure can result in adverse effects to aquatic biota either by direct toxicity, or by increasing residues of bioaccumulative chemicals both directly and through the food chain. Where remedy activities are focused on deeper deposits that may have higher contamination and/or different geochemistry than surficial deposits, increases in water column exposure have the potential to be more severe if resuspension is not controlled.

TABLE 1. Residual and implementation risks.

Type of Risk	Source of Risk	Measure of Risk
Implementation Risks		
Temporary impacts of sediment suspension on water quality and sediment recovery (Figure 4)	Dredging or capping	Predicted dispersion and deposition rates; water quality testing, post-remediation monitoring
Construction or transportation accidents	Construction and remediation operations	Fatal and non-fatal accident rates
Air emissions	Dredging; dewatering and stabilization; transportation and disposal	Health risk to sensitive receptors or odors
Public quality of life issues	Restricted use of a resource; vehicle traffic; odors; noise	Time for loss of amenity; Truck volumes on local roads; air quality monitoring
Temporary loss of habitat (Figure 4)	Physical removal of benthic macro-invertebrate community and habitat	Habitat recovery time
Residual Risks		
Long term chemical residual risks (Figure 5)	Dredging and background recontamination	Residual deposition away from remediation area (potentially even chemicals unrelated to target compounds) and recontamination of cleaned area from background sources

Even best-management practices to control particulate releases cannot contain 100% of chemical contamination, particularly once desorbed and partitioned into the aqueous phase. However, short-term effects are defined by the exposure period and remediation can be strategically implemented to minimize impacts by, for example, avoiding activities such as dredging during wildlife breeding seasons or during high water flow conditions. While this may seem to be an obvious risk management consideration, the U.S. and Europe have felt the need to codify this behavior in regulatory guidance such as the U.S. Sustainable Fish Act and the E.U. Water Framework Directive to require the consideration of ecological factors that may influence the implementation of construction and remediation projects. Another alternative is to minimize dredging by focusing on hot-spots or by employing in situ remedies, such as capping or MNR, particularly of those remedies achieve the same level or greater risk reduction than dredging.

Post-remedy risks are determined primarily by changes in the exposure field resulting from residuals and redistributed sediments. These are often referred to as longer term risk considerations. The ability to predict changes in bioaccumulation and other risks depends not only on the ability to describe the post-remedy exposure field, but also the degree to which pre-remedy modeling accurately reflected the relationship between sediment/water/food chain exposure and actual bioaccumulation. Because many sediment projects identify both short-term and long-term risk reduction goals, it is important for risk evaluations to consider both short- and long- term environmental changes and subsequent risks.

4. Approaches to Evaluation of Sediment Risks

Several paradigms have been used that incorporate risk analysis or environmental benefits analysis into environmental management [11,14,4]. At present, two different risk assessment paradigms offer the clearest direction on how best to merge quantitative risk assessment with economic analysis and societal valuation of potentially affected natural resources. Both cases [21, 8] demonstrate that the extent to which environmental risks and benefits can be quantitatively included in economic analysis is largely dependent on the use of detailed risk assessment methods [7].

Suter et al. [21] proposed a model for organizing the results of a risk assessment for evaluation of health and ecological risks before and after remediation, as well as evaluation of the impacts posed by different remedy options. The approach is illustrated in Figure 2.

Suter et al. [21] proposed three categories of human and ecological risk. The first category is *de manifestis* risks, which are those risks that require

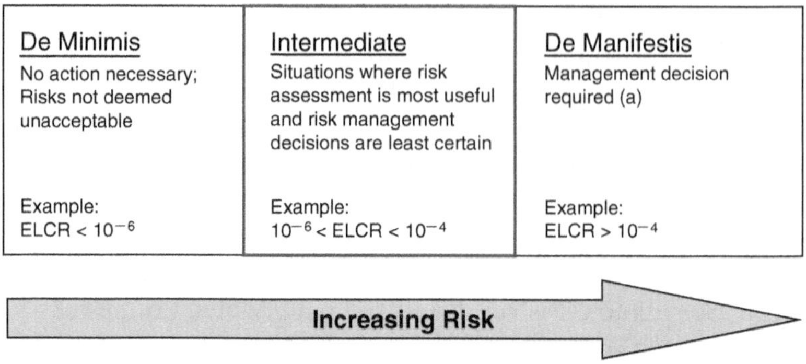

(a) The management decision does not necessarily presume sediment removal (e.g., dredging). A comparative analysis of remedy alternatives may lead to a determination that intrusive action is not warranted. For example, the management decision may be the collection of more data, the refinement of risk assessment approaches, and consideration of in situ remedies such as monitored natural recovery or contaminant isolation via in situ capping.

Figure 2. De manifestis, de minimis, and intermediate risks.

remediation, unless the remedial action conflicts with the protection of human health. De manifestis ecological risks often include risks imposed on threatened or endangered species, wetlands, and ecological components with extraordinary local or ecological value. De manifestis human health risks are defined by an excess cancer risk greater than or equal to 1 in 10,000 (i.e., 10^{-4}) or a hazard quotient (HQ) or hazard index (HI) greater than or equal to a value of 1 for any individual contaminant or for combined exposures across contaminants of similar toxicological effects.

The second category is *de minimis* risks. De minimis ecological risks do not normally require remediation because the risks are considered so low that they do not impact populations in a measurable way. Suter et al. [21] defined *de minimis* ecological risk based on regulatory precedents as a less than 20% reduction in the abundance or production of a population within suitable habitat, the loss of less than 20% of the species in a community, or the loss of less than 20% of the habitat area. Suter et al. [21] defines *de minimis* human health risk as excess cancer risk less than or equal to 1 in 1,000,000 (i.e., 10^{-6}) or an HQ or HI less than a value of 1 for any individual contaminant or for combined exposures across contaminants of similar toxicological effects.

The third category of risk are intermediate health and ecological risks, which fall between *de manifestis* and *de minimis* risks and generally require further evaluation and cost-benefit analyses to determine the need for remediation. It is often within this intermediate range that risk assessment tools can make the most significant contribution, by supporting comparative risk analysis of multiple remedial alternatives.

The approach by Suter et al. [21] does not account for the ecological recovery of a site, the different recovery rates for different species, or the relative abundance of the species in the larger ecosystem, each of which can be critical decision-making components. Although Suter et al. [21] argue there is legal precedent for using the 20% criterion when evaluating a remediation project and acute impacts, they acknowledge this criterion is a generalization with considerable uncertainty. The appropriate benchmark for pass-fail (either below or above 20%) should be evaluated on a site-specific basis.

A different approach proposed by Efroymson et al. [8] uses net environmental benefit analysis (NEBA) to quantify the gains in environmental services or other ecological properties attained by remediation or ecological restoration, as well as the environmental injuries caused by the same actions. The approach is illustrated in Figure 3.

The approach relies on the use of reference sites to establish baseline conditions before and after remediation. It also provides a framework to include ecological improvement activities that do not necessarily focus directly on contaminant isolation or removal and may or may not even occur at the same location as the contamination. Ecological improvement activities may include wetland construction, reef construction, or other types of habitat mitigation measures that can be quantified either economically or based on some index of value to society.

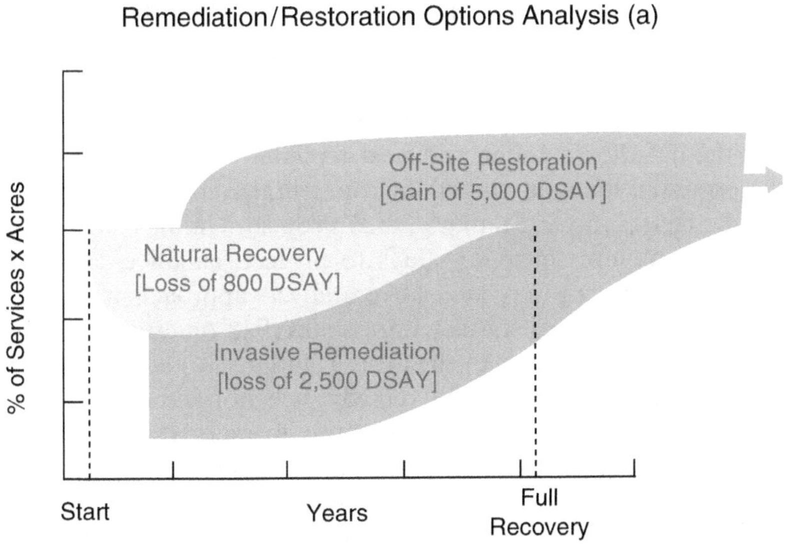

Figure 3. Net environmental benefits analysis. NEBA results show that offsite restoration can yield greater days of service acre years (DSAYS) than invasive remediation.

According to Efroymson et al. [8], a remedy may provide no net benefit or may cause more harm than good if it is ineffective and does not substantially reduce risk, or if the action causes environmental injuries greater than the damage associated with baseline conditions. The underlying challenge of this approach is clear understanding of baseline and post-remediation conditions. Equally important challenges are comparisons made solely on the basis of chemical contamination or societal values that may bias decision making toward remediation, and comparisons made solely on the basis of cost, which may bias decision making toward inaction.

5. Risk Assessment Applied to Sediment Management

Risk assessment prior to remediation establishes baseline conditions, from which:

- A determination is made about whether risks are such that corrective action must be considered (i.e., *de manifestis*, *de minimis*, or intermediate risks).
- Remedy alternatives can be systematically compared to identify the greatest risk reduction with the least amount of impact from the remedy itself (i.e., greatest net benefit).

Where predictive models and bench-scale testing support conclusions that the implementation of a remedy will cause more environmental benefit than harm, recent USEPA guidance on sediment remediation at contaminated sites indicates that so-called active monitoring or in situ remediation may be the preferred approach [29]. In the U.S., ecological risk assessment guidance emphasizes that ecological risk reductions associated with a remedy must be balanced against the potential impacts of the remedy itself [26]. Guidance from Australia, New Zealand, Canada, and the Netherlands also specifies the need for baseline and post-remediation risk analyses to ascertain the potential for unintended environmental consequences and adverse biological effects [2, 5,16]. Similarly, the European Union Water Framework Directive [10] requires member states to protect, enhance, and restore all controlled waters using a comparative analysis approach with the aim of achieving good ecological status within a specified timeframe [10].

In the context of human health, a failure to adequately evaluate implementation risks during the remedy selection process can result in unanticipated injuries (or even fatalities) to workers and nearby residents during cleanup. The consequences may also include costly delays associated with substantial remedy modifications or abandonment of an incomplete remedy [6]. In the context of ecosystem restoration, the U.S. Army Corps of Engineers [23] increasingly includes non-monetary metrics to evaluate the accomplishments or results of environmental projects, judge cost effectiveness, and identify both short- and long-term success. Thus, short-term implementation risks, such as those posed by extensive dredging activities and the transportation of waste materials and both capping and backfill materials, must be evaluated carefully and weighed against any long-term risk reduction that may be achieved.

The timeframe to evaluate different remedy alternatives (e.g., the required time to meet remediation goals for the site) is an important consideration [29, 10]. USEPA and EU guidance generally requires remediation efforts to meet cleanup goals within a "reasonable timeframe," leaving a large degree of subjectivity when defining what is "reasonable" for remediation [28, 29, 10]. At many sites, restoration may require years or decades. In the absence of imminent risk to human health and the environment, it may be acceptable to allow contaminants to remain in place and allow ongoing natural processes to contribute to long-term ecosystem health.

Figure 4 illustrates the importance of considering time-scale in sediment management decision making. In Figure 4, changes relative to baseline, pre-remedy conditions show that dredging impacts may alter ecological conditions for longer period of time relative to other less intrusive remedies; however, the impacts are often perceived as short-term in comparison to the long-term benefits of remediation.

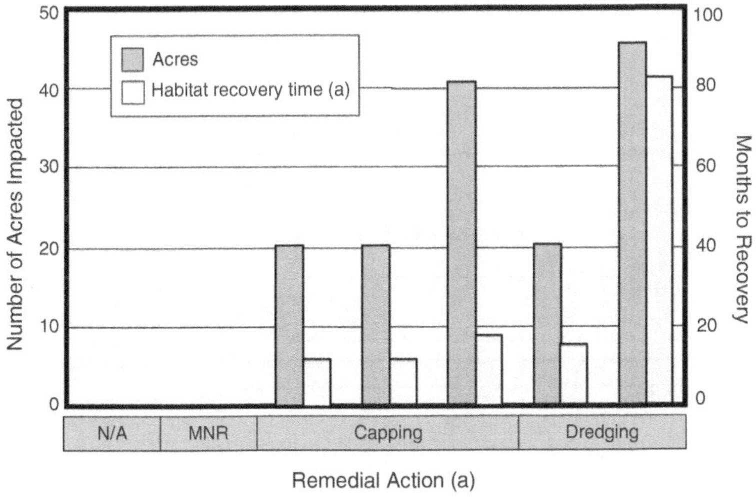

(a) Changes relative to baseline, pre-remedy conditions; results show that dredging impacts the largest area for the longest period of time. However, these impacts are often perceived as short-term in comparison to the benefits of remediation.

Figure 4. Comparative analysis of habitat recovery time for different sediment management options.

For sediment sites, residual risks should be considered as part of the decision making process, including post-remedy potential risks to benthic organisms, wildlife, and humans resulting from exposure to chemicals in sediment, fish tissue, and surface water. The effects of different sediment management options on both chemical and physical stressors should be quantified to the extent practicable by the state of the science and available site-specific information. By doing so, the risk assessment becomes a useful tool to quantify, compare, and communicate residual risks associated with remedy alternatives. Post-remedy risks require an understanding of short-term and long-term residual risks after remedy implementation. Though it may be reasonable to assume negligible short-term residual risks after sediment capping, due to the creation of a clean sediment surface via capping, background contaminants common such as PAHs or metals are likely to recontaminate the sediment surface after remediation. At a minimum, long-term residual risks are likely to resemble background risks associated with background contaminants, and may be greater than expected due to uncontrolled contaminant sources.

For example, Figure 5 shows the results of a comparative risk assessment using the U.S. Army Corps of Engineers Trophic Trace model to evaluate the effects of six different sediment management options on chemical exposures to two species of water birds. The analysis also considered the anticipated long-term residual risks.

The analysis illustrates the level of information available to decision makers faced with selecting the most ecologically protective or least

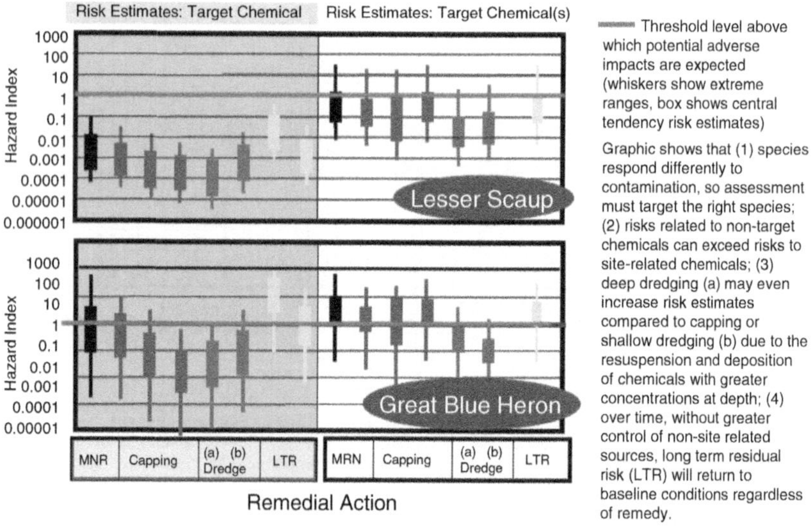

Figure 5. Post-remedy comparative risk analysis and long-term residual risks of chemical exposure for two species of water birds, Lesser Scaup and Great Blue Heron.

damaging outcome. In the example shown above, an in situ sediment cap appears to provide similar or better overall risk reduction as compared to other options.

Comparisons of baseline risks with risk reduction measures and evaluation of the time to achieve risk-based goals require relatively accurate ecological and human health risk projections for each remedy alternative. These projections have to be modeled using site-specific hydrology, hydrodynamic, and hydrogeology data; sediment geophysical and contaminant properties; fate and transport mechanisms; and ecological inputs. The models or assessment tools used will depend on the type of risk evaluated. For example, prediction of the risk of exceeding species-specific critical body residues might involve the prediction of surface water concentrations and the use of a food chain model.

In some cases, current available modeling tools may not be sophisticated enough to make the necessary calculations to forecast risk into the future, particularly with respect to integrating contaminant fate and transport with risk. Research is needed to develop modeling capabilities up to the task of making long-term predictions of sediment contaminant concentrations, in situ or technology-related risks, ecological state, and costs to permit comparisons of remediation alternatives based on short- and long-term risk reduction and risk management.

6. Summary

The management of contaminated sediments is one of the largest environmental challenges facing commercial shipping ports. The primary goals of a sediment management program are to:

1. Reduce the risks of sediment contaminants such that human health and the environment are protected
2. Restore ecosystems damaged by accidental or historical discharges and potential future catastrophic releases due to sabotage or natural disasters, if port security measures are circumvented.

Figure 6 shows a sediment risk-management decision framework. The framework considers baseline risks, implementation risks, and residual risks in the decision analysis and remedy selection process.

The success of applying net environmental benefit analysis to sediment management decision making will require greater reliance on human health and ecological risk assessment, in conjunction with the evolution of multivariate decision-making methods such as multi-criteria decision analysis (MCDA) and comparative risk assessment (CRA). The extent to which benefits can be quantitatively included in an economic analysis is largely determined by the choice of risk assessment methods [7].

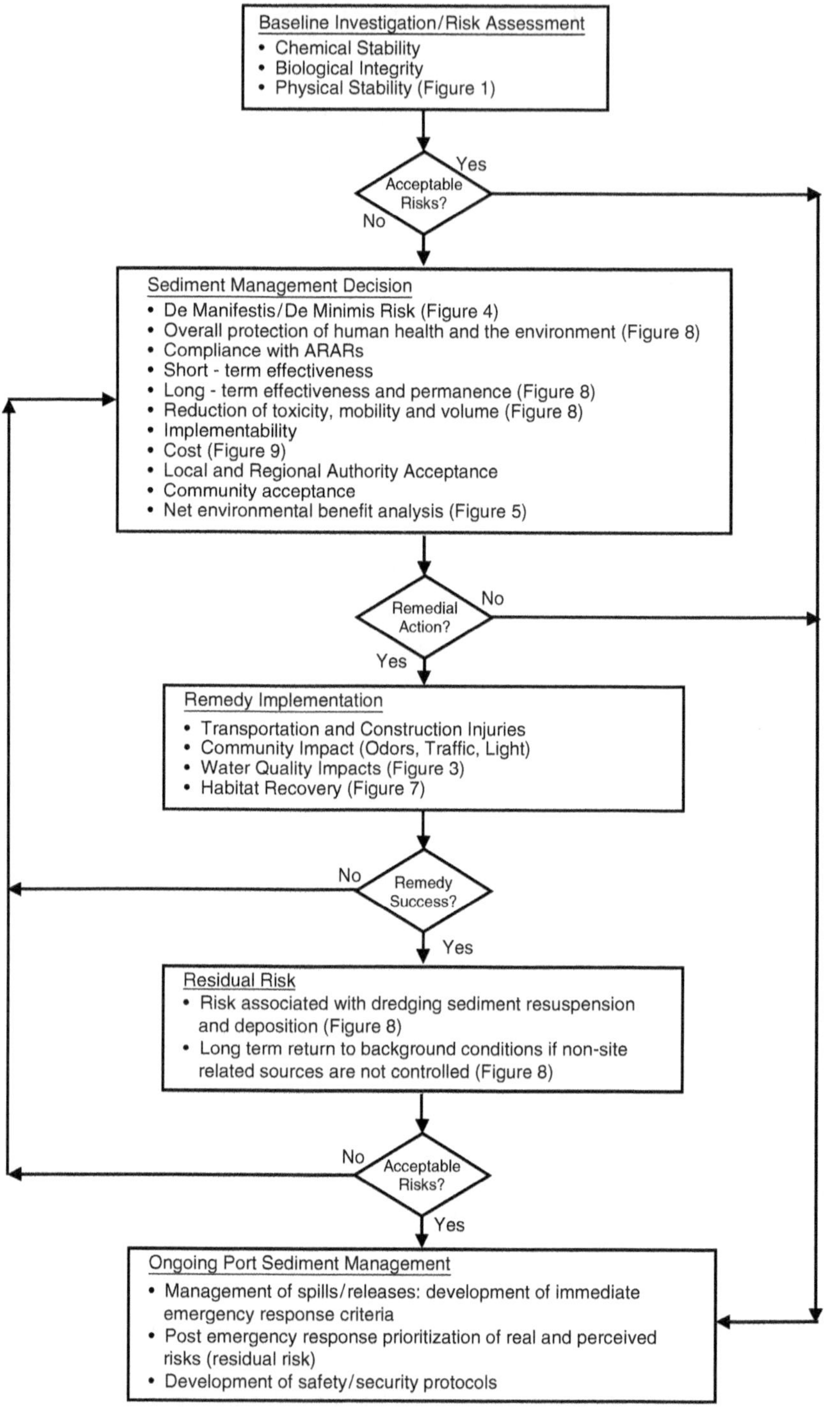

Figure 6. Sediment risk assessment and risk management framework.

Interdisciplinary collaboration between engineers, economists, regulatory specialists, community stakeholders, and experts in risk assessment-related disciplines will be critical to further development of objective, quantitative remedy alternatives analysis.

At present, advanced assessment tools such as MCDA and CRA are emerging as integral components of both risk evaluation and decision making. A review of MCDA and CRA methods is available elsewhere [13]. The approach merges probabilities associated with different threats (either manmade or natural), various consequences of threats, and the consequences of different response actions and preventive measures on successful mitigation or prevention of threats posed by manmade events (e.g., terrorism) and natural disasters (e.g., extreme weather events). MCDA and CRA also have the additional advantage of highlighting differences among threat scenarios, sediment management options, as well as tradeoffs among conflicting decision criteria and their uncertainties when evaluating possible response actions and preventive measures.

7. References

1. Apitz SE, Davis JW, Finkelstein K, Hohreiter DL, Hoke R, Jensen RH, Kirtay VJ, Jersac J, Mack EE, Magar VS, Moore DW, Reible D, Stahl R. Assessing and managing contaminated sediments: part I, developing an effective investigation and risk evaluation strategy. Integrated Environmental Assessment and Management 2005; 1:2-8.
2. Australian and New Zealand Environment and Conservation Council/National Health and Medical Research Council (ANZECC/NHMRC). Australian and New Zealand guidelines for the assessment and management of contaminated sites. 1992.
3. Bailey D, Plenys T, Solomon GM, Campbell TR, Feuer GR, Masters J, Tonkonogy B. Harboring pollution, the dirty truth about U.S. ports. San Francisco: Natural Resources Defense Council; 2004. Available at http://www.nrdc.org/air/pollution/ports1/contents.asp. Accessed on 5 March 2007.
4. Bojorquez-Tapia LA, Sanchez-Colon S, Florez A. Building consensus in environmental impact assessment through multicriteria modeling and sensitivity analysis. Environ Manage. 2005; 11(Epub 11 August).
5. Canadian Council of Ministers of the Environment. Guidance document on the management of contaminated sites in Canada, PN1279. Winnipeg; 1997.
6. Church BW. Remedial actions: the unacknowledged transfer of risk. Environ Sci Pollution Res. Special . 2001; 1:9-24.
7. Dockins C, Griffiths CW, Owens N, Simon NB, Axelrad DA. Linking economics and risk assessment. J Toxicol Environ Health A. 2004; 67(8-10):611-620.
8. Efroymson RA, Nicolette JP, Suter II GW. A framework for net environmental benefit analysis for remediation or restoration of petroleum-contaminated sites, ORNL/TM-2003/17. Oak Ridge: Oak Ridge National Laboratories, Environmental Sciences Division; 2003.
9. Environment Canada. Environmental assessments of priority substances under the Canadian Environmental Protection Act: guidance manual version 1.0, PS/2/CC/3E. Chemicals Evaluation Division; 1997.
10. European Union. European water framework directive (Directive 2000/60/EC); 2000.

11. Hauger MB, Rauch W, Linde JJ, Mikkelsen PS. Cost benefit risk – a concept for management of integrated urban wastewater systems. Water Sci Technol. 2002; 45(3):185-193.
12. International Maritime Organization. IMO's global programme on maritime and port security - the work continues. 2004. Available at: http://www.imo.org/Newsroom/mainframe.asp?topic_id=848&doc_id=3656. Accessed on 22 July 2005.
13. Kiker GA, Bridges TS, Varghese A, Seager TP, Linkov I. Application of multicriteria decision analysis in environmental decision making. Integ Environ Assess Manage. 2005; 1(2):95-109.
14. Knapp KC, Weinberg M, Howitt R, Posnikoff JF. Water transfers, agriculture, and groundwater management: a dynamic economic analysis. J. Environ Manage. 2003; 67(4):291-301.
15. Martello LB, Sorensen MT, Fuchsman PC, Wenning RJ. Chromium geochemistry and bioaccumulation in sediments from the lower Hackensack River, New Jersey. Archives of Environmental Contamination and Toxicology. In press.
16. Ministry of Housing, Spatial Planning and the Environment. Environmental quality objectives in the Netherlands: a review of environmental quality objectives and their policy framework in the Netherlands. Bilthoven: Risk Assessment & Environmental Quality Division, Directorate for Chemicals, External Safety & Radiation Protection; 1994.
17. National Research Council. Contaminated sediments in ports and waterways: cleanup strategies and technologies. Washington: National Academy Press; 1997.
18. National Research Council. A risk-management strategy for PCB-contaminated sediments. Washington: National Academy Press; 2001.
19. Sorensen MT, Gala WR, Margolin JA. Chemical-specific risk characterization – selecting the right tool for the job. Human and Ecological Risk Assessment. 2004; 10:245-269.
20. Sorensen, MT, Conder JM, Fuchsman PC, Martello LB, Wenning RJ. Using a sediment quality triad approach to evaluate benthic toxicity in the lower Hackensack River, New Jersey. Archives of Environmental Contamination and Toxicology. In press.
21. Suter II GW, Cornaby BW, Hadden CT, Hull RN, Stack M, Safran FA. An approach for balancing health and ecological risks at hazardous waste sites. Risk Anal. 1995; 15(2): 221-231.
22. Urban Harbors Institute. America's green ports, environmental management and technology at U.S. ports. Boston: University of Massachusetts Press; 2000. Available at http://www.uhi.umb.edu/pdf_files/greenports.pdf. Accessed on 5 March 2007.
23. United States Army Corps of Engineers. Improving environmental benefits analysis in ecosystem restoration planning, IWR report 03-PS-3. Alexandria: Institute for Water Resources; 2003. Available at: http://www.iwr.usace.army.mil/iwr/ecosystem/envirobenefits.htm. Accessed on 22 March 2005.
24. U.S. Environmental Protection Agency. Guidelines for ecological assessment, EPA/630/R-95/002FA. Washington: Office of Research and Development; 1998.
25. U.S. Environmental Protection Agency. Contaminated sediment management strategy, EPA/823/R-98/001. Washington: Office of Water; 1998.
26. U.S. Environmental Protection Agency Guidelines for ecological risk assessment, EPA/630/R-95/002F. Federal Register. 1998; 63(93):26846-26924.
27. U.S. Environmental Protection Agency. Risk assessment guidance for Superfund: volume 1, human health evaluation manual, part D, standardized planning, reporting and review of Superfund risk assessments, Final. 2001.
28. United States Environmental Protection Agency. Principles for managing contaminated sediment risks at hazardous waste sites, OSWER Directive 9285.6-08. Washington: Office of Solid Waste and Emergency Response; 2001.
29. United States Environmental Protection Agency. Contaminated sediment remediation guidance for hazardous waste sites, OSWER Directive 9355.0-85. Washington: Office of Solid Waste and Emergency Response; 2005.

30. Wenning RJ, Sorensen M, Magar VS. Importance of implementation and residual risk analyses in sediment remediation. Integrated Environmental Assessment and Management. 2006; 2:00-00.
31. Wenning RJ, Apitz SE, Baba A, Citron M, Elliott K, Al-Halasah N, Kapustka L, Lazarescu M, Mohamed M, Rutjes R. Understanding environmental security at ports and harbors. In: Linkov I, Kiker G, Wenning RJ, editors. Understanding environmental security at ports and harbors. Dordrecht: Springer; 2007. In press.
32. Wenning RJ, Della Sala S, Magar VS. Role of risk assessment in environmental security planning and decision-making. In: Linkov I, Kiker G, Wenning RJ, editors. Environmental security and environmental decision-making. Dordrecht: Springer; 2007. In press.
33. Wenning RJ, Apitz SE, Belluck DA, Chiesa S, Figueria J, Filip Z, Ignatavieius G, Kiker MK, Lahdelma R, MacDonell M, Mannini A, Raissouni B, Sinka D, Xenidis Y. Environmental security: approaches and tools. In: Linkov I, Kiker G, Wenning RJ, editors. Environmental security and environmental decision making. Dordrecht: Springer; 2007. In press.

CHAPTER 12

HARMFUL CYANOBACTERIAL BLOOMS

Identifying Data Gaps and the Need for a Management Framework

I. LINKOV

Intertox, Inc., 83 Winchester Street, Suite 1, Brookline, MA 02446 USA

A. FRISTACHI

Oak Ridge Institute for Science and Education assigned to U.S., Environmental Protection Agency, Office of Research and Development, National Center for Environmental Assessment, Cincinnati, OH USA

F.K. SATTERSTROM, A. SHIFRIN

Cambridge Environmental, Inc., 58 Charles Street, Cambridge, MA 02141 USA

J. STEEVENS

U.S. Army Corps of Engineers, Engineer Research and Development Center, Vicksburg, MS USA

G.A. CLYDE, JR.

U.S. Army Corps of Engineers, Tulsa, OK USA

G. RICE

U.S. Environmental Protection Agency, Office of Research and Development, National Center for Environmental Assessment, Cincinnati, OH USA.

I. Linkov et al. (eds.), Managing Critical Infrastructure Risks, 207–242.
© 2007 *Springer*.

Abstract: Harmful algal blooms (HABs) are frequently reported in many countries around the world. Even though this phenomenon has been known for a long time[1], the causes of HABs and toxicity releases are still not understood. Effective management of HABs of cyanobacteria requires an understanding of both the environmental factors associated with their formation and the effectiveness of available management alternatives for a given body of water. HABs can affect the quality of water bodies, thereby disrupting drinking water supplies, impacting recreational uses, and releasing toxins that can harm ecosystems and humans. HABs present a growing problem to water resource managers. HAB-related management decisions are complex and involve:

- Selecting the appropriate management interventions.
- Identifying important knowledge and technology gaps.
- Allocating funding for appropriate research or management projects.
- Communicating with stakeholders.

This chapter discusses the uncertainties associated with environmental factors linked to HAB formation and toxin release and proposes a management decision framework using multi-criteria decision analysis (MCDA). MCDA links management strategies with decision criteria and weightings elicited from decision makers or the public, allowing visualization and quantification of the tradeoffs between many social and technical factors involved in the decision-making process.

1. Background

Freshwater cyanobacteria periodically accumulate, or bloom, in water bodies worldwide. These blooms, also known as harmful algal blooms (HABs), can lead to a reduction in the number of individuals who engage in recreational activities in lakes and reservoirs [1] and to the degradation of aquatic habitats [2]. Cyanobacteria may also release toxins (i.e., cyanotoxins) into water bodies during or immediately following a HAB. The release may cause adverse human and ecological health effects. Deaths of fish, cattle, horses, pigs, sheep, and dogs have been associated with cyanotoxin exposures [3, 4, 5]. Human exposures to these toxins through recreational contact and tap water consumption have been associated with acute gastroenteritis [6, 7, 8]. The use

[1] The Bible may contain the earliest description of HABs "... all the waters that were in the river were turned to blood. And the fish that was in the river died; and the river stank, and the Egyptians could not drink of the water of the river" (Exodus 7: 20—21).

of cyanotoxin-contaminated water in dialysis equipment in Brazil resulted in human fatalities [9,10].

This paper is prompted by the needs of environmental managers within the U.S. Army Corps of Engineers (USACE), which oversees more than 400 U.S. freshwater reservoirs, to assess the HAB problem. We interviewed six USACE managers from the Louisville, Norfolk, Buffalo, Portland, Jacksonville, and Tulsa districts who have recently experienced HABs. The interviewees reported that HABs commonly resulted in ecological (e.g., fish kills), economic (e.g., reduced recreational use or increased water treatment) or possible human health impacts (e.g., drinking water contamination) [11]. In recent years, they noted marked increases in the recognition of HABs, in the resources allocated to address them, and in public concerns about HABs, prompting congressional inquiries on HABs and USACE operations [12]. Even though HABs pose many different problems, USACE management actions to address them have generally been restricted by the following factors:

- Relatively few personnel dedicated to HABs.
- Organizational focus on flood control and water supply (i.e., higher-level reservoir management objectives preclude management actions to reduce HABs).
- Ineffectiveness of short-term interventions for preventing HABs.
- Lack of advisory criteria and guidance on HAB management.

The occurrence and public awareness of HABs has steadily been increasing in the U.S., while sound HAB management methods remain undeveloped, prompting Congress, in 2004, to reauthorize the 1998 Harmful Algal Bloom and Hypoxia Research and Control Act (HABHRCA) [13]. HABHRCA authorizes funding for the many federal agencies responsible for managing risks associated with HABs. With this funding, these agencies will conduct research that increases the scientific understanding of the environmental factors associated with HABs, leading to a greater ability to detect, assess, predict, control, and mitigate HAB events. Effective HAB management requires an understanding of the following:

- Environmental factors associated with HABs.
- Economic, human health, and ecological impacts of HABs.
- Effectiveness of available management alternatives for reducing the frequency or intensity of existing or future HABs.

This paper reviews available information on environmental factors associated with HAB formation and cyanotoxin release in fresh waters. Since most water management practices can result in broad ecosystem impacts, this review focuses on studies reporting multiple factors that have been associated with HAB formation. Alternative practices that managers can use to control HABs are also discussed. The paper describes how the application of specific MCDA methods could improve HAB management, focusing on a generalized framework encompassing multiple lines of evidence for prioritization of future HAB research. Finally, we illustrate the framework with an example based on experiences at a USACE reservoir in Kansas.

1.1. MARION RESERVOIR, KANSAS: A CASE STUDY

Marion Reservoir (Figure 1) is a 25.13×10^6 m^2 (6,210-surface-acre) reservoir located at river mile 126.7 in Marion County, Kansas, on the Cottonwood River, a tributary of the Grand-Neosho River, about 46 miles north-northwest of Wichita, Kansas.

The reservoir was formed to provide flood control, water supply, water quality control, recreation, irrigation, and fish and wildlife enhancement. The Marion Reservoir watershed covers 243 square miles with land use comprising approximately 52% cropland and 46% grassland, with minimal urban or woodland uses.

On 5 June 2003, Marion Reservoir experienced an extensive HAB extending along the dam from the Hillsboro Cove Public Recreation Area to the Marion Cove Public Recreation Area (area 4 to area 2 in Figure 1). Marion

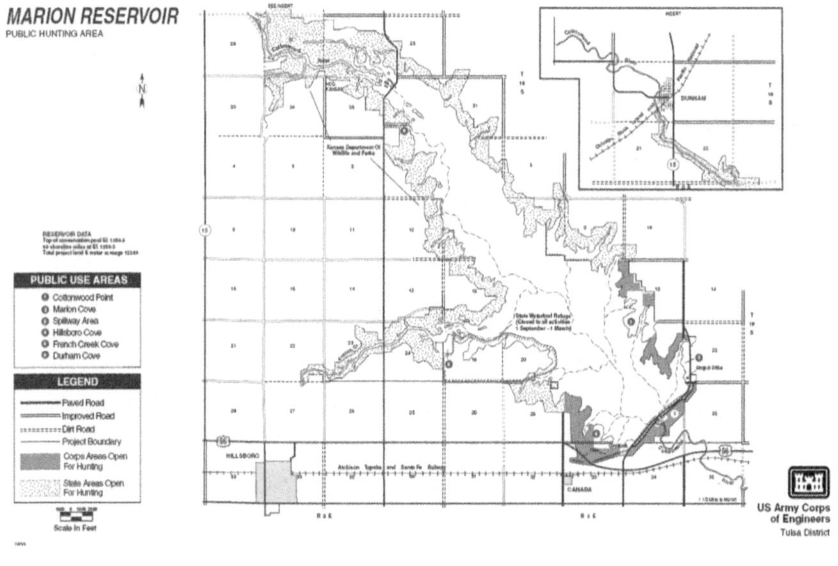

Figure 1. Map of Marion Reservoir, Marion County, Kansas.

Project personnel and local officials in Marion County, Kansas, began coordinating with personnel in the Tulsa District Office as well as the Kansas Water Office (KWO), Kansas Department of Wildlife and Parks (KDWP), and Kansas Department of Health and Environment (KDHE) on 6 June 2003. The next day, as the extent of the cyanobacteria bloom was assessed, the City of Hillsboro stopped using Marion Reservoir for water supply and began trucking water into the community, pending a more complete assessment. Additional precautions were taken by local officials and USACE personnel at the Marion Project Office to close the reservoir to public recreation. On 18 June 2003, in a letter from the Director of the Bureau of Water, KDHE provided requested recommendations to local officials:

- The reservoir may be reopened for recreational use based upon decreasing algal densities with following caveat:

...individuals choosing to recreate in and on the reservoir should avoid direct contact with visible surface scums and not swim in visibly affected waters.

- Fish consumption from Marion Reservoir may be resumed.
- Treatment of drinking water from the reservoir follows the commonly accepted practices for treatment of algae and algal toxins.

The City of Hillsboro resumed using the Marion Reservoir as a source of drinking water on 27 June 2003 after implementing the WHO and KDHE recommendations for drinking water treatment. Those changes included ceasing the use of potassium permanganate at the reservoir pump house, moving the point of chlorination prior to filtration and after the upflow clarifier, and adding 20 ppm powered activated carbon into the upflow clarifiers.

USACE and KDHE personnel conducted a site visit to Marion Reservoir on 12 June 2003 to assess the extent of the HAB. Phytoplankton samples were taken at two established USACE water quality sampling sites at the dam and at a mid-lake site. These samples were processed for identification and enumeration by Dr. Russell G. Rhodes, Department of Biology, Missouri State University. The 2003 bloom was dominated primarily by *Aphanizomenon*, with *Anabaena* and *Microcystis* present (Table 1).

Additional samples for cyanotoxin analyses were obtained on 11 and 17 June 2003. Samples were tested for microcystin, nodularian, saxitoxin, and neosaxitoxin using ELISA assay methods. All toxin analyses for the City of Hillsboro were performed by Dr. Wayne W. Carmichael, Wright State University. Toxins were analyzed from samples taken at the surface of the reservoir at the water intake for the City of Hillsboro, at the lake surface in Hillsboro Cove and at various points in the water treatment process. The only toxin present in the reservoir was microcystin, in concentrations ranging from 20.4 ug/L to 60.9 ug/L. Results for all toxin analyses are presented in Table 2.

TABLE 1. Algal density of selected taxa sampled at two locations on 12 June 2003 at Marion Reservoir, Marion Country, Kansas [64]

Location	*Anabaena*		*Aphanizomenon*		*Microcystis*		*Cryptomonas*	
Dam Site	Raw Count	Units/mL	Raw Count	Units/mL	Raw Count	Units/mL	Raw Count	Units/mL
Surface	9	570.6	31	1965.4	0	0	5	317
1 m	7	443.8	29	1838.6	1	63.4	0	0
3 m	11	697.4	28	1775.2	4	253.6	0	0
7 m	6	980.4	13	824.2	6	380.4	1	63.4
9 m	4	253.6	1	63.4	4	253.6	0	0
Mid Lake Surface	1	63.4	56	3550.4	5	317	2	126.8
1 m	4	253.6	61	3867.4	5	317	2	126.8
3 m	2	126.8	51	3233.4	2	126.8	1	63.4
7 m	0	0	34	2155.6	0	0	1	63.4

In May 2004, the Tulsa District began monitoring Marion Reservoir through the District's Water Quality Monitoring Program, focusing primarily on collecting chemical and biological data in order to provide algal densities and chlorophyll-*a* concentrations to local decision makers in coordination with the KWO and KDHE. In July 2004 Marion Reservoir again experience a substantial algae bloom, causing beach closures and an issuance of warnings to the public. Again, as in 2003, all decisions to close recreation areas at Marion Reservoir were made by local decision makers in the affected communities and Marion Project personnel. In the absence of state and national guidance, USACE reached a consensus to use WHO public health guidance and criteria to issue beach closures and public health warnings.

In 2005, the Tulsa District continued monitoring activities at Marion Reservoir. In collaboration with Dr. Robert A. Lynch, University of Oklahoma Health Science Center, College of Public Health, microcystin analysis was incorporated into the monitoring activities. Between May and September 2005, microcystin was present in Marion Reservoir on all sampling dates. Microcystin concentrations ranged from 0.6 ug/L to 16.4 ug/L and did not reach concentrations observed in 2003. In 2006, monitoring of algal toxins present in Marion Reservoir has continued and includes Anatoxin-a, (+/−)—beta-N-methylamino-L-alanine (BMAA), cylindrospermopsin, domoic acid (DMCA), microcystin-LA, microcystin-LF, microcystin-LR, microcystin-LW, microcystin-RR, and microcystin-YR. A HAB again covered a large extent of the reservoir on 28 June 2006, resulting in beach closures along the east shore during the 4 July holiday weekend.

The factors contributing to these repeated cyanobacterial blooms at Marion Reservoir are, at present, not well understood. KDHE has identified Marion Reservoir as being impaired due to eutrophication and has an approved TMDL for the Upper Cottonwood River in Marion and McPherson Counties, Kansas. Currently the total phosphorus (P) loading into Marion Reservoir is 133,299 lb/year and the total P concentration at the dam is 81.3 ug/L [14]. TMDL reduction goals include a 75% reduction in reservoir P loading and a 52% reduction in P concentration at the dam. In Kansas, the Watershed Restoration and Protection Strategy (WRAPS) is the preferred means for implementation of TMDL goals. In the Marion Reservoir watershed, the Marion Reservoir WRAPS has implemented conservation practices on 68% of the total acres of cropland and grassland within the watershed. The conservation practices being implemented within Marion County generally include best management practices (BMPs) implemented through the conservation reserve program (CRP) managed by the Marion County Conservation District. The most common BMPs used

TABLE 2. *Microcystin* concentrations present in Marion Reservoir surface waters and in City of Hillsboro, KS, finished drinking water on 11 and 17 June 2003

Sample Location	Date	Sample volume (L)	Sample dry wt. (g)	Aliquot dry wt (g)	Aliquot Conc.		Sample Conc.	
					ug/L	±Std(n)	ug/L	ug/g
Dam surface at intake structure	6/11/03	1.75	19.87	0.491	2639	150(3)	60.98	5.37
Surface at Hillsboro Cove	6/11/03	0.1	0.188	0.188	4168	160(3)	41.68	22.17
Finished water after filters	6/11/03	0.17	0.067	0.067	1.25	0.07(6)	0.007	0.019
Dam surface at intake structure	6/17/03	0.15	0.089	0.089	3061	493(4)	20.41	34.47
Two feet below dam surface at intake structure	6/17/03	1.75	0.924	0.924	1925	518(3)	1.16	2.20
Finished water after filter #1	6/17/03	1.75	1.073	1.073	5.61	0.68(5)	0.003	0.006
Finished water after filter #2	6/17/03	1.75	1.115	1.115	10.29	0.50(4)	0.006	0.009

in Marion County have targeted erosion and sediment transport and include buffer strips, filter strips, and terraces. In 2003, the reservoir was reported to be substantially less turbid as a result of the success of the Marion Reservoir WRAPS program in controlling erosion within the watershed, decreasing sediment loads into Marion Reservoir, and increasing light availability, and the depth of euphotic zone in Marion Reservoir. WRAPS activities, however, were not successful in decreasing P loading, with the majority of P loading appearing to result from sources external to the reservoir; however, the sources of external P loading (anthropogenic versus naturally occurring) cannot be determined at present.

2. Factors Influencing HABs

The factors that lead to HAB formation and induce toxin production are incompletely understood [15]. Many factors are thought to be associated with HAB occurrence, but successful bloom management requires further understanding of how the factors are related to one another. Although multiple studies have looked at the empirical relationship between environmental variables and toxin concentrations, few have measured the many different variables necessary to compile a comprehensive dataset for analysis [16], and even fewer have analyzed the data through sound statistical methods recommended by EPA and other agencies [8]. Temporal and spatial patterns of HAB formation appear complex, and the mechanisms of bloom progression are largely unexplained. Additionally, HABs can include both toxic and nontoxic strains, and many strains can be present during HABs in specific lakes [18,19,20]. A major source of variation in toxin concentration appears to be the strain-dependent influence of environmental factors on toxin release [21,22] and successive replacement of toxic and nontoxic species during the progression of a bloom [23]. No single modeling study discussed in this paper has attempted to predict the dynamics of toxin releases and/or toxin levels, reflecting the limited understanding of the environmental factors that influence toxin release.

To examine the relationships between different environmental factors associated with HABs in freshwater ecosystems, we conducted a literature review using the Science Citation Index (SCI) Expanded Database, the Google search engine for gray literature, and federal and state reports and references cited in review and original papers. Search terms included *HABs*, *algal bloom*, *cyanobacteria*, *microcystin*, *Microcystis*, *statistics*, *regression*, *model*, *environmental*, and *factors*. Of the several hundred candidate papers returned from our searches, we reviewed approximately 60 papers, focusing our efforts on experimental studies that used:

- Statistical models (i.e., regression models describing observed relationships between environmental factors and bloom formation/progression).
- Mechanistic models (i.e., coupled differential equations designed to predict bloom formation from basic biological principles).
- Fuzzy-logic models that combine expert knowledge on bloom formation with mechanistic or statistical data.

Several recent reviews [24, 25, 26, 27, 28, 29, 30] provide a comprehensive discussion of individual environmental factors that may be responsible for HABs. Figure 2 shows the main factors associated with bloom formation and proliferation that are amenable to management control.

Physical factors include temperature, light availability, and meteorological conditions. Hydrologic factors include alteration of water flow, turbidity, and vertical mixing. Chemical factors include nutrient loading (principally nitrogen (N) and P), pH changes, and trace metals, such as Cu, Fe, and Zn.

Table 3 lists the articles we classified as experimental studies and Table 4 lists articles we classified as modeling studies based on their primary objective and focus.

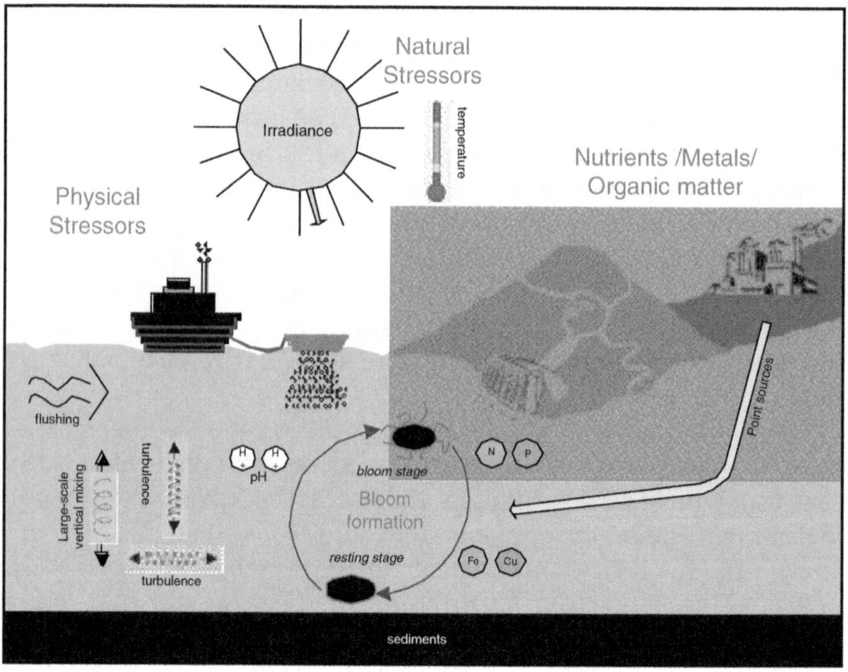

Figure 2. Environmental factors affecting HABs.

TABLE 3. Experimental studies that discuss factors associated with HABs. The factors are (1) vertical mixing, water flow, and light availability, (2) nutrient input, (3) water temperature and pH, and (4) toxin release data. Small boxes indicate factors identified as minor; large boxes indicate major factors.
ABBREVIATIONS: FW = freshwater, APH = Aphanizomenon, N = nitrogen, TN = total N, P = phosphorous, TP = total P.

Year	Toxin/Species	Ecosystem	Analytical Methods	Factors 1 2 3 4	Findings
1997	*Oscillatoria*	Great Ouse River, England	Time series, regression	▪	Examined possible causes for crash in spring population of cyanobacteria and chlorophyceae. Nine regions of the river were examined over five years, measuring phytoplankton chlorophyll-a, cell number, and underwater light and water turbidity/flow. While nutrient limitation or poor light did not seem to be the cause for the drop in population, a connection was established between increased phytoplankton populations, especially *Oscillatoria*, and decreased water turbidity/flow, suggesting that restricted flow and/or drier conditions favor phytoplankton growth [55].
1997	*Microcystis*, *APH*	Lake Mendota, Wisconsin.	Time series analysis, correlation analysis	■▪	Daily samples from the lake taken June-October and measured for chlorophyll-a, surface scum, and Secchi depth. The goal was to correlate scum formation with these factors and weather conditions. Physical factors and weather conditions proved to have more effect on biomass than nutrients and zooplankton grazing [24].
1998	*Anabaena* (4 strains), *Microcystis*	Meta-analysis of 72 lakes in Finland	Correspondence analysis	■▪▪	Nutrients (dissolved P and N) are the most significant factors determining which strain dominates in a given bloom. Neurotoxic blooms are associated with low P and high N. Hepatotoxic blooms are associated with lowest P. Nontoxic blooms are associated with somewhat higher phosphate concentrations [22].

(continued)

TABLE 3. (continued)

Year	Toxin/Species	Ecosystem	Analytical Methods	Factors 1	2	3	4	Findings
1998	*Microcystis*, *Anabaena*	Lake in France	Correlation analysis	■			▪	Microcystin-LR concentration was positively correlated with inorganic and organic C and chlorophyll- a; Microcystin-LR' was positively correlated with dissolved P and negatively correlated with inorganic C; Microcystin-LR" was positively correlated with N and chlorophyll-a, and negatively correlated with solar radiation. Total Microcystin was positively correlated with chlorophyll-a and negatively correlated with solar radiation [18].
2000	*Microcystis*	Lake in Washington	Time series, regression analysis	▪	■	▪	▪	A comparison of lake conditions over two summers, one with bloom occurrence and the other without, revealed lake characteristics that favor blooms: high N: P, decreased water transparency, high water column stability, high surface water temperature, high water pH, decreased lake flushing. Additionally, low N:P and abundance of ammonium-N—and not nitrate-N—favored the bloom of *Microcystis* over other species. Toxin production, separate from cyanobacteria growth, may be regulated by P [20].
2000	*Microcystis*; microcystin-LR	Eutrophic and hypereutrophic lakes, Alberta, Canada	Correlation coefficients, sometimes using long-transformed data; regression	■	▪	▪		Bacteria and toxins were significantly related to measures of TP. Bacteria concentration was related to chlorophyll concentration, as expected, but toxin concentration was not. Despite lack of linear relationship, toxin concentration was maximal when inorganic N was low. Toxin was most clearly related (inversely) to total N: P. Cyanobacteria biomass and toxin concentration were also inversely related to Secchi depth, an indicator of light availability. This result may not hold, as the bacteria can rapidly change their vertical location [37].

2000	*Cylindrospermopsis*	Multiple tropical and subtropical reservoirs in Australia	Summary statistics	■ ■ ■	HABs were recorded in most of the reservoirs under study. High pH, water temperature between 28 and 32°C, and stable stratified water with hydraulic retention time of more than 2.5 yrs are associated with increased *C. raciborskii* populations. Toxins measured up to six weeks after decrease in cell count were below detection [45].
2001	Cyanobacteria (unspecified)	99 lakes around the world	Correlation and regression analysis	■	This study argues that N:P is not the most important factor in causing blooms and that variations in TP, TN, and biomass are far more important factors in predicting freshwater blooms [36].
2001	*Cylindrospermopsis*	Reservoir in Australia	Linear regression analysis	■ ■	Reduced rates of vertical mixing, reduced mixed:euphotic depth ratios, and high temperature (>25°C) result in more cyanobacteria blooms [25].
2002	*Microcystis*	Laboratory conditions	Multivariate regression analysis, mathematical modeling	■ ■	Examined growth of *Microcystis*, toxic and nontoxic, in batch culture with variable nutrient (TN and TP) concentrations. High nutrient concentrations promoted growth of the toxic over the nontoxic strain. Specifically, high TN seemed more important than high TP for growth of all *Microcystis* strains [21].
2002	*Microcystis*	Hypereutrophic Lake Suwa, Japan	Regression analysis	■ ■	Higher water temperatures promote toxic blooms although toxin concentrations within bivalves increase only once the water temperature drops (as in summer to autumn) [2].
2003	*Microcystis*	San Roque Reservoir, Cordoba, Argentina	Discriminant analysis	■ ■ ■	Bloom occurrence is promoted by inorganic P and is reduced by carbonates. Toxin releases were favored by high temperatures (>23°C), while high concentrations of NH_4 and Fe resulted in decreased toxic releases [46].

(continued)

TABLE 3. (continued)

Year	Toxin/Species	Ecosystem	Analytical Methods	Factors 1 2 3 4	Findings
2003	*APH*, *Anabaena*, *Microcystis*	Artificial enclosures within Steele Lake, Alberta, Canada	Regression, summary statistics	■	Lowering the N:P ratio (5:1) promotes *Microcystis* growth. Higher ratios (20:1) favor *Anabaena* blooms. *Aphanizomenon* growth is not sensitive to N:P [40].
2004	*Microcystis*, *Anabaena*, *Coelo-sphaerium*, *Oscillatoria*	241 FW lakes in Missouri, Iowa, Kansas and Minnesota	Nonparametric Spearman-Rank correlation, linear and nonlinear regression analysis	■ ■ ■	Lakes were sampled between May and September, measuring Secchi depth, surface temperature, TP, TN, suspended solids, and chlorophyll. Microcystin concentration in the samples correlated with latitude, nutrients, TN:TP, chlorophyll, chlorophyll:TP, suspended solids, Secchi depth, and surface temperature. The highest correlations were with latitude, TN, and TP [39].
2004	*APH*, *Plankothrix* *Anabaena*	FW Dam reservoir in Poland	Concentration estimation; correlation	■ ■ ■	Influence of environmental factors on relative abundance of cyanobacteria species and toxin releases was studied. Increase of N:P results in replacement of *Aphanizomenon* by *Planktothrix*. *Anabaena* was co-dominant at lower temperatures. *Anabaena* was responsible for toxin production [19].

TABLE 4. Modeling studies that examine factors associated with HABs

Year	Toxin/ Species	Ecosystem	Analytical Methods	Factors 1	2	3	4	Findings
1997	*Microcystis, Oscillatoria*	Lake Kasumigaura, Japan	Predictive modeling using Artificial Neural Network	■		■		Developed a predictive model for blooms and validated it for five species of blue-green algae. *Microcystis* proves less affected by change in nutrient conditions than by change in light, water transparency, and temperature. *Oscillatoria* is highly affected by changes in nutrient and zooplankton conditions. Management options for preventing blooms are discussed [42].
1997	*Microcystis, Oscillatoria, Phormidium*, and two diatoms	Eutrophic lake, Japan	Artificial neural network to predict phyto-plankton blooms	■				The ANN predicted total chlorophyll-a and *Microcystis* concentrations with a 7-day lead-time, but did not predict other species well. The most influential factors for chlorophyll-a were transparency, pH, and PO_4 [44].
1998	None in particular	Lake Mendota, WI	Exploration of a P-dependent bloom model	■				21 years of measurements of inputs and outputs of P into this lake system were used to predict the change in probability of a bloom if P loadings were decreased by 50%. No other variables were incorporated into this lake-specific model [41].
1999	Diatoms, cyano-bacteria (generic)	Lake Vortsjärv, Estonia	Model of phyto-plankton dynamics and water quality	■	■			The contribution of water level was found to be very significant for cyanobacteria and diatom biomass. Nutrient loading has a minimal effect on cyanobacteria population but a greater effect on the population of diatoms and species diversity [56].
2001	Microcystin	Farmoor Reservoir, Oxfordshire, UK	Generic model of microcys-tin growth and movement	■			■	Generic model relating blooms and environmental factors (light, temperature, and turbulent lake mixing) is reported. The population movements are compared with published observations from field and laboratory studies [32, 57].

(continued)

TABLE 4. (continued)

Year	Toxin/ Species	Ecosystem	Analytical Methods	Factors 1	2	3	4	Findings
2001	*Microcystis, Phormidium, Synedra, Oscillatoria*	Lake Kasumigaura, Japan	Artificial neural network model applied to actual site	■	■	■		Effects of environmental factors and their combinations were studied for four cyanobacteria species. pH was found to be the most important factor for proliferation. TP was found to be important for *Microcystis, Synedra*, and *Oscillatoria*, while *Phormidium* demands more N [43].
2002	*Microcystis*	Man-made lake, France	1D vertical model of phytoplanktonic succession	■			■	Generic model shows that vertical stratification and temperatures above 24°C lead to cyanobacteria dominance in the lake [38].
2003	*Microcystis*, APH	Lake in Netherlands	Predictive modeling, fuzzy logic	■				Temporally and spatially explicit predictive model for bloom occurrence. Cell buoyancy and stability of the water column increase the likelihood of blooms [35].
2003	*Microcystis*	River and Lake ecosystem in South Korea	Predictive modeling using genetic algorithms, validation using extensive empirical dataset	■				Increased water turbidity and water flow result in increased *Microcystis* concentrations [31].
2004	*Microcystis*	Lake in Netherlands	Predictive model for species abundance in water column	■				Model predicts how changes in turbulent mixing affect competition for light between buoyant and sinking phytoplankton species. Model results confirmed by experiment with artificial mixing of an entire lake. Buoyant *Microcystis* dominated at low turbulence; diatoms and green algae dominated at high turbulence [34].

The literature was reviewed for information on three factors that typically have been associated with the formation of freshwater HABs:

1. Water column stability and light availability.
2. Increased nutrient loading.
3. pH and temperature.

The tables also show whether toxin data were reported. Other factors (e.g., trace metals, organic carbon, and biological factors) may be important, but they are less amenable to management control. Individual discussion of these factors follows.

2.1. WATER COLUMN STABILITY AND LIGHT AVAILABILITY

Water column stability, including turbulence, flushing, vertical mixing, and water flow, affects the amount of light available to cyanobacteria and plays an important role in controlling the formation and persistence of HABs [10]. Stable, thermally stratified water columns favor HAB [25]. Water column stability affects the ability of cyanobacteria to regulate their buoyancy, which is important in the formation of scums, a characteristic of HABs [24]. Water mixing hinders the ability of cyanobacteria to locate in the optimal strata for light and nutrients, reducing their capacity to compete for energy resources and nutrients in the environment.

Water column mixing was a common parameter in predictive models for HABs that we reviewed. A statistical regression model by Jeong et al. (2003) places turbidity as the second most important factor (after biovolume) for *Microcystis* blooms in Korea [31]. A mechanistic model by Howard and Easthope (2002) allows modeling of vertical movement of *Microcystis* in response to different lake mixing regimes [32]. Coupling of a vertical mixing model with phytoplankton succession modeling for a lake in France suggests that vertical stratification influences *Microcystis* development [33]. Huisman et al. (2004) concluded that turbulent water mixing and light availability are the most important factors controlling cyanobacteria growth and bloom formation in the mechanistic model that they developed [34]. Using fuzzy-logic algorithms, Ibelings et al. (2003) found water level to be a key factor in regulating the trophic status of the lake and thus HABs [35].

2.2. NUTRIENTS

N and P enrichment are considered cyanobacteria bloom precursors. Except for *Microcystis*, most cyanobacteria are N-fixers, making N a limiting factor in competition over other phytoplankton species. Although nutrient influence is a widely studied factor associated with HABs, conclusions are still controversial. Downing et al. (2001) conducted regression analysis using data from approximately 100 temperate lakes and demonstrated that total N (TN) and total P (TP) are better predictors for HABs than N:P ratio [36]; yet N:P was the most important factor in a regression model for 13 lakes in Canada [37]. Dissolved P was found to be the main factor positively correlated with microcystin levels in a lake in France [38]. Graham et al. (2004) found that TN and TP are important but also that their ratio shows a statistically significant correlation with microcystin concentration [39]. Studies in lakes and lagoons show the importance of interactive N and P effects on cyanobacteria growth—one study showed that a relatively low N:P ratio (5:1) promoted *Microcystis* growth, while higher ratios (20:1) favored *Anabaena* blooms [40]. *Aphanizomenon* growth was insensitive to the N:P ratio [40].

Dissolved N and P may also be important for discriminating between toxic and nontoxic blooms [22]. Neurotoxic *Anabaena* blooms were associated with low P and high N concentrations, while high P concentrations were associated with hepatotoxic *Microcystis* blooms and nontoxic *Anabaena* blooms [22]. Vézie et al. (2002) found that toxin-releasing strains of *Microcystis* needed more nutrients for their growth than other strains [21].

Modeling studies suggest the importance of nutrient loading on improving water quality and reducing HAB probability and severity. Lathrop et al. (1998) predicted that Lake Mendota in Wisconsin should experience proportional reduction in bloom probability per unit reduction in P loading [41]. Nutrient conditions were found to be an important factor in the succession of cyanobacteria species in artificial neural network (ANN) models [42]. An ANN model study predicted that *Microcystis*, *Oscillatoria*, and *Synedra* levels would decrease with increases in TN, while *Phormidium* levels would increase [43]. An ANN model on data for a lake in Japan predicted dissolved N to be the factor most related to cyanobacteria biomass and dissolved P to be associated with chlorophyll concentration [44].

2.3. WATER TEMPERATURE AND pH

Temperature and pH, in combination with other factors, are important in HAB formation and proliferation [20, 26]. High pH is usually associated with depletion of dissolved CO_2, which cyanobacteria use very efficiently at low concentrations [20]. Reservoirs with high pH had increased levels of *Cylindrospermopsis* [25]. In several lakes in Canada, pH was found to correlate well with *Microcystis* biomass but not with microcystin [24, 37]. ANN models have also predicted pH to be an important factor in HAB formation [43, 44].

Temperatures above 20°C, usually observed during summer, tend to favor HAB formation [45], with optimal temperatures varying by study and species:

- 28–32°C for *Cylindrospermopsis* [25,45]
- 21–24°C for *Aphanizomenon*, *Anabaena*, and *Plankothrix* [19]
- 8.3–38.9°C for *Microcystis*, *Anabaena*, *Coelosphaerium*, *Oscillatoria* [39]
- 8.7–29.1°C for *Microcystis* [46]

Temperature change may also influence species dominance, as temperature decrease was associated with a shift from *Aphanizomenon* to *Plankothrix* as the dominant species in a lake in Poland [19].

Temperature was an important parameter for several mechanistic models of bloom development. Howard and Easthope (2002) incorporated temperature into a cyanobacteria growth model to predict response to climate change [32]. Temperature was also used as a variable in a vertical mixing model [38]. An ANN modeling study predicted that *Microcystis* would be sensitive to temperature conditions and *Oscillatoria* would be insensitive to temperature [42], although *Oscillatoria* is less adapted to higher temperatures than *Microcystis*, *Phormidium*, and *Synedra* [43].

3. HABs: Data/Analysis Needs

Table 3 lists many factors that may be associated in some way with HAB occurrence, but successful management of algal blooms or prevention of future blooms requires an understanding how these multiple factors are related to one another and the bloom. Relationships between these factors have been demonstrated primarily in essentially descriptive studies. We identified a few studies where regression analysis and modeling (primarily mechanistic) were attempted. Given the limited number of samples used in these papers and the often suboptimal quality of the data, linear regression methods are not optimal for HABs predictions, where multiple factors and nonlinearity is a key [16].

To be useful to management decision makers, HAB monitoring data must be analyzed using statistical tools to deduce relationships among the measured variables. The crucial issue in using advanced statistical methods is availability of high-quality datasets. In 1997, Kneale and Howard reviewed HAB monitoring data available for statistical analysis and concluded that the data was of limited value [16]. They called for devising a suitable sampling strategy to monitor water bodies. In the U.S., even though no regulatory strategy has been implemented specifically for HAB monitoring, interviews with USACE personnel [11] and discussion with various state and federal environmental protection personnel reveal that a large volume of water quality monitoring data and some measurements of algal toxins are currently available for statistical analysis.

For example, for the past 10 years, the Portland District of the USACE has taken comprehensive measurements at three projects: the Willow Creek Reservoir (which often experiences algal blooms), Lost Creek Lake, and Applegate Dam. Sample analysis consists of organism ID, cell counts, cell density, and biovolume, as well as environmental factors (including temperature, pH, turbidity, dissolved oxygen, and dissolved solids). The Louisville District collects data from each one of its reservoir projects, from the inflows and from the tail waters. USACE collects water for analysis from different depths at tail waters, dam sites, and major tributaries.

As discussed above, HAB studies often measure algae levels and construct modest datasets, but stop short of comprehensive statistical analysis. We believe that cluster analysis methods may be useful to advance the understanding of factors responsible for HABs, specifically as related to HAB management. In large studies, statistical analysis becomes more complex as the number of variables increases [17]. Different cluster analysis methods group multiple factors into similar categories or clusters based on more than one ecological indicator (such as measures of HABs occurrence). Methods of ordination analysis are frequently used to reduce many variables to fewer variables by grouping together similar variables and treating them as one [17]. In principal components analysis, for example, the dimensionality of the data is reduced when new variables are introduced as linear combinations of original variables. Principal components analysis is linear regression extended into multiple dimensions [17], and it has been used in multivariate biological and chemical studies [47]. Discriminant analysis is a statistical method used to place data into predefined categories, and it has been used in multiple studies [46,48]. Multidimensional scaling is another statistical method that helps investigators visualize relationships between measured variables [49]. From a two-dimensional table of the correlations between all measured variables, it creates a spatial map of relative distances between the various parameters.

Cluster analysis has been applied to address problems similar in scope to HABs. Stewart et al. (1999) used cluster analysis in their investigation of whether land use patterns affect algal communities since such communities are often used as water quality indicators [50]. Cluster analysis was used to group sampling sites by their diatom assemblage patterns. The sites were largely grouped into their three geographic watersheds. Exceptions occurred when cluster analysis identified sites in different geographic groups that were more similar to one another than to others in the group. For example, algae patterns were quite similar at two sites in different groups that were each near a road, free-flowing, with a sandy bottom. Hartwell and Claflin (2005) grouped contaminated sites in Delaware Bay geographically, and they used cluster analysis to group sites according to their benthic species profiles as well [51]. Sites were primarily grouped along a salinity gradient. Clusters included upper estuary, depositional estuary, and tributary sites. Species groupings reflected such distinctions as whether the species was found in tidal freshwater or not and whether it lived on a muddy or sandy bottom. Habitat blocks were outlined based on clusters of species being found abundantly in certain site clusters. Gasiunaite et al. (2005) defined seasonal succession patterns in the Baltic Sea by using cluster analysis to identify eight phytoplankton community types [52]. These community types were each associated with seasonal physicochemical monitoring data, including temperature, salinity, and eutrophication. Samples used to determine community types were also distinguished by biomass percentages of each of several taxonomic groups of phytoplankton, relative to the biomass of the sample as a whole.

We performed cluster analysis (as well as multidimensional scaling) on a limited dataset from the USACE Portland District (see Supplemental Information). The analysis evaluated nine variables:

1. Oxidation reduction potential
2. Dissolved oxygen
3. pH
4. Specific conductance @ 25°C
5. Water temperature
6. Dissolved organic carbon
7. Chlorophyll a levels
8. Turbidity
9. Total algal biovolume

Cluster analysis found that total algal biovolume fluctuated most closely with turbidity and chlorophyll-a levels. Because cyanobacteria contain chlorophyll-a and organic matter such as cyanobacteria causes the turbidity, these factors are likely associative rather than causal.

4. HAB Management Options

As noted previously, decisions regarding HABs are made despite high levels of uncertainty. This uncertainty results from a limited understanding of the factors that cause HABs and the variable (or unknown) responses of a watershed to specific management practices that target HABs [11]. Tables 5 and 6 show the factors, effects and impacts, likelihood of impact, and efficiency of

TABLE 5. Management actions controlling physical factors thought to affect HABs

Factors	Effects and impacts	Likelihood of impact	Implementation Efficiency
Flushing/altered water residence time	May disrupt cyanobacteria filaments, colonies, aggregates, and mutualistic ssociations with other microflora and microfauna.	Water residence time for cyanobacteria is one of the most important factors associated-with HAB formation. Flushing will also reduce nutrient loading and thus decrease bloom formation likelihood.	Easy to implement in many reservoirs. Managers have the ability to alter-water flow from a dam, but other-factors (such as needs of endangered fish species) may take precedence.
Large-scale vertical mixing and small-scale turbulence	Counteracts near-surface accumulations of buoyant bloom populations. Forces competition for light and nutrients with more desirable, nonbuoyant eukaryotic taxa.	Mixing is reported as one of the most important factors influencing cyanobacterial bloomsand is likely to be very efficient for small reservoirs.	Relatively difficult to implement and-costly. Can be implemented only atsmall water bodies. Large-scale mixing may be damaging to other ecosystem components.
Shading (reduced surface irradiance)	Can alter phytoplankton communitycomposition and can negatively affect cyanobacteria surface bloomtaxa.	Light availability is the major-driver for HABs. Likely to be efficient.	Very expensive. Would only work on very small bodies of water.
Temperature	Generally, temperatures in excess of20°C accompanied by stratificationand high nutrient loading can pro-mote blooms for some species.	HABs have been observed in a wide range of temperatures. Nevertheless, the likelihood and direction of the effect is controversial. Most likely temperature will shift species composition, but effect on toxin releases and HAB magnitude is not clear.	Relatively expensive. In some instances, managers can release water from different depths behind a dam, thus changing temperature regime for the lake.

TABLE 6. Management actions controlling chemical factors thought to affect HABs

Factors	Effects and impacts	Likelihood of impact	Implementation Efficiency
Reduction of nutrient (N and P) inputs	Long-term (months, years)reductions in both N and P inputs may be effective in reducing HABs.	Nutrient loading is one of the crucial factors control-ling blooms and its imple-mentation should be veryefficient. Low N:P loading(<20) often caused byexcessive P loading canenhance bloom potential.	Several methods available for nutrient reductionwithin lake (chemical precipitation, sediment dredging and biological removal). In general, these measures will work only in combinationwith reduction of external inputs.
Algaecides	Algaecides (e.g., $CuSO_4$) kill-cyanobacteria.	Efficient, but should be usedas emergency measure only.Significant ecological dam-age is likely.	Effective, but side effects are significant. Efficient only when applied at early stages ofbloom development. Other ecological recep-tors can be severely affected. Water reservoirsshould be isolated to allow degradation of toxins, cyanobacteria, and odor.

two broad categories of management alternatives that can affect the factors associated with HAB formation. These categories are reactive treatments designed to reduce concentrations of cyanobacteria after a HAB has formed and preventive treatments including watershed management and other long-term measures designed to improve ecosystem health.

In general, management practices can exert a greater influence on physical rather than chemical factors. Many of the actions are directly related to water management alternatives. In many instances, water flow can be altered, which decreases residence time in a reservoir or affects vertical mixing and nutrient loading in the water column. Similarly, the ability to release water from different depths behind dams could affect HABs by changing the temperature profile of water downstream. Consistent forceful mixing could prevent cyanobacteria from maintaining optimal water depth, slowing their growth.

Algaecides, such as copper sulfate ($CuSO_4$), can control HABs after they occur. Managers using algaecides need to be mindful of ecosystem impacts and the potential for the algaecide to cause rapid toxin release. Cu is an essential trace element in plant and animal nutrition, yet $CuSO_4$ is relatively toxic to trout and other fish, especially in soft or acid waters [53, 54]. Mismanagement of $CuSO_4$ application may cause decreases in populations of aquatic invertebrates, plants, and fish [55].

5. Proposed HAB Management Framework

The task of deciding among alternative practices to prevent or mitigate HABs for a given body of water is challenging. Effective decision making requires consideration of the environmental, ecological, technological, economic, and sociopolitical factors relevant to evaluating existing HAB management alternatives. Additional data on the physical and chemical factors that influence HAB formation may also help to identify new management practices in the future. The decision-making process would benefit greatly from a structured, adaptive framework to enable the decision maker to examine the impact of a given alternative across multiple objectives [56].

Figure 3a illustrates the decision-making process and the tradeoffs associated with typical HAB management decisions. While risk assessment is a critical input for environmental managers, alone it fails to provide information that compares risks, costs, political considerations, and uncertainties associated with alternative management practices. Integrative decision-analytic processes have been implemented in some federal agencies to evaluate alternatives and understand the differences among competing risk management practices. Decision analysis connects the technical aspects of a

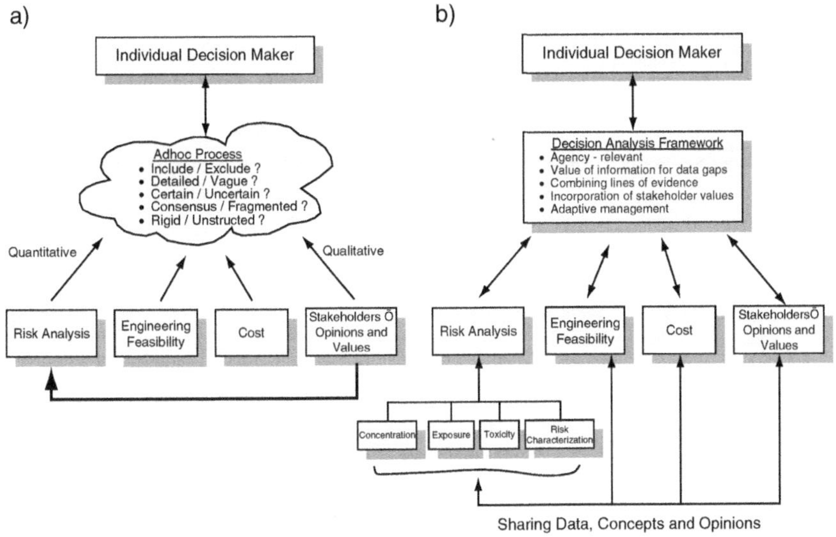

Figure 3. (a) Unstructured Decision Making; (b) Decision Making with Multi-criteria Decision Analysis.

decision process with the social sciences through the integration of subjective values into the analysis. Decision-analytic methods, such as comparative risk assessment (CRA) or MCDA, can highlight the tradeoffs among competing project objectives and help compare alternatives that are dramatically different in their potential impacts or outcomes. Decision-analytic methods integrate a wide variety of information into a structured approach (Figure 3b) [57]; this facilitates both the identification of missing or highly uncertain data and an evaluation of the importance of obtaining such data.

Previous papers by Linkov et al. introduce a general framework for management of environmental challenges based on MCDA [58, 59]. This framework has also successfully been focused on contaminated sediment management [57].

The systematic decision framework shown in Figure 4 provides a generalized road map to the decision-making process for HAB management.

Having the right combination of people is an essential element in the decision process. The activity and involvement levels of HAB managers and decision makers, scientists and engineers, and the general public are symbolized by dark lines for direct involvement and dotted lines for less direct involvement. The same for Risk Assessment Modeling and MCDA tools used in the decision-making process. Each group may have a unique perspective on the HAB issue, its own method of envisioning solutions, and its own societal responsibility. Policy and decision makers spend most of

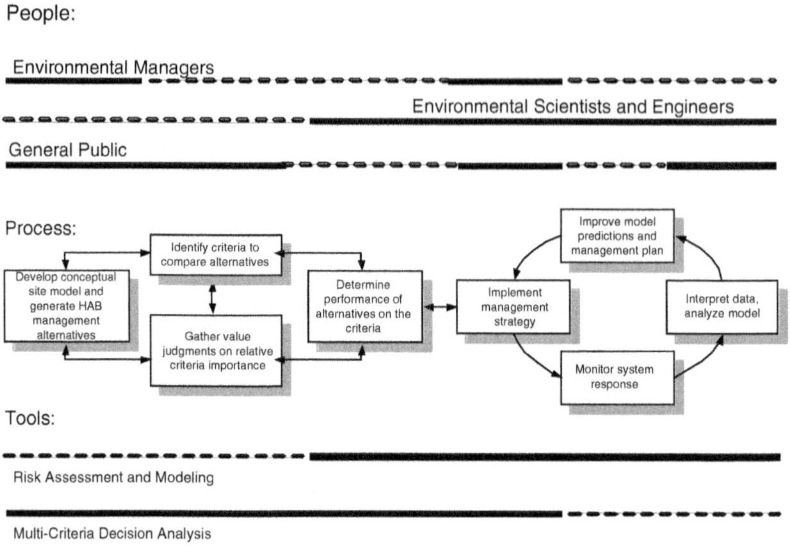

Figure 4. General MCDA framework. (Solid lines symbolize direct group involvement; dashed lines symbolize less direct involvement.)

their effort defining the problem's context and the overall constraints on the decision. In addition, they may have responsibility for final policy selection and implementation. Stakeholders may provide input towards defining the problem, but they contribute most in helping to formulate performance criteria and to make value judgments for weighting the various success criteria. Depending on the problem and regulatory context, stakeholders may have some responsibility in ranking and selecting the final option. Scientists and engineers provide the measurements or estimations of the desired criteria that determine the success of various alternatives. They also may take a secondary role as stakeholders or decision makers.

The process depicted in Figure 4 follows two basic themes:

1. Generating management alternatives, success criteria, and value judgments.
2. Ranking the alternatives by applying value weights.

The first part of the process generates and defines choices, performance levels, and preferences. The latter section methodically excludes nonfeasible alternatives by first applying screening mechanisms (for example, overall cost, technical feasibility, or general societal acceptance) and then ranking in detail the remaining options by decision analysis techniques, such as *Analytical Hierarchical Process* (AHP), *Multi-Attribute Utility Theory* (MAUT) or *Outranking*, that use various criteria levels generated by environmental tools, monitoring, or stakeholder surveys [57]. A detailed analysis of the theoretical foundations of different MCDA methods and their comparative strengths and weaknesses for environmental applications is provided in our previous paper [59]. While it is reasonable to expect that the process may vary in specific details among regulatory programs and project types, emphasis should be given to designing an adaptive management structure. In contrast to traditional "command and control" approaches, adaptive management uses adaptive learning as a means for incorporating changing decision priorities or new knowledge from monitoring programs into HAB management strategy selection or change [57]. Chorus (2005) stressed the importance of adaptive approaches for HAB management [60].

As shown in Figure 4, the methods used within group decision making and scientific research are essential elements of the overall decision process. The applicability of decision analysis methods, symbolized by solid lines (direct or high-utility) and dotted lines (indirect or lower-utility), helps to generate and map stakeholder preferences and individual value judgments into organized structures that can be linked with other technical tools from risk analysis, modeling, monitoring, and cost estimation. Decision analysis software can also provide useful graphical techniques and visualization methods to express the results in understandable formats. When new information becomes available or changes occur in the requirements of the decision process, decision analysis methods can respond efficiently to reprocess and iterate with the new inputs. This integration of decision analysis with scientific and engineering methods allows each to have a unique and valuable role in the decision process without attempting to apply either type of tool beyond its intended scope.

The MCDA framework links heterogeneous information on causes, effects, and risks of HABs with decision criteria and weightings elicited from decision makers, allowing visualization and quantification of the tradeoffs involved in the decision-making process. The proposed framework can also be used to prioritize research and information-gathering activities and thus can be useful for information analysis.

6. Illustrative Example

For an illustrative example of AHP, consider a manager choosing among three alternative HAB management strategies:

- Physical intervention (flushing).
- Chemical intervention (algaecides).
- No action.

Given the complex choices and numerous criteria to consider, the manager decides to use AHP as a decision aid. In the first step, the manager determines the criteria by which the alternative strategies will be evaluated. These criteria could include human health impacts, environmental impacts, and social impacts. In the second step, the decision maker weights the importance of these criteria. Although in this simple scenario it would be possible to assign weights directly, in many instances it could be difficult because of the multitude of criteria and sub-criteria that the manager may wish to consider.

In the AHP method, the decision maker does not give importance weightings directly, but rather derives the category weights from a series of relative judgments. In this example, there are three relative judgments, in the form of weightings ratios. From these relative weightings, AHP derives normalized weightings of 0.655, 0.250, and 0.095 for human health, environmental, and social impacts, respectively. Main criteria categories may be too broad to be used directly in evaluating management alternatives; therefore, sub-criteria within each of these categories need to be developed (see Table 7).

For example, within the human health impacts category, drinking water quality, recreational impacts, and inhalation effects may be considered. Sub-criteria are compared and weighted in a pairwise manner similar to that for the main criteria. Table 8 shows the importance weightings for each of the human health sub-criteria.

Similarly, Tables 9 and 10 show the importance weightings for environmental and social impacts.

Once relative weightings have been derived for each of the sub-criteria, normalized weightings for use in scoring different HAB management alternatives may be calculated by finding the eigenvector corresponding to the highest eigenvalue of the (symmetric) weighting matrix. The goal of the

TABLE 7. Sub-criteria for each main criteria category

Main criteria category	Sub-criteria
	Drinking water quality
Human Health Impacts	Recreational impacts
	Inhalation effects
	Effects on fish
Environmental Impacts	Effects on birds
	Effects on mammals
Social Impacts	Cost
	Public acceptability

TABLE 8. Importance weightings for human health impacts sub-criteria

Sub-Criteria	Drinking water quality	Recreational impacts	Inhalation effects
Drinking water quality		2.0	5.0
Recreational impacts			3.0
Inhalation effects			

TABLE 9. Importance weightings for environmental impacts sub-criteria

Sub-Criteria	Effects on fish	Effects on birds	Effects on mammals
Effects on fish		1/3	1/7
Effects on birds			1/3
Effects on mammals			

TABLE 10. Importance weightings for social impacts sub-criteria

Sub-Criteria	Cost	Public acceptability
Cost		1/3
Public acceptability		

weighting process is to set absolute weights that reflect as closely as possible the relative ratings input by the user.

The next step is to measure or assess the relative performance of each management option on each criterion. Again, the decision maker inputs a relative ranking—only now it is an evaluation of how well the alternatives perform for given sub-criteria. If a quantitative answer is not given, a qualitative statement may be transformed into a numerical value through a standardized system (e.g., the numbers 1, 3, 5, 7, and 9 correspond to the judgments "equally important," "moderately more," "strongly more," "very strongly more," and "extremely more," respectively). Once the decision maker gives inputs for each alternative under each sub-criterion, he may use the previously obtained weightings to calculate scores for each main criterion, followed by an overall score for each alternative. As Table 11 shows, algaecides score highest on human health impacts in this hypothetical example.

However, their advantage in this category is outweighed by the better performance of flushing in the environmental and social impact categories. When criteria weights are taken into account, flushing scores as the best option overall. The highest scoring alternative is, according to the rankings and preferences given by the decision maker throughout the analytic hierarchy process, the best strategy for the situation.

TABLE 11. Score breakdown

Management Response	Flushing	Algaecides	No Action
HUMAN HEALTH IMPACTS	0.403	0.496	0.100
Drinking water quality	0.226	0.674	0.101
Recreational impacts	0.627	0.280	0.094
Inhalation effects	0.719	0.166	0.115
ENVIRONMENTAL IMPACTS	0.705	0.173	0.122
Effects on fish	0.729	0.050	0.221
Effects on birds	0.707	0.070	0.223
Effects on mammals	0.702	0.226	0.073
SOCIAL IMPACTS	0.642	0.246	0.112
Cost	0.637	0.258	0.105
Public acceptability	0.644	0.242	0.114
OVERALL SCORE	0.502	0.392	0.107

7. Acknowledgements

This work was supported in part by EPA through Dynamac, Inc., via EPA contract number 68-D-01-005, with thanks to Kathy Martin, Tony Olsen, and Barbara Cook; and by USACE. Mr. Fristachi was supported through an appointment to the Research Participation Program at the EPA National Center for Environmental Assessment administered by the Oak Ridge Institute for Science and Education through an Interagency Agreement between the U.S. Department of Energy and the EPA. We would like to thank Sarah Armstrong of Cambridge Environmental for editing draft versions of this paper. Although this work was reviewed by the EPA and USACE and approved for publication, it does not necessarily reflect official Agency policy. The authors have no conflicts of interest to report.

8. Supplemental Information

8.1. CLUSTER ANALYSIS FOR AN EXAMPLE

To illustrate how cluster analysis can be used for HAB data, we performed an example cluster analysis on the subset of the data available for the Willow Creek reservoir in the Portland District of the USACE mentioned above previously. The dataset is in general representative of monitoring programs and their drawbacks: it contains widely differing numbers of measurements for different variables, and not all variables were measured at the same times. Although more than 1,600 samples have been taken on the reservoir since 1996, many parameters—such as total N P and total iron (Fe)—were measured less than 10% of the time. More than 40 different parameters are reported in the Willow Creek , but only six—oxidation reduction potential, dissolved oxygen, pH, specific conductance @ 25C, water temperature, and turbidity—were measured more than half the time. Additionally, measurements of water temperature and flow were often taken on alternate days when other variables were measured (Figure 5).

Cluster analysis of nine variables shows that total algal biovolume fluctuates most closely with turbidity and chlorophyll-a levels. Because cyanobacteria contain chlorophyll, one would expect a close relation between the two. Cluster analysis helps investigators visualize such relationships as well as others that might not be expected. It shows us that studies that measure both variables may be able to easily collapse a dimension by combining the two in order to perform more efficient statistical analysis.

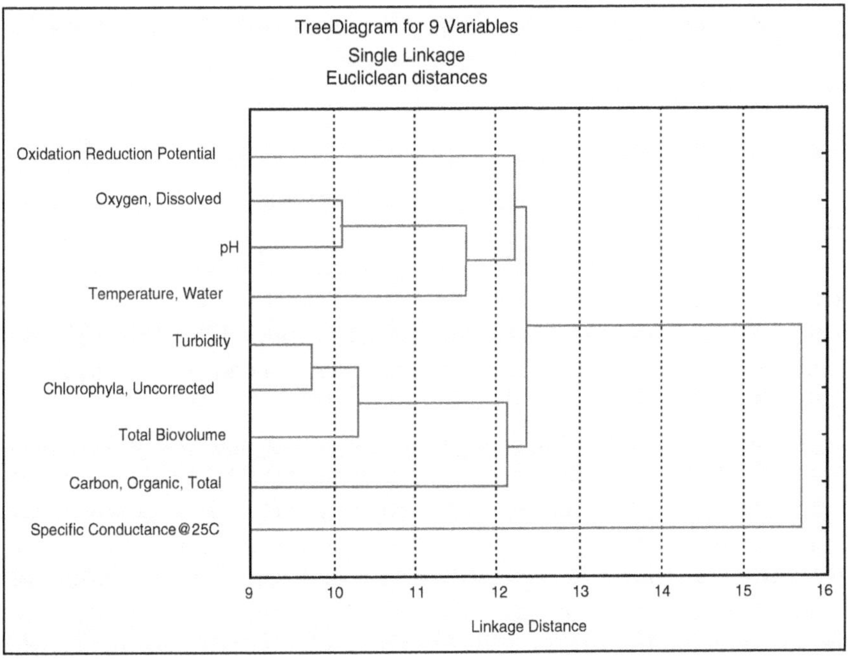

Figure 5. Cluster analysis of nine variables measured at Willow Creek Lake, Oregon. (Note: This analysis is done on a dataset where it's still not clear what "site 3" refers to with the algae data.)

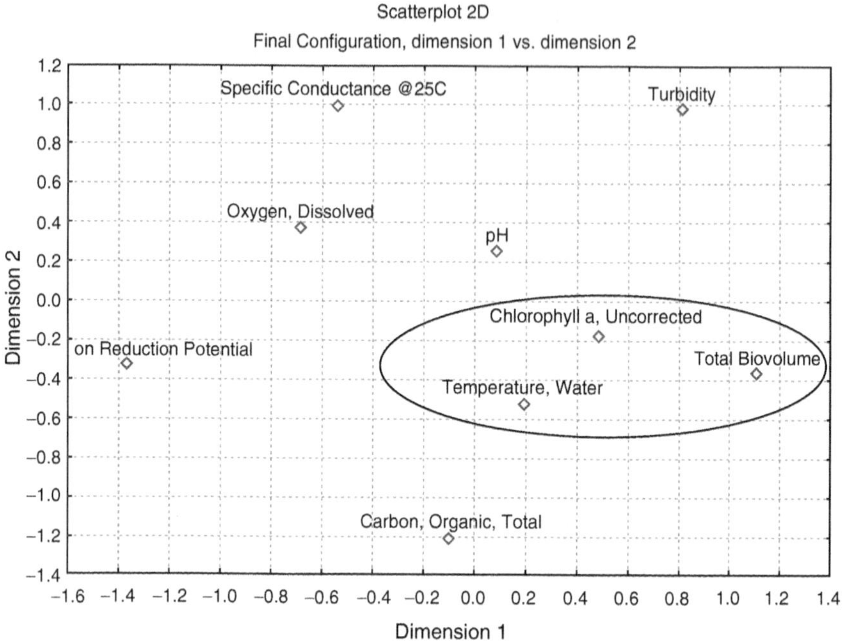

Figure 6. Multidimensional scaling output with nine variables.
(Note : This analysis is done on a dataset where it's still not clear what "Site 3" refers to with the algae data.)

Cluster analysis is, however, subject to such inconsistencies as the switching of pairings when variables are added or removed from the analysis. For instance, in the example above, pH groups most closely with dissolved oxygen, whereas it groups with water temperature when chlorophyll-a and total organic carbon levels are removed from the analysis (data not shown). Nonetheless, in both cases oxidation reduction potential, dissolved oxygen, pH, and water temperature are all more closely related to one another than they are to the other variables (Figure 6).

Multidimensional scaling is another statistical method that helps users visualize relationships between measured variables [58]. From a two-dimensional table of the correlations between all measured variables, it creates a spatial map of relative distances between the various parameters. In the above graph, a curved line can be drawn that separates oxidation reduction potential, dissolved oxygen, pH, and water temperature from the other variables, showing that the multidimensional scaling output is in accordance with the cluster analysis results. Both methods help to make obvious the relationships between variables.

9. References

1. Anderson DM, Hoagland P, Kaoru Y, White AW. Estimated annual economic impacts from harmful algal blooms (HABs) in the United States, technical report WHOI-2000-11. Woods Hole: Woods Hole Oceanographic Institution; 2000.
2. Yokoyama A, Park HD. Mechanism and prediction for contamination of freshwater bivalves (unionidae) with the cyanobacterial toxin microcystin in hypereutrophic Lake Suwa, Japan. Environ Toxicol. 2002; 17:424-33.
3. Bury NR, Eddy FB, Codd GA. The effects of the cyanobacterium microcystis aeruginosa, the cyanobacterial hepatotoxin microcystin-lr, and ammonia on growth rate and ionic regulation of brown trout. J Fish Biol. 1995; 46(1668):1042-54.
4. Rodger HD, Turnbull T, Edwards C, Codd GA. Cyanobacterial (blue-green algal) bloom associated pathology in brown trout, salmo trutta l. in Loch Leven, Scotland. J Fish Dis. 1994; 17(736):177-81.
5. Carmichael WW. A status report on planktonic cyanobacteria (blue green algae) and their toxins. Cincinnati: Environmental Monitoring Systems Laboratory, Office of Research and Development, US Environmental Protection Agency; 1992.
6. Lippy EC, Erb J. Gastrointestinal illness at Sewickley, Pa. J Am Water Works Assoc. 1976; 68(462):606-10.
7. Tisdale ES. Epidemic of intestinal disorders in Charleston, W. Va., occurring simultaneously with unprecedented water supply conditions. Am J Pub Health. 1931; 21(972):198-200.
8. Veldee MV. Epidemiological study of suspected waterborne gastroenteritis. Am J Pub Health. 1931; 21(1008):1227-35.
9. Azevedo SM, Carmichael WW, Jochimsen EM, Rinehart KL, Lau S, Shaw GR, Eaglesham GK. Human intoxication by microcystins during renal dialysis treatment in Caruaru—Brazil. Toxicology. 2002; 181-182(661):441-46.
10. Carmichael WW, Azevedo SMFO, An JS, Molica RJR, Jochimsen EM, Lau S, Rinehart KL, Shaw GR, Eaglesham GK. Human fatalities from cyanobacteria: chemical and biological evidence for cyanotoxins. Environmental Health Perspectives. 2001; 109(6448):663-68.
11. Satterstrom FK, Linkov I, Shifrin A, Steevens JA. The impact of harmful algal blooms on USACE operations, ERDC TN- ANSRP-Exx. Vicksburg: US Army Engineer Research and Development Center; 2006.
12. Voinovich GW. Statement of George V. Voinovich. 2002. Available at: http://epw.senate.gov/107th/voi_080502.htm.
13. NOAA. Harmful algal bloom and hypoxia research and control reauthorization. Public Law 108-456, 01/06. 2004.
14. KDHE. Neosho river basin total maximum daily load, Marion Lake, Marion County, Kansas. Kansas Department of Environment and Health. 2004.
15. Chorus I, Bartram J. Toxic cyanobacteria in water: a guide to their public health consequences, monitoring and management. London: E & FN Spon; 1999.
16. Kneale PE, Howard A. Statistical analysis of algal and water quality data. Hydrobiologia. 1997; 349:59-63.
17. Office of Wetlands, Oceans, and Watersheds. Statistical analysis methods for biological assessment. Washington, DC: US Environmental Protection Agency; 2003. Available at: http://www.epa.gov/owow/monitoring/tech/appdixe.html.
18. Vezie C, Brient L, Sivonen K, Bertru, G, Lefeuvre J-C, Salkinoja-Salonen M. Variation of microcystin content of cyanobacterial blooms and isolated strain in Lake Grand-Lieu (France). Microbial Ecol. 1998; 35(2978):126-35.
19. Pawlik-Skowronska B, Skowronski T, Pirszel J, Adamczyk A. Relationship between cyanobacterial bloom composition and anatoxin-a and microcystin occurrence in the eutrophic dam reservoir (SE Poland). Polish J Ecology. 2004; 52:479-90.
20. Jacoby JM, Collier DC, Welch EB, Hardy FJ, Crayton M. Environmental factors associated with a toxic bloom of microcystis aeruginosa. Can J Fish Aquat Sci. 2000; 57:231-40.

21. Vezie C, Rapala J, Vaitomaa J, Seitsonen J, Sivonen K. Effect of nitrogen and phosphorus on growth of toxic and nontoxic microcystis strains and on intracellular microcystin concentrations. Microbial Ecol. 2002; 43(492):443-54.
22. Rapala J, Sivonen K. Assessment of environmental conditions that favor hepatotoxic and neurotoxic anabaena spp. strains cultured under light limitation at different temperatures. Microbial Ecol. 1998; 36(8038):181-92.
23. Kardinal WEA, Visser PM. Dynamics of cyanobacterial toxins. In: Huisman J, Matthijs HCP, Visser PM, editors. Harmful cyanobacteria, aquatic ecology series. Dordrecht: Springer; 2005.
24. Soranno PA. Factors affecting the timing of surface scums and epilimnetic blooms of blue-green algae in a eutrophic lake. Can J Fish Aquat Sci. 1997; 54:1965-75.
25. Saker ML, Griffiths DJ. Occurrence of blooms of the cyanobacterium cylindrospermopsis raciborskii (woloszynska) seenaya et subba raju in a North Queensland domestic water supply. Mar Freshwater Res. 2001; 52 (320):907-15.
26. Paerl HW. Microscale physiological and ecological studies of aquatic cyanobacteria: macroscale implications. Microscopy Res Technique. 1996; 33(3658):47-72.
27. Landsberg JH. The effects of harmful algal blooms on aquatic organisms. Rev Fish Sci. 2002; 10(505):113-390.
28. Huisman J, Hulot FD. Population dynamics of harmful cyanobacteria. In: Huisman J, Matthijs HCP, Visser PM, editors. Harmful cyanobacteria, aquatic ecology series. Dordrecht: Springer; 2005.
29. Codd GA, Morrison LF, Metcalf JS. Cyanobacterial toxins: risk management for health protection. Toxicology and Applied Pharmacology. 2005; 203:264-72.
30. Carmichael WW, Sirenko LA, Klochenko PD, Shevchenko TF. 2001 A comparative assessment of the toxicity of algae and cyanobacteria in water bodies of Ukraine. Phycologia. 2005; 40(10338):15.
31. Jeong KS, Kim DK, Whigham B, Joo GJ. Modelling microcystis aeruginosa bloom dynamics in the Nakdong River by means of evolutionary computation and statistical approach. Ecol Model. 2003; 161:67-78.
32. Howard A, Easthope MP. Application of a model to predict cyanobacterial growth patterns in response to climatic change at Farmoor Reservoir, Oxfordshire, UK. Science of the Total Environment. 2002; 282(423):459-69.
33. Jann-Para G, Schwob I, Feuillade M. Occurrence of toxic planktothrix rubescens blooms in Lake Nantua, France. Toxicon. 2004; 43(13128):279-85.
34. Huisman J, Sharples J, Stroom JM, Visser PM, Dardinaal WEA, Verspagen JMH, Sommeijer B. Changes in turbulent mixing shift competition for light between phytoplankton species. Ecology. 2004; 85:2960-70.
35. Ibelings BW, Vonk M, Los HFJ, van der Molen DT, Mooij WM. Fuzzy modeling of cyanobacterial surface waterblooms: validation with NOAA-AVHRR satellite images. Ecol Appl. 2003; 13(10328):1456-72.
36. Downing JA, Watson SB, McCauley E, Frisk T, Bilaletdin A, Kaipainen H, Malve O, Mols M. Predicting cyanobacteria dominance in lakes: modelling phytoplankton dynamics of the eutrophic Lake Vortsjarv, Estonia. Can J Fish Aquat Sci. 2001; 58:1905-08.
37. Kotak BG, Lam AKY, Prepas EE, Hrudey SE. Role of chemical and physical variables in regulating microcystin-LR concentration in phytoplankton of eutrophic lakes. Can J Fish Aquat Sci. 2000; 57(127):1584-93.
38. Bonnet MP, Poulin M. Numerical modelling of the planktonic succession in a nutrient-rich reservoir: environmental and physiological factors leading to microcystis aeruginosa dominance. Ecol Model. 2002; 156:93-112.
39. Graham JL, Jones JR, Jones SB, Downing JA, Clevenger TE. Environmental factors influencing microcystin distribution and concentration in the Midwestern United States. Water Research. 2004; 38:4395-404.
40. Ghadouani A, Pinel-Alloul B, Prepas EE. Effects of experimentally induced cyanobacterial blooms on crustacean zooplankton communities. Freshwater Biol. 2003; 48:363-81.

41. Lathrop RC, Carpenter SR, Stow CA, Soranno PA, Panuska JC. Phosphorus loading reductions needed to control blue-green algal blooms in Lake Mendota. Can J Fish Aquat Sci. 1998; 55(12738):1169-78.
42. Recknagel F. Anna – artificial neural network model for predicting species abundance and succession of blue-green algae. Hydrobiologia. 1997; 349:47-57.
43. Wei B, Sugiura N, Maekawa T. Use of artificial neural network in the prediction of algal blooms. Water Research. 2001; 35:2022-28.
44. Yabunaka K, Hosomi M, Murakami A. Novel application of a back-propagation artificial neural network model formulated to predict algal bloom. Water Science and Technology. 1997; 36:89-97.
45. McGregor GB, Fabbro LD. Dominance of cylindrospermopsis raciborskii (nostocales, cyanoprokaryota) in Queensland tropical and subtropical reservoirs: implications for monitoring and management. Lake Reservoir. 2000; 5:195-205.
46. Amé MV, Díaz MP, Wunderlin DA. Occurrence of toxic cyanobacterial blooms in San Roque Reservoir (Cordoba, Argentina): a field and chemometric study. Environ Toxicol. 2003; 18(762):192-201.
47. Andren C, Eklund B, Gravefors E. A multivariate biological and chemical characterization of industrial effluents connected to municipal sewage treatment plants. Environmental Toxicology and Chemistry. 1998; 17:228-33.
48. Wunderlin DA, Díaz María del Pilar AMV, Pesce SF, Hued AC, Bistoni M. Pattern recognition techniques for the evaluation of spatial and temporal variations in water quality: a case study: Suquía River Basin (Córdoba–Argentina). Water Research. 2001; 35:2881-94.
49. Kruskal JB, Wish M. Multidimensional scaling. Newbury Park: SAGE Publications; 1978.
50. Stewart PM, Butcher JT, Gerovac PJ. Diatom (bacillariophyta) community response to water quality and land use. Nat Area J. 1999; 19:155-65.
51. Hartwell SI, Claflin LW. Cluster analysis of contaminated sediment data: nodal analysis. Environmental Toxicology and Chemistry. 2005; 24:1816-34.
52. Gasiunaite ZR, Cardoso AC, Heiskanen AS. Seasonality of coastal phytoplankton in the Baltic Sea: influence of salinity and eutrophication. Estuar Coast Shelf S. 2005; 65:239-52.
53. Pimentel D. Ecological effects of pesticides on nontarget species. Washington, DC: Executive Office of the President's Office of Science and Technology; 1971.
54. Tucker CS, Hargreaves JA. Copper sulfate to manage cyanobacterial off-flavors in pond-raised channel catfish. In: Rimando AM, Schrader KK, editors. Off-flavors in aquaculture, acs symp. ser. Washington, DC: American Chemical Society/Oxford Univ. Press; 2003.
55. EPA. Guidance for reregistration of pesticide products containing copper sulfate; fact sheet no 100. Washington, DC: US Environmental Protection Agency, Office of Pesticide Programs; 1986.
56. Linkov I, Satterstrom FK, Kiker G, Bridges T, Benjamin S, Belluck D. From optimization to adaptation: shifting paradigms in environmental management and their application to remedial decisions. Integr Environ Assess Manage. 2006; 2:92-98.
57. Linkov I, Satterstrom FK, Seager TP, Kiker G, Bridges T, Belluck D, Meyer A. Multi-criteria decision analysis: comprehensive decision analysis tool for risk management of contaminated sediments. Risk Anal. 2006; 26:61-78.
58. Linkov I, Varghese A, Jamil S, Seager TP, Kiker G, Bridges T. Multi-criteria decision analysis: framework for applications in remedial planning for contaminated sites. In: Linkov I, Ramadan A, editors. Comparative risk assessment and environmental decision making. Amsterdam: Kluwer; 2004.
59. Kiker G, Bridges T, Varghese AS, Seager TP, Linkov I. Application of multi-criteria decision analysis in environmental management. Integr Environ Assess Manage. 2005; 1:49-58.
60. Chorus I. Water safety plans. In: Huisman J, Matthijs HCP, Visser PM, editors. Harmful cyanobacteria, aquatic ecology series. Dordrecht: Springer; 2005.
61. Marker AFH, Collett GD. Spatial and temporal characteristics of algae in the River Great Ouse. I Phytoplankton Regul River. 1997; 13:219-33.

62. Frisk T, Bilaletdin A, Kaipainen H, Malve O, Mols M. Modelling phytoplankton dynamics of the eutrophic Lake Vortsjarv, Estonia. Hydrobiologia. 1999; 414:59-69.
63. Howard A. Modeling movement patterns of the cyanobacterium, microcystis. Ecol Appl. 2001; 11(563):304-10.
64. Rhodes RG. Blue green algae in Marion Reservoir, Marion County Kansas, sampled on June 12, 2003. Available at: http://biology.missouristate.edu/faculty_pages/Rhodes/marion_reservoir/blue_green_algae_in_marion_reser.htm.

PART 4

DECISION-MAKING AND RISK ASSESSMENT METHODS, TOOLS, AND APPLICATIONS FOR CRITICAL INFRASTRUCTURE AND PORT SECURITY

CHAPTER 13

DECISION ANALYSIS TOOLS FOR SAFETY, SECURITY, AND SUSTAINABILITY OF PORTS AND HARBORS

G.S. PARNELL

U.S. Military Academy at West Point
and Innovative Decisions Inc.
USA

J.R. FIGUEIRA

Instituto Superior Tecnico
Technical University of Lisbon
Portugal

S. BENNETT

National Biodefense Analysis and Countermeasures Center
USA

N. BOBYLEV

St. Petersburg State Polytechnic University
Russia

M. DEL PUP

Comunita Porti Adriatici
Italy

J. GANOULIS

Aristotle University of Thessaloniki
Greece

N. HARUVY

Netanya Academic College
Israel

I. Linkov et al. (eds.), Managing Critical Infrastructure Risks, 245–260.
© 2007 *Springer*.

S. MENONI

Politecnico di Milano
Italy

F. PERUZZO

Direzione Sicurezza e Ambiente (Safety and Environment Dept.)
Italy

O. SALVI

INERIS (National Institute for Industrial Risks and Environment Protection)
France

V. SARGSYAN

Yerevan State Institute of Economy
Armenia

U. SCHLINK

UFZ - Centre for Environmental Research
Germany

D. SCHNELLE

DTRA/CBT
USA

S. EL SHELTAWI

Cairo University
Egypt

Abstract: Industrial ports and harbors face significant safety, security, and sustainability challenges. Decisions must be made to assess risk and allocate risk management resources. Decision analysis has nearly 40 years of successful application to decision challenges. We present a comprehensive decision analysis framework that captures the concepts and tools of decision analysis. We use four illustrative examples to demonstrate the potential use of decision analysis:

1. Management of hazardous material at ports.
2. Dredging of contaminated material.
3. A general terrorist attack at ports.
4. A specific terrorist attack, the detonation of a radiological dissemination device (a "dirty bomb") in a port.

We show how decision analysis concepts could be applied in each case to make risk management decisions. We conclude with some illustrative results of the analysis.

1. Introduction

Two small watercraft enter the wet dock of the harbor undetected. They approach a tank that stores 20 tons of chlorine. A major metropolitan area is located 3 km from the chlorine storage in the harbor. Two boats increase probability of success; each boat transports 1 ton of explosive material, enough to cause total loss of containment of the chlorine. After the two explosions, employees within a 200 m radius are dead or injured, and the installations within a 130 m radius are destroyed, including the alarm system with the sirens. The chlorine begins to disperse to the surrounding area…

The above scenario is an example of one of many threats facing port authorities. Industrial ports and harbors consist of water, natural environment, buildings, equipment, infrastructure, organizations, and people. Ports and harbors exist for two important economic purposes. First, ports and harbors provide a transportation node to support national and international trade. Import shipments arrive by boat and are placed on another means of transportation to convey the goods to the interior of the nation or to another nation. Export shipments arrive at the port by a variety of means and placed on ships to transport to another port, usually, in another country. Second, private commercial port and harbor operators need to make a profit to continue operations.

Industrial ports and harbors face significant safety, security, and sustainability challenges. The safety challenges arise from the operation of large equipment by many people in a large industrial enterprise. Potential health and environmental issues can arise from the accidental or intentional release of hazardous materials. In the past, security issues have focused on the illegal transport of people or materials through the ports. In the current international security environment, terrorist groups may attack an industrial port as a symbol of global trade or to destabilize a government opposed to their objectives. The security risks are especially challenging since they involve intelligent, resourceful, and adaptive adversaries. The sustainability challenges for the continued operation of the port and harbor are social, environmental/ecological, and economic.

These challenges involve international, national, local, and private stakeholders. These stakeholders expect port and harbor operations to assess all of the safety, security, and sustainability risks. Decisions must be made to assess risk and allocate risk management resources. Governmental and commercial organizations need to spend risk assessment and risk management funds wisely.

Decision analysis has about 40 years of successful application to decision challenges. Here, we present a comprehensive decision analysis framework that identifies the concepts and tools of decision analysis. We use four illustrative examples to discuss the potential use of decision analysis:

1. Management of hazardous material at ports.
2. Dredging of contaminated material.
3. A general terrorist attack at ports.
4. A specific terrorist attack, the detonation of a radiological dissemination device (a "dirty bomb") in a port.

We show how decision analysis concepts could be used in each case to inform risk management decisions, and conclude by providing some illustrative displays of decision analysis results.

2. Decision Analysis Framework

In order to assess the potential use of decision analysis to meet decision challenges for the safety, security, and sustainability of ports and harbors, we need a clear understanding of the concepts and tools of decision analysis. Figure 1 provides a decision analysis framework to describe the major concepts of decision analysis [24]. The framework is comprised of dimensions, branches and levels. The dimensions are the decision (single and multiple alternative branches); the decision maker (single and multiple decision maker branches) and stakeholder interaction; the value preference (time preference, objectives, and value preference branches); and the uncertainty with risk preference (value and utility) branches. The levels on each branch represent increasing complexity as we move away from the origin. The gray levels are not recommended.

There are many other useful decision analysis frameworks and typologies. One typology characterizes decisions in relation to external conditions of uncertainty and conflict [11], which include:

- Decision under certainty.
- Decision under risk.

- Decision under uncertainty.
- Decision under conflict (this includes *game theory* [8]).

In Figure 1, we focus on the number of decision makers and their interaction with the stakeholders. An alternative would be to classify decision-making methods by the number of objectives and the number of decision makers. This classification might include the following:

- Optimization methods for one objective and one decision maker.
- Team theory for one objective and two or more decision makers.
- Multicriteria modeling for multiple objectives and one decision maker.
- Game theory for multiple objectives and multiple decision makers.

In this chapter, we will use the decision analysis framework in Figure 1.

2.1. OPPORTUNITY DEFINITION

We begin our description of the decision framework at the center of the framework with the opportunity definition. It is essential to describe the opportunity or problem that requires a decision. We choose the word opportunity as more general than problem, since we believe that "every problem is an opportunity." Developing a clear one-sentence description of

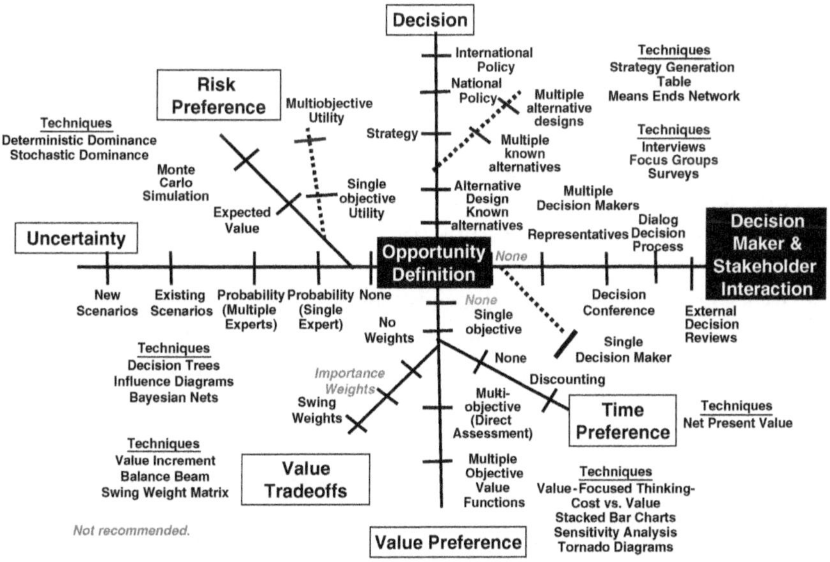

Figure 1. Decision analysis framework.

TABLE 1. Four illustrative industrial port and harbor decisions

Example Decisions	Short Description
1. Management of hazardous material at ports	1. The port authorities must choose actions to reduce the risk of hazardous materials at ports and in the harbors.
2. Dredging of contaminated material	2. The port authorities must decide the best alternative to dredging of contaminated material.
3. General terrorist attacks at ports	3. The port authorities must determine the best set of risk management actions to reduce the likelihood and consequences of a terrorist attack.
4. Specific terrorist attack: "dirty bomb" in a port	4. The port authorities must determine the best set of risk management actions to reduce the likelihood and consequences of a "dirty bomb" in a port.

the decision opportunity is a very useful first step. This sentence will need to be refined as our understanding of the opportunity or problem increases.

Table 1 provides a list of the four illustrative decisions that we will consider in this chapter to demonstrate the use of decision analysis.

2.2. DECISION

There are three important distinctions about decisions. The first is the number of decisions. We may be selecting one alternative from a list of alternatives (e.g., when we purchase a home to live in) or we may be selecting many alternatives from a list of alternatives (e.g., when we develop our investment portfolio of stocks, bonds, mutual funds, and cash). Multiple decisions increase the complexity. The second distinction is the type of decision—alternative selection or design. For example, we can purchase an existing home or we can design a home. The decisions involving design are more complex. The third distinction is the level of the decision. The decision level may be local, regional, national, or international. Two useful techniques are the strategy generation table [18] and the means-ends network [16]. Table 2 describes the four illustrative decisions using the three decision distinctions. Only the dredging decision involves one decision. Most of the alternatives involve decision selection and may involve design. All of the decisions could involve decision approvals above the local level.

2.3. DECISION MAKER AND STAKEHOLDER INTERACTION

It is almost impossible to have a clear understanding of the decision opportunity and the potential decisions without decision maker and stakeholder interaction. We believe the lack of a plan for involving decision makers and

TABLE 2. Four illustrative industrial port and harbor decisions

Example Decisions	Number of Decisions	Type of Decision	Decision Levels
1. Management of hazardous material at ports	Several	Alternative selection and design	Local and regional
2. Dredging of contaminated material	One	Alternative selection	Local and regional
3. General terrorist attacks at ports	Several	Alternative selection and design	Local, regional, and national
4. Specific terrorist attack: "dirty bomb" in a port	Several	Alternative selection and design	Local, regional, and national

stakeholders will result in the decision analysis not being accepted. There are two important distinctions: the number of decision makers and the level of review of the decision. In some cases there is one decision maker. In many cases however, there are multiple decision makers. The number of stakeholders typically depends on the decision and the level of the decision. The level of review of the decision can be internal and external to the organization. The more reviews and the higher the level of review, the more decision makers and stakeholders are involved.

Two processes have proven very successful: decision conferences [32] and the dialog decision process [1]. Decision conferencing uses single or multiple meetings to make the key decisions. Each decision conference is attended by key decision makers or their representatives. The dialog decision process plans for regular interaction with key decision makers. The interaction is usually a presentation to the decision maker or decision board. There are several techniques for decision maker and stakeholder interaction: interviews, focus groups, and surveys [28].

We identified the following as potential decision makers and stakeholders for industrial ports and harbors. Our list was developed from a list presented in Steve Rowan's presentation at the workshop [26]. This list clearly illustrates the complexity of decision making for industrial ports and harbors.

Using the three categories of decision makers and stakeholders, it is clear that our four decision examples involve many of the stakeholders in Table 3. Table 4 shows the likely involvement of decision makers and stakeholders for each of the four examples. The terrorist attack studies would involve all the stakeholders.

TABLE 3. Industrial port and harbor decision makers and stakeholders

Operators	Regulatory	Other
Port Operators	Department of Transport	Citizenry
Port Security	Health and Safety Agencies	Professional Organizations
Shipping Companies	Environmental Agencies	Consumer Organizations
Port/Harbor Employees	Railway Regulators	Local Industry
Funders and Insurers	Customs and Revenue Service	Nongovernmental Organizations
Intermodal Facilities	Homeland Security	Port Investors
Rail/Truck (ground transportation)	International Trade Organizations	Local Politicians
Ship Operators Fire Brigade/	Local Port Authorities	Regional Politicians
Emergency Services	Local Police	National Politicians
Coast Guard	Local Nature Conservation Bodies	International Trade Organizations
Civil Defense Military		

TABLE 4. Decision maker and stakeholder involvement

Example Decisions	Operators	Regulators	Other
1. Management of hazardous material at ports	Most	Safety and environmental	Some
2. Dredging of contaminated material	Some	Environmental	Some
3. General terrorist attacks at ports	All	All	All
4. Specific terrorist attack: "dirty bomb" in a port	All	All	All

2.4. VALUE PREFERENCE

The third dimension is value preference. Thinking about our preferences is critical to good decision making. Value preference has three branches: objectives, time preference, and value tradeoffs. The objectives can be single or multiple objectives. The most common single objective is profit [25]. Many complex problems with multiple stakeholders, such as industrial ports and harbors, involve multiple objectives. Value-focused thinking [16] involves using our values to evaluate our decisions and to create better decisions. Time preference means that we value attaining an objective sooner rather than later. Net Present Value is a common technique to use a discount rate to discount the value of funds received in future years. We use value tradeoffs to compare conflicting objectives. We typically use weights to compare measures of objectives. A common mistake is to use importance weights. The weights depend on the importance and the variation of the measure [16]. Swing weights are the appropriate technique to use [18]. A new technique, the swing weight matrix, helps decision makers and stakeholders assign the appropriate swing weights and document the rationale for the weight assessment [7]. The technique also helps analysts explain the weights to decision makers and stakeholders who did not participate in the weights assessment. Other weighting techniques include balance beam and value increment. Two useful techniques to identify and display to the most significant parameters of the analysis are sensitivity analysis and tornado diagrams [18]. Two useful techniques for summarizing the results are stacked bar charts and cost versus value charts (see section 2.6).

There are alternatives to value and utility. For example, multi-criteria decision analysis (MCDA) uses other methods, including distance-based techniques, which seek to find a solution as "close" as possible to an ideal point, such as the Compromise Programming Method [30] and outranking techniques, which compare alternatives pairwise and reflect preferences between alternatives. The Analytical Hierarchy Process is another approach that uses pairwise comparisons [27].

For industrial ports and harbors, we identified several important objectives for the decision makers and stakeholders to consider in the evaluation of the decision alternatives.

- **Safety**. The large equipment and movement of materials at an industrial port create the potential for life-threatening accidents.
- **Public Health**. The shipment of hazardous materials poses the risk of release of materials that would be hazardous to humans.
- **Security**. The shipment of illegal items or a terrorist attack could create major security risks.
- **Sustainability:**
- **Social Issues**. The industrial ports must be socially acceptable to the local community stakeholders.
- **Environmental Protection**. The operations of the port and harbor should not endanger the environment.
- **Economic Issues**. The port operators must make money to stay in operation. Stakeholders are also concerned with economic equity issues.

Table 5 provides an assessment of the importance of the objectives for the four example decisions. The meaning of the words in the cells is the following: major importance to the example, minor importance to the example, and not applicable to the example. Most of the objectives are important in every example. Security is focused on the last two examples.

2.5. UNCERTAINTY AND RISK

The final dimension is uncertainty and risk. Decision making for industrial ports and harbors involves significant future uncertainties due to natural and manmade causes. Severe weather, such as a hurricane, can dramatically impact port operations; e.g., the impact of Katrina on the Port of New Orleans [34]. The future event of a hurricane striking a port or a terrorist attack at a port is uncertain. Risk is the likelihood and consequences (usually adverse or severe consequences) of an uncertain event. The consequences of an uncertain event can be a dramatic change in each of the objectives shown in Table 5.

There are several approaches to representing and managing uncertainty in the decision analysis process. The decision maker may choose to ignore uncertainty or consider it explicitly. In decision analysis, probability and scenarios are primarily used to model uncertainty. If all of the potential outcomes can be identified, probability distributions can be assigned to an uncertain event or events. It is easier to obtain a probability distribution from an expert than a group of experts. Two useful techniques for analyzing uncertainty are influence diagrams [33] and Bayesian Nets [31]. Two useful techniques for examining decisions under uncertainty are decision trees and influence diagrams [5]. If we cannot identify all of the potential outcomes, we may use scenarios that span the future to address potential uncertainties [21]. Scenarios take significant time and resources to develop. Therefore, existing scenarios are easier than new scenarios.

TABLE 5. Relative importance of objectives for four examples

					Sustainability	
Example Decisions	Safety	Health	Security	Social	Environ-mental	Economic
1. Management of hazardous material at ports	Major	Major	N/A	Minor	Major	Minor
2. Dredging of contaminated material	Major	Major	N/A	Minor	Major	Minor
3. General terrorist attacks at ports	Major	Major	Major	Major	Major	Major
4. Specific terrorist attack: "dirty bomb" in a port	Major	Major	Major	Major	Major	Major

The U.S. Department of Homeland security has created 15 "planning scenarios" for use in national, federal, state, and local homeland security planning activities [15].

Risk preference can be approached in two ways. The analyst can assess utility functions (single or multiple attribute) from the decision maker(s). This requires lotteries of uncertain attribute levels [18]. Lotteries are difficult to obtain from decision makers for many different attributes. An alternative approach is to use expected value or to use Monte Carlo simulation to show the impact of the uncertainties in the scores of alternatives on alternative value [19]. The output of the Monte Carlo simulation is a probability distribution of the value of each alternative. A tornado diagram can be used to graphically identify the uncertainties that contribute the most to the variance of the alternative value. Two useful techniques for comparing uncertain alternatives are deterministic and stochastic dominance [5].

Table 6 identifies the sources of uncertainty for each of the four examples. The table also identifies the potential to use probabilities and scenarios. In addition, the table lists the major risks.

2.6. ILLUSTRATIVE DECISION ANALYSIS RESULTS

In this section, we will show some illustrative decision analysis results for single-decision (e.g., dredging of contaminated materials) and multiple-decision problems (e.g., the other three illustrative examples). The multiple-decision problem is sometimes called the resource allocation problem or the portfolio decision problem. In each case, the alternatives, scores, values, and costs are notional. Our purpose is to illustrate the types of analysis results available using decision analysis techniques.

2.6.1. Single Decision Results

For our illustration, we used six objectives and one measure for each objective. We used the additive value model [18] to calculate the value of each alternative using value functions and weights. Figure 2 shows an alternative evaluation using a stacked bar chart. We are easily able to see the value contribution for each measure. Our baseline approach only provides a value of 10 out of a possible 100 (the ideal alternative). Alternative A is the best alternative with a value of 77. By comparing alternative A with the ideal alternative, we can see opportunities for Value-Focused Thinking to improve the value of alternative A, especially measures 1, 2, and 3.

Figure 3 displays the cost versus value plot. For this figure, we can see that alternatives B and D are dominated by alternative A, which has higher value and lower cost than B and D. The efficient frontier is composed of the

TABLE 6. Sources of uncertainty and use of probabilities and scenarios

		Appropriate uncertainty technique		
Example Decisions	Uncertainties	Probabilities	Scenarios	Risk
1. Management of hazardous material at ports	Types of materials Amounts of each type Sources of accidents Port response Consequences	May be reasonable to assign probability distributions.	Might be useful if we were considering too many materials to analyze separately.	Environmental, public health, and social consequences
2. Dredging of contaminated material	Type of material Location of material Effectiveness of dredging Sources of accidents Consequences	May be reasonable to assign probability distributions.	Might be useful if we were considering too many materials to analyze separately.	Environmental, public health and social consequences
3. General terrorist attacks at ports	Terrorist group Attack objectives Type of attack Magnitude of attack Time of day Port and community response Consequences	Very. difficult to identify outcome space	Selection and analysis of bounding scenarios may be best method.	Many significant consequences
4. Specific terrorist attack: "dirty bomb" in a port	Terrorist group Attack objectives Type of bomb Magnitude of bomb Time of day Weather Port and community response Consequences	Very difficult to identify outcome space.	Selection and analysis of bounding scenarios may be best method.	Many significant consequences

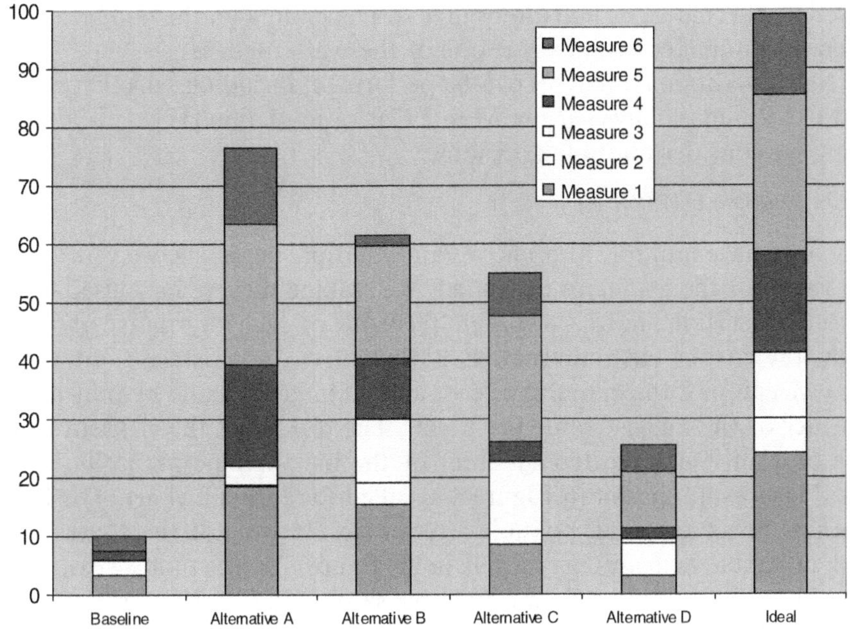

Figure 2. Alternative evaluation using a stacked bar chart.

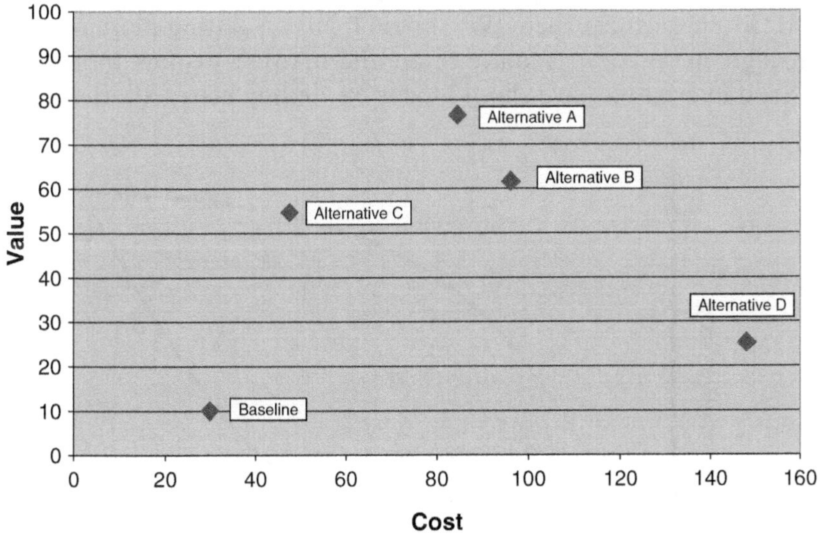

Figure 3. Alternative evaluation using a cost versus value plot.

baseline, alternative C, and alternative A. Depending on the budget available, the decision maker would select one of these alternatives.

Many additional analyses can be performed, including sensitivity analysis and uncertainty analysis using Monte Carlo simulation [19].
Next, we consider multiple decisions.

2.6.2. Multiple Decision Results

When we have multiple alternatives and multiple decisions, we want to show the value for the resources expended. We assume that we have used multiple objective decision analysis to obtain the value of each alternative and that we know the cost of each alternative. If the values and costs are independent, then we can plot the cumulative cost and cumulative value by plotting them in order of the highest value per dollar. The top line is the efficient frontier. The bottom line is plotted by order of the lowest-value-per-dollar alternatives. The resulting plot in Figure 4 is called the football chart. The football bounds the space of all possible alternatives. If we plot the chart, we first look at our current strategy (*Today* in the figure). We see that we can increase our value for the same cost or obtain the same value for less cost. In fact, any point in the region between the dotted lines and the efficient frontier would be an improvement. Once we get our portfolio on the efficient frontier, we might find that we are low or high on the curve. If we are low on the curve (require additional funding), we have many good alternatives but we lack the resources to implement them. We should focus on getting more resources. If we are high on the curve (require better alternatives), we have more resources than good alternatives. We should focus on finding better alternatives.

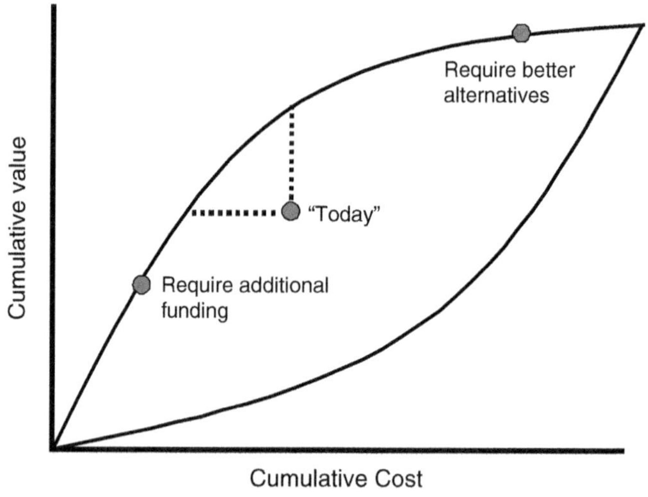

Figure 4. Benefit cost analysis summary of alternatives.

If we have dependent values or costs, optimization with constraints can be used to obtain the best portfolio [18].

2.7. SUMMARY

Decision analysis offers a systematic methodology for assessing conflicting objectives, considering uncertainties, and evaluating complex alternatives. It is essential to begin with a clear understanding of the decision opportunity or problem. The identification of stakeholders and their objectives is essential to the understanding of the problem, the development of alternatives, the evaluation of alternatives, and the implementation of the decision. Uncertainty is an important consideration in industrial and port security. The use of scenarios will be an important technique to understand the uncertainties and assess the risk. The integration of risk analysis and decision analysis will be required to make wise risk management decisions.

To achieve acceptance in the industrial ports and harbors domain, we will need to perform studies and develop decision analysis case studies that demonstrate the value of decision analysis to decision makers responsible for ports and harbors. We are ready to begin!

3. References

1. Barabba V. Meeting of the minds. Watertown: Harvard Business School Press; 1995.
2. Bodily SE, editor. Special issue: decision & risk analysis. Interfaces. 1992; 22(6):1-166.
3. Brown RV, Ulviva J. Decision analysis comes of age. Harvard Business Review. 1982; 60(5):130-41.
4. Buckshaw DL, Parnell GS, Unkenholz WL, Parks DL, Wallner JM, Saydjari OS. Mission oriented risk and design analysis of critical information systems. Military Operations Research. 2005; 10(2):19-38.
5. Clemen R. Making hard decisions, 2nd edition. Belmont: Duxbury Press; 1996.
6. Corner JL, Kirkwood CW. Decision analysis applications in the operations research literature, 1970-1989. Operations Research. 1991; 39(2):206-219.
7. Ewing P, Tarantino W, Parnell G. Use of decision analysis in the army base realignment and closure (BRAC) 2005 military value analysis. Decision Analysis Journal. 2006; 3(1):33-49.
8. Fraser NM, Hipel KW. Conflict analysis: models and resolutions. Amsterdam: Elsevier; 1984.
9. Ganoulis J. Water resources management and environmental security in Mediterranean transboundary river basins, In: Environmental security and environmental management: the role of risk assessment. Dordrecht: Springer; 2006.
10. Ganoulis J. Evaluating alternative strategies for wastewater recycling and reuse in the Mediterranean area. J. of Wat. Sci. Tech. Water Supply. 2003; 3(4):11-20.
11. Ganoulis J. Engineering risk analysis of water pollution: probabilities and fuzzy sets. Weinheim: Wiley; 1994.
12. Grelk BJ, Kloeber JM, Jackson JA, Deckro RF, Parnell GS. Quantifying CERCLA using site decision maker values. Remediation. 1998; 8(2):87-105.
13. Howard RA. Decision analysis: practice and promise. Management Science. 1988; 34(6):679-695.

14. Howard RA, Matheson JE, editors. The principles & applications of decision analysis [two volumes]. Menlo Park: Strategic Decisions Group; 1983.
15. Howe D. Planning scenarios: executive summaries. White House Homeland Security Council; 2004.
16. Keeney RL. Value-focused thinking: a path to creative decisionmaking. Cambridge: Harvard University Press; 1992.
17. Keeney RL, Raiffa H. Decision making with multiple objectives: preferences and value tradeoffs. New York: Wiley; 1976.
18. Kirkwood CW. Strategic decision making: multiobjective decision analysis with spreadsheets. Belmont: Duxbury Press; 1997.
19. Kwinn MJ Jr, Parnell GS. Decision making. In: Parnell et al., editors. Systems decision making for systems engineering and management. New York: John Wiley & Sons Inc.; 2007.
20. Merrick J, Parnell GS, Barnett J, Garcia M. A multiple objective decision analysis of stakeholder values to identify watershed improvement needs. Decision Analysis. 2005; 2(1):44-57.
21. Parnell GS, Jackson J, Burk R, Lehmkuhl L, Engelbrecht J. R&D concept decision analysis: using alternate futures for sensitivity analysis. Journal of Multi-Criteria Decision Analysis. 1999; 8:119-127.
22. Parnell GS, Frimpon M, Barnes J, Kloeber J, Deckro RJ, Jackson JA. Safety risk analysis of an innovative environmental technology. Risk Analysis. 2001; 21(1):143-155.
23. Parnell GS, Dillon-Merrill RL, Bresnick TA. Integrating risk management with homeland security and antiterrorism resource allocation decision-making. In: David Kamien, editor. The McGraw-Hill handbook of homeland security. New York: McGraw-Hill; 2005.
24. Parnell GS. Decision analysis tools for risk management of industrial ports and harbors. Risk Management Tools for Port Security, Critical Infrastructure, and Sustainability, NATO Advanced Research Workshop, Venice, Italy. 16–19 March 2006.
25. Raiffa H. Decision analysis: introductory lectures on choices under uncertainty. Boston: Addison-Wesley; 1968.
26. Rowan S. Preventing & responding to major environmental incidents in the UK: an overview. Risk Management Tools for Port Security, Critical Infrastructure, and Sustainability, NATO Advanced Research Workshop, Venice, Italy. 16–19 March 2006.
27. Saaty TL. The analytic hierarchy process. New York: McGraw-Hill; 1980.
28. Trainor T, Parnell GS. Problem definition. In: Parnell et al., editors. Systems decision making for systems engineering and management. Mississauga: John Wiley & Sons Inc.; 2006.
29. Van Winterfeldt D, Edwards W. Decision analysis & behavioral research. Cambridge: Cambridge University Press; 1986.
30. Zeleny M. Multiple criteria decision making. New York: McGraw-Hill; 1982.
31. Bayesian network. Available at: http://en.wikipedia.org/wiki/Bayesian_network; accessed July 13, 2006.
32. Decision conferencing. Available at: http://www.lse.ac.uk/collections/decision Conferencing/; accessed July 13, 2006.
33. Influence diagram. Available at: http://en.wikipedia.org/wiki/Influence_diagram; accessed July 13, 2006.
34. Port of New Orleans recovery message. Available at: http://www.portno.com/message.htm; accessed July 13, 2006.

CHAPTER 14

A MULTI-CRITERIA DECISION ANALYSIS APPROACH FOR PRIORITIZATION OF PERFORMANCE METRICS

U.S. Government Performance And Response Act And Oil Spill Response

I. LINKOV, F.K. SATTERSTROM, A. TKACHUK, A. LEVCHENKO

Intertox, Inc.
83 Winchester Street
Brookline, MA 02446 USA

T.P. SEAGER

Social and Environmental Research Institute
278 Main Street, Room 404
Greenfield, MA 01301 USA

J.R. FIGUEIRA, T. TERVONEN

CEG-IST (Center for Management Studies,
Instituto Superior Tecnico)
Technical University of Lisbon
Tagus Park, Av. Cavaco Silva
2790-990 Porto Salvo, Portugal

Abstract: The Government Performance and Results Act (GPRA) of 1993 requires federal agencies to establish objective, quantifiable goals describing the outcomes of agency programs. The purposes of the Act are fairly straightforward: to improve confidence, performance, accountability, feedback, decision making and management in federal agencies. However, implementation of the Act in different agencies has been complicated by several factors including a lack of expertise in strategic planning, difficulty reconciling multiple missions or divergent perspectives, a lack of structured approaches to engaging relevant stakeholder groups, and the presence of multiple confounding factors that obscure the relationship between agency efforts and outcomes. Currently, there is no reliable, structured approach to establishing GPRA metrics that can be transferred from one agency to the next. Successful case studies have largely

I. Linkov et al. (eds.), Managing Critical Infrastructure Risks, 261–298.
© 2007 *Springer*.

resulted from ad hoc efforts within agencies with different degrees of expertise related to performance measurement, public or stakeholder participation, and decision support. Consequently, the requirements of the Act remain a source of concern to many federal agency managers. This paper hypothesizes that many of the challenges posed by the Act are characteristic of the types of problems that multi-criteria decision analysis (MCDA) is appropriate for and proposes an MCDA framework for establishing a suite of GPRA metrics for gauging the effectiveness of oil spill response efforts. The key advantage of this framework is the visualization and quantification of the uncertainties involved in planning or decision-making processes and the potential to model multiple stakeholder responses. Moreover, we expect an MCDA approach to be generalizable to multiple agencies engaged in environmental protection missions and/or crisis response.

1. Introduction

Government agencies can rarely be characterized by a single, focused mission. Most government agencies work towards multiple objectives and serve many "customers" with divergent goals. Alternatively, a single purpose (e.g., protecting human health) may be served by multiple agencies. Consequently, agencies are often required to partner with each other and nongovernmental organizations to accomplish goals that seem to always be increasing in complexity. No single outcomes metric can reliably capture the efficacy of any single agency [10]. Nonetheless, the Government Performance and Results Act (GPRA) of 1993 requires federal agencies to set objective, measurable performance goals for the purpose of assessing the effectiveness of individual agency programs. Ostensibly, program budgets are revised in accordance with their perceived effectiveness in relation to the Act. A more efficient allocation of government resources is expected to result. However, in reality, implementing both the spirit and letter of the Act has been complicated by a surfeit of challenges related to strategic planning, performance measurement, stakeholder engagement, and disentangling the relationship between individual agency effectiveness and measurable program outcomes.

In retrospect, passage of the Act may seem like the inevitable intersection of two important sociopolitical trends of the early 1990s: the increasing adoption of measurable performance management systems in industry and the Clinton-Gore administration's "reinventing government" reform initiative [4]. In industry during the early 1990s, there was an increasing movement towards adoption of standardized systems for managing quantitative information (as exemplified by International Standards Organization specifications for quality and environmental management) and application of quantitative management systems to foster continuous quality improvements (such as the Six Sigma approach [7]). There is no doubt that the framers of the Act envisioned making government more businesslike by increasing responsiveness to "customers," improving efficiency, reducing bureaucratic procedures (i.e., "red tape"), and fostering a culture of accountability. However, unlike business units that receive continuous feedback and capital from financial markets, government agencies operate in an environment comparatively isolated from market forces (relative to political forces). While connecting budgets and performance assessment may seem consistent with the goal of building a businesslike culture in government, in reality government agencies operate in a different milieu. Consequently, federal agencies must develop unique approaches to both information management systems and performance assessments.

In this paper, we argue that many of the challenges posed by the GPRA are amenable to multi-criteria decision analytic approaches such as those that have been adopted in some instances for social problems such as environmental decision making (Table 1). Multi-criteria decision analysis refers to a group of methods used to impart structure to the decision-making process. Generally, these decision analysis methods consist of four steps:

1. Creating a set of criteria relevant to the decision at hand, for use in evaluating the decision alternatives.
2. Weighting the relative importance of the criteria.
3. Scoring how well each alternative performs on each criterion.
4. Combining scores across criteria to produce an aggregate score for each alternative or pairwise comparisons of alternatives.

The goal of the process is often to select the single best alternative, but ranking alternatives is very useful. Most multi-criteria decision analysis (MCDA) methodologies share similar steps 1 and 3, but diverge in their processes for steps 2 and 4 [22]. A detailed analysis of the theoretical foundations of different MCDA methods and their comparative strengths and weaknesses is presented in [1].

2. Performance Metrics for Oil Spill Response

Measuring the effectiveness of an oil spill response is extremely challenging. Until very recently there has been little research on what constitutes "good" environmental performance metrics within the context of environmental crisis, different types of metrics, how these could apply to oil spill response, and how

TABLE 1. Comparison of GPRA challenges and MCDA capabilities

GPRA	MCDA
Requires expertise in strategic planning	Facilitates structured examination of objectives, alternatives
Must balance multiple missions and perspectives	Criteria result from multiple objectives and may be weighed differently by different groups
Must engage stakeholders and be responsive to feedback	Structures and facilitates stakeholder value elicitation and non-expert knowledge
Requires expertise in performance measurement	Accommodates multiple sources and types of information including qualitative and semi-quantitative

they might relate to stakeholder or public attitudes, expectations, or objectives. Incident-specific strategies must be identified early and on a case-by-case basis that usually must be implemented by an ad hoc association of many organizations that may be unfamiliar with working with one another [5]. To further complicate matters, little guidance is available to responders on how to incorporate stakeholder or public views into the initial assessment of priorities. Consequently, methods for setting objectives, tracking progress, and communicating or determining success depend largely on the experience of the on-scene coordinator and the level of interaction with state, local, or other nonfederal government groups outside the command structure, including the media. Even when the response is closely coordinated among agencies and planning documents are scrupulously adhered to, public perception can be that the response has failed—partly because it is not apparent what normative standards of success should be applied or how the measures of success employed by decision makers will be interpreted by the public or intermediaries (such as journalists or nongovernmental organizations [6, 2]). In short, there currently is no comprehensive consensus approach to assessing the success of response efforts [18].

2.1. DIFFICULTIES IN PERFORMANCE MEASURES

Oil spill response, like many emergency or crisis response situations, displays many characteristics of a government function that is extremely difficult to assess with respect to the GPRA:

- Every oil spill is unique. Although planning efforts are typically intense with regard to pre-positioning of equipment, establishing communication and leadership protocols, and personnel training, the response *tactics* are virtually impossible to plan ahead of time. Optimal alternatives are highly dependent upon the circumstances of individual spills.

Therefore, compliance with the GPRA—in which strategic planning is an essential aspect—is complicated by the inability to establish a universal vision of how to achieve a successful response.

- Even with generalizable goals for oil spill response in place, performance measurement requires a benchmark to determine the sensitivity of outcomes to agency efforts. In the case of oil spills (and other crisis response situations), it may be impossible to say what the level of measurable outcomes would be under alternative response scenarios. The unique nature of the crisis precludes comparative assessments to control or "normal" circumstances. Moreover, the time-sensitive nature of the crisis typically prohibits dynamic modeling of alternative scenarios in real time. Consequently, decision outcomes must be judged without the benefit of context.
- Oil spills inevitably engage multiple stakeholder groups including industry (e.g., fishing, tourism), nongovernmental organizations, local government agencies, journalists, and the greater public. In many cases, the goals of these groups may be at odds, complicating the process of incorporating key stakeholder concerns into GPRA measures.
- Effective oil spill response requires close coordination between multiple public and private agencies (such as the responsible party and response contractors). The perceived success or failure of a response is a function of the collective efforts of the wide-ranging ad hoc administrative structure. Disentangling the contribution of each of these agents to the eventual outcome may be impossible, if not explicitly counterproductive to the extent that organizations may fail to cooperate effectively if they become preoccupied with their own performance assessment at the expense of overall system effectiveness.

2.2. REVIEW OF PERFORMANCE METRICS USED FOR OIL SPILLS

To examine performance metrics in oil spill response, we interviewed key stakeholders to study performance metrics used to assess two recent oil spill responses: the *Bouchard-120* spill response that began on April 27, 2003, as the tugboat *Evening Tide* ran aground and released No. 6 home heating fuel oil just at the entrance to Buzzards Bay, Massachusetts; and the Chalk Point spill response that began on April 7, 2000, when an intrastate pipeline that transports oil to the Potomac Electric Power Company's (PEPCO) Chalk Point electrical generating facility in Prince George's County released No. 2 and No. 6 home heating fuel oil into Swanson Creek and subsequently into the Patuxent River. Those interviewed represented a diverse group involved in the spill response, including federal, state, and local government officials, NGO representatives, and local community members. The federal on-scene coordinator and state on-scene coordinators from each response were included in the interview samples.

We found that the stakeholders were collectively able to generate a large number of performance metrics. A sampling is listed in Table 2. Each metric can be characterized by where in the organization the metrics applies, how (and how well) the information is expressed, and what type of information is collected (Table 3) according to the typology of environmental performance metrics described in previous work [18].

The HOW columns in Table 3 characterize the mathematical form and quality of the metric. Mathematically, all metrics could be classified as either quantitative (e.g., cardinal measurement), semi-quantitative, or qualitative. In the GPRA, a clear preference is expressed for quantitative measures—although they may not always be available. Regarding quality, different metrics may have multiple attributes to different degrees. Therefore, a judgment regarding the quality of any metric may depend upon the relative importance of each attribute such as cost or verifiability to the decision maker.

The WHERE columns describe the relationship of a metric to both organizational structure and sequence events in crisis response. A different suite of metrics may be applicable at the strategic level of thinking than at

TABLE 2. Example oil spill response metrics by type of information measured

Economic	Thermo-dynamic	Environmental	Ecological	Human Health	Socio-political
Clean up costs.	Volume of oil spilled, recovered, destroyed, or contained.	Chemical concentration and toxicity.	Wildlife deaths or populations, fecundity, and recovery rates.	Threatened population	Newspaper column inches, minutes TV coverage, web hits.
Property and ecosystem damage.		Habitat suitability; e.g., acres shellfish bed.	Biodiversity.	Quality-adjusted-life-years (QALYS).	Volunteerism.
Ecosystem damages or lost services.	Slick area and thickness.	Length of oiled shoreline.	Catch sizes.	Disability-adjusted-life-years (DALYS).	Public meeting attendance.
Lost marginal profits.	Mass of clean up wastes generated.	Degradation rates.	Plantings, seedings.	Life expectancy.	Critical sites protected.
Volunteer opportunity costs.	Volume cleaning agent deployed.	Residual risk.	Habitat suitability.	Injuries.	Historic sites protected.

TABLE 3. Characterization of performance metrics

How		Where		What	
Mathematical	Quality	Decision Level	Causal Chain	Type	
quantitative	verifiability	strategic	inputs	planning	economic
semi-quantitative	cost	tactical	processes	response	environmental
qualitative	communicability	operational	outputs	recovery	ecological
	sensitivity		outcomes	restoration	socio-political
	credibility				human health
	scalability				thermo-dynamic
	relevancy				

the tactical or operational. Similarly, different metrics are applicable for an accounting of resources, processes, or outcomes depending upon whether the concern is spill response, systems recovery, or restoration. In some cases, outcome metrics for response (e.g., rescued birds) may be viewed as input metrics for recovery (e.g., viable breeding population) or restoration. For example, an overarching strategy may be to reduce the severity of oil spills (e.g., as measured by wildlife deaths). One tactical approach may be to contain and remove slicks. In pre-spill planning, a resource metric might be to measure capital equipment expenditures for purchase and pre-positioning of additional equipment. In the event of an oil spill, the on-scene coordinator might track tactical measures such as the time required to deploy the pre-positioned equipment during the response. At the operational level, the effectiveness of the deployed equipment must be tracked to ensure success of the overall strategy. The importance of tracking resource and process (or output) measures must be emphasized in cases where outcomes are significantly delayed or disconnected in time from the actual decisions. In many instances, mid-course corrections must be made before final outcomes can be tallied.

It has often been written that what gets measured will be managed. Therefore, it is essential to understand the relationship of any metric to the multiple objectives of the spill response. Table 3 proposes a fairly comprehensive—albeit far from exhaustive—list of different dimensions by which oil spill response success might be gauged. However, where two different dimensions are incompatible with one another (e.g., additional expense is required to remove additional oil from the environment, although the ecological benefits may be dubious) disagreement may exist about which measures are most important.

The ecological dimension refers to biological aspects of living organisms such as populations, health, or incidence of injury or disease. Presumably, human populations could be lumped together with wildlife in this category. However, the preeminence of human health protection in oil spill response likely justifies a separate category specifically tracking human health measures, such as worker injuries or exposures.

The environmental dimension refers to the chemical or physical characteristics of the impacted area, such as oil or toxic chemical concentrations. These are distinguished from ecological measures because the relationship between environmental conditions and biological endpoints may be complex, indirect, or difficult to reveal. Moreover, environmental endpoints (such as presence of oil in shoreline soils) that may have limited ecological relevance may nonetheless be held in high regard by some stakeholder groups.

The economic dimension may itself have several important aspects, such as the direct costs to the responsible party and the indirect costs to volunteers, impacted industries, or even potential beneficiaries of the spill expenses such as contractors hired to participate in response.

The thermodynamic dimension characterizes all the mass and/or energy measures that may be relevant process outputs (or resource inputs). For example, the mass of oiled litter recovered from beaches, the volume of oil recovered from shipboard, or the total gallons (barrels) of oil spilled into the environment.

The sociopolitical dimension describes the quality of human relationships and how they are impacted by the spill. For example, the incidence of litigation may be one symptom of a deterioration of trust. Criminal proceedings may be an indication of an ethically bankrupt culture. However, high incidence of volunteerism may be indicative of the quality or effectiveness of community outreach programs.

Ideally, an agency will populate the entire WHERE and WHAT dimensions of the typology with high-quality metrics that are applicable to every level of decision making within the organization. For example, one important strategic goal listed in *A Guide for Spill Response Planning in Marine Environments* is:

> …minimize adverse environmental and socioeconomic effects by coordinating all containment and removal activities to carry out a timely, effective response [17].

This single statement cuts across several dimensions of the environmental performance metrics typology. First, the notion of environmental "effects" is a qualitative, outcomes-based, environmental and ecological metric that should inform strategic decision making—while the notion of socioeconomic effects is also qualitative, outcomes-based, and strategic. It seemingly suggests aggregating the economic and sociopolitical information that are treated separately in the typology of Table 3. Second, specific suggestions are provided to tactical decision makers thinking about achieving the strategic goal, namely containment and removal. In this regard, the emphasis is on tactical processes, rather than strategic outcomes and the type of information to be gathered would presumably be thermodynamic (i.e., volume of oil contained or removed). Lastly, criteria for judging the spill response are suggested: specifically, the response should be "timely" and "effective." However, it should also be apparent that this statement alone is insufficiently detailed to allow responders to gauge the effectiveness of their efforts. For example, how should containment operational effectiveness bye measured? How would timeliness be measured quantitatively? A more comprehensive elicitation process such as interviews, literature review, and consultation with experts is called for. This exercise will allow identification of gaps that may concern managers or stakeholder groups. For example, they may discover a paucity of strategic or quantitative metrics, which may cause them to return to the step of metrics elicitation or invent new metrics to fill the gaps they consider important.

Following elicitation, the agency must reorganize the information in multiple directions to discover whether a suitable cross-section of different types of feasible, high-quality metrics are available for each organizational objective (strategic, tactical, operational) or at each step in the agency's chain of influence (inputs, process outputs, and outcomes) or for each type (economic, environmental, ecological, sociopolitical, human health, and thermodynamic). Once again, significant deficiencies may be corrected by further elicitation or invention.

In any case, it is unlikely that all stakeholders (or agency managers) agree on which metrics (and their quality) should be employed to determine the relative success or the quality of any response. Overall, a four-step process that precedes MCDA is called for:

1. First, metrics must be *elicited* from the salient stakeholder groups.
2. Then, they must be *characterized* with regard to the typology in Table 3.
3. Where important gaps are identified, a *return to elicitation* may be called for—or perhaps entirely new metrics must be devised.
4. Lastly, the quality of the metrics must be must be *assessed.*

Only then should value-based weights be assigned to any set of sub-characteristics to allow aggregation to a higher-level assessment of which metrics should be adopted for any given situation. This prioritization framework could be easily coded in one of the MCDA platforms. Even then, the disaggregated information should remain available to agency managers.

5. Finally, decision makers should *deliberate* about the results of the overall process to select a final suite of metrics most relevant for a particular region or problem.

2.3. USE OF MCDA TO RESOLVE DIFFICULTIES

Multi-criteria decision support processes can overcome many of the challenges of performance measurement faced by federal agencies. The typical approach to an environmental MCDA involves identifying feasible alternatives and the criteria by which they should be judged and assessing the performance of each alternative relative to the salient criteria. While alternatives may be generated by experts, stakeholders, and/or public groups, the decision criteria and their relative importance may justifiably be the purview of affected parties (such as stakeholder groups), who may or may not also be experts. The performance assessments are typically performed by experts. For example, in considering the problem of managing contaminated sediments, stakeholder groups may emphasize the importance of environmental quality, preservation of ecological habitat, or economic development. All of the alternatives generated by expert and stakeholder groups must then be assessed relative to measurable criteria that capture each of these performance criteria—although different groups may value each criterion differently. The results are called a *performance table* that is subsequently analyzed with different mathematical approaches including multi-attribute utility theory, outranking, pairwise comparison, or others. The advantage of MCDA is that alternatives can be ranked or prioritized for multiple decision criteria, tradeoffs can be elucidated, different types of information including semi-quantitative or uncertainty information can be handled in a structured way, and conflicts or opportunities for compromise can be observed that may not otherwise have been discovered [13, 3]. The principal disadvantage of MCDA approaches is that they can be time-intensive, especially when multiple stakeholder perspectives are important. In crisis response, there may not be time to engage in structured decision making [21, 9]. Therefore, the typical MCDA approach to alternatives assessment is not helpful.

In the case of crisis response (such as oil spills), existing MCDA processes are at a distinct disadvantage. However, we propose a unique approach for the purposes of the GPRA. In this case, the *metrics* are the alternatives. Usually, metrics are associated with assessment criteria and relate to the underlying objectives or goals of the organization. However, the GPRA requires agencies to select a limited number of metrics from a theoretically infinite universe of potential outcomes metrics. Not all metrics would be cost-effective for the agency to track, nor can any single metric perfectly fulfill the requirements of the GPRA. Therefore, agencies need a process for generating a set of potential metrics, vetting these with key internal and external stakeholder groups, and prioritizing among them to establish a concise suite of metrics that capture agency performance. As in a typical environmental MCDA, some agency customers may value certain metrics within the suite more highly than others. (Some may even place zero weighting on certain metrics.) Consequently, despite a common set of assessment metrics, it is possible that agency performance could be viewed positively by one group and negatively by others—a situation that may be all too familiar to government managers. The critical difference between the process we propose for GPRA and a typical MCDA is that the decision alternatives are *which metrics to use*. The criteria that represent the other half of the performance table are characteristics that agency and key stakeholder feel the selected metrics should have. Thus, the performance table is a guide to determining which metrics are better than others for any particular program.

3. Decision Analysis Example

As an illustrative example of MCDA application, consider the problem of prioritizing performance metrics for oil spill response according to specified management objectives. As a hypothetical example, we consider 14 alternative metrics in bold type in Table 2. These metrics represent different phases of the response (response, recovery, restoration) and the different types of metrics (economic, thermodynamic, environmental, human health, ecological, sociopolitical). Many of these performance metrics were mentioned by stakeholders in the case studies we examined [20].

3.1. SMAA METHODOLOGY

One of the most time-intensive, problematic aspects of MCDA for group decision making is the elicitation of weights that represent the values of each decision maker. The larger the group and the greater the number of relevant criteria (or objectives), the more burdensome the problem of value elicitation becomes. Moreover, the process is complicated by the fact that decision makers themselves may be uncertain about weights, preferences may evolve or change throughout the decision-making process, or the weights elicited may be unreliable in the sense that they depend upon the method of elicitation. Moreover, in some case the results of an MCDA may be highly sensitive to treatment of uncertainty in the weights. However, stochastic multiobjective acceptability analysis (SMAA) circumvents the difficulties of value elicitation and expedites analysis of uncertainty by using Monte Carlo simulation to explore a wide range of the weight spaces. The results can provide decision makers with a probabilistic characterization of preferred alternatives even prior to value elicitation. The results are especially helpful for understanding multiple-decision-maker problems by identifying when some alternatives are likely over a wide range of potential weighting schemes [19,11].

3.2. CRITERIA

For the purposes of illustration, 11 of the criteria from Table 3 have been chosen for an example in the use of SMAA and are depicted in the rows of Figure 1, which is an input screen captured from a software program known as SMAA-III [19,12].

The criteria are:

- Cost (e.g., of obtaining the data required to track a specific metric)
- Ecological (Rel_Ec)
- Environmental (Rel_Env)
- Human health (Rel_HH)
- Sociopolitical (Rel_SP)
- Verifiability (ver)
- Communicability (com)
- Causality (causal)
- Credibility (cred)
- Scalability (scal)
- Sensitivity (sens)

The analyst must specify different aspects of these criteria, such as whether ascending or descending ordering is preferred (i.e., more or less of the characteristic), the minimum precision with which two estimates on the scale of this criteria may be distinguished (the indifference threshold) and the difference between two estimates that would result in a complete preference for one estimate over another (the preference threshold; between indifference and preference, a decision maker may be uncertain about the extent to which one number is preferred over another). The SMAA-III software can work with different scales (including poorly defined ordinal scales), does not require precise preference information and is particularly adequate for situations where "big" uncertainties are present; these uncertainties referred to imperfect knowledge of basic data and technical and preference parameters. (In addition, SMAA-III software allows the definition of other model specifications that are not used in this example).

3.3. ASSESSMENT

Assessment of the performance of different metrics relative to the 11 listed criteria in general requires consultation with decision makers and technical experts. For example, the total volume of oil recovered from the environment may be very difficult to verify (because the liquids removed will be an oil-water mixture), or it may be expensive to determine. However, this measure may also be easily communicated and understood by the public. In any case, a determination must be made about the performance of a metric such as "total volume of oil recovered" relative to each criterion. It may be helpful to elicit performance estimates from a wide range of experts or stakeholders; these estimates need not agree. They may be expressed as a probabilistic distribution, or alternative analysis can be conducted for different performance assessments and the results compared. The quantities used in this example have been furnished by the authors for illustrative purposes only. Fourteen candidate metrics listed in bold in Table 2 are assessed:

- Threatened human populations (TP)
- Residual risk (RR)
- Habitat suitability index (HSI)
- Habitat suitability (HS)
- Clean up costs (CLC)
- Property damage (PD)
- Volume of oil (VO)
- Mass of waste (WM)
- Length oiled shoreline (LOS)
- Chemical concentration (CHC)
- Injuries (IN)
- Media attention (MA)
- Critical sites (CUS)
- Public meeting attendance (PUB)

Cost is measured in US dollars required to collect information for quantification. For example, we assumed a cost of $50,000 to track changes in the TP. For those criteria with the "Rel" prefix (all from the "What" column of Table 2) assessments of relevance to the criteria are gauged on a scale of 1-10 (10 being highly relevant and 1 having little relevance). All the candidate measures are recognized as being relevant to more than one of the "what" criteria, which is to say that these criteria are not completely independent. Metrics performance for other criteria (e.g., credibility, communicability) is assessed as merely High, Medium, and Low (represented by 3, 2, and 1 in Figure 2). For example, the TP metric is scored high on all criteria because it is a robust measure that is easy to communicate. Even though the MA metric is easy to communicate and somewhat sensitive to public perception, here it is listed as having low credibility and assessed low on other factors. The assessments are summarized in Figure 2. It bears repeating that these assessments

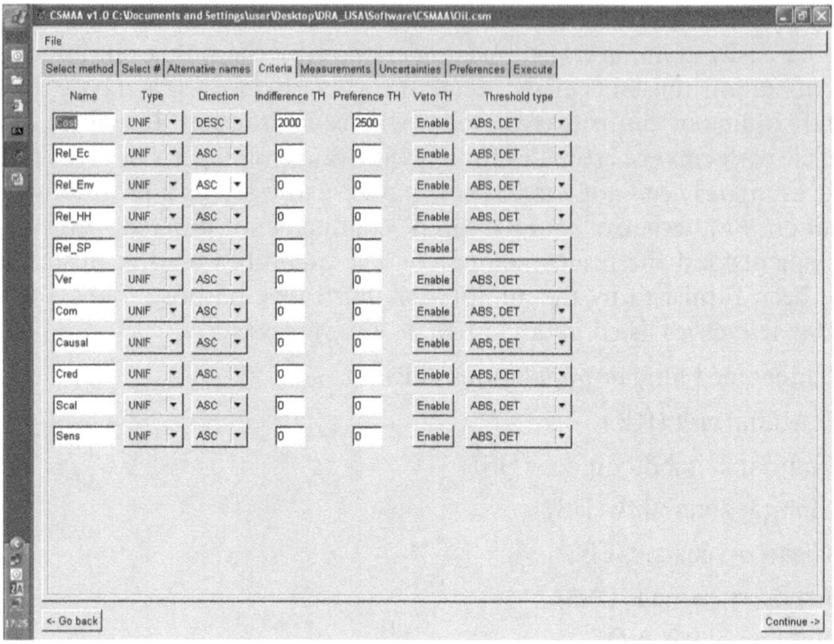

Figure 1. Input criteria types screen from SMAA-III software.

are purely for the mathematical convenience of the authors. A more robust assessment would entail a wide elicitation of expert and stakeholder views.

Unlike some other MCDA tools, SMAA-III also allows defining the indifference and preference thresholds. Indifference thresholds could be used by decision makers to acknowledge that a small difference in assessments could be interpreted as indifference between two alternatives. In our example, indifference and preference thresholds are defined for cost criteria only. The indifference threshold of $2,000 defined in our example means that two alternatives with price difference of $2,000 are considered as indifferent (i.e., this amount of money is too small to worry about in spill response). However, differences of greater than $2,500 are considered to be important, which is to say that a decision maker would clearly prefer a metric that was at least $2,500 less costly to track. Between $2,000 and $2,500 savings, the decision maker may or may not consider the cost difference important. Therefore, the preference is considered to be incomplete.

4. Preference Analysis

SMAA-III is based upon an outranking approach to preference determination rather than a utility function approach [18,15]. SMAA-III offers several advantages compared with other MCDA approaches:

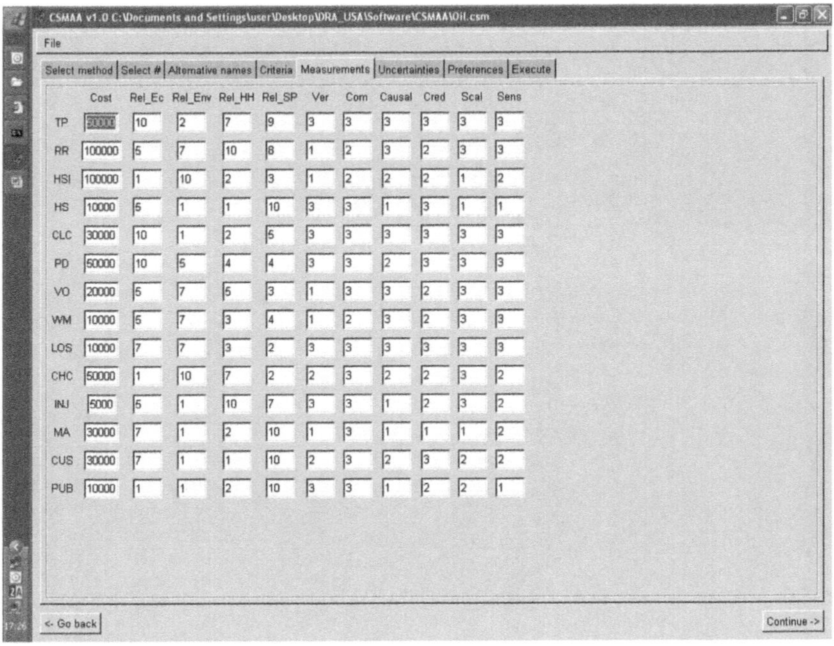

Figure 2. Hypothetical assessment of candidate metrics relative to representative criteria.

- As depicted in Figure 1, SMAA-III works with the original and natural scales for different assessment criteria, including cardinal (e.g., cost), semi-quantitative, or qualitative (e.g., High, Medium, Low). There is no need to translate assessments into a single-dimensional utility or value function.
- Low time requirements and cognitive efforts from decision makers.
- Uncertainty can be incorporated in every aspect, including assessments and weightings.

In our example, we assumed that the decision maker cannot assign weights to the criteria a priori but believes that all of them are important and none dominates. This ignorance about the exact criteria importance is modeled by using imprecise weights (Figure 3). In an actual case, decision makers may assign their own bounds and different analyses may be conducted for individual bounds. The results may be important for identifying opportunities for decision makers with different views to compromise on alternative(s) that are nonetheless highly ranked by all.

4.1. RANK ACCEPTABILITY ANALYSIS

The rank acceptability index measures the share of feasible weights that grant each metric specific rank in the order of importance by simultaneously

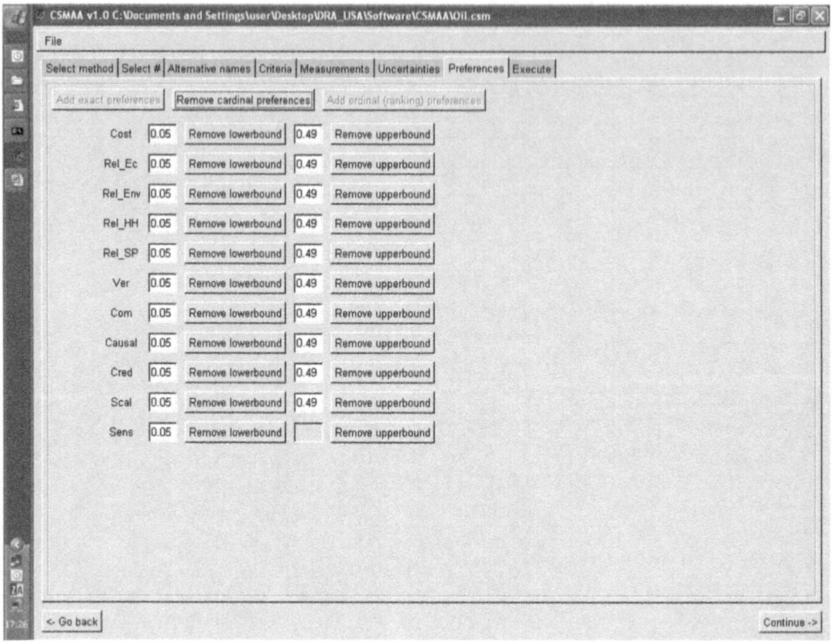

Figure 3. The range of weights tested for each assessment criterion range from a lower bound of 5% to an upper bound of 49%.

taking into account uncertainty in all parameters and criterion evaluations. It represents the share of all feasible parameter combinations that make the alternative acceptable for a particular rank, and it is most conveniently expressed as a percentage. The most acceptable ("best") metrics are those with high acceptabilities for the best ranks. These are the metrics that are most likely to be valued by decision makers with a random assignment of criteria weights between the lower and upper bounds. That is, they may be the alternatives most likely to result in consensus. The results of the rank acceptability analysis for our case are presented at Figure 4. The left-most box represents the likelihood of the alternative metric being ranked as the most important, whereas the right-most indicates the likelihood of being ranked last. The top metrics include TP, LOS, CLC, and PD.

4.2. PAIRWISE ALTERNATIVE COMPARISON

Alternatively, it is possible to compare the likelihood that one metric may be preferred to one other metric in a pairwise comparison rather than an overall ranking. Figure 5 is a pairwise comparison table. The alternative metrics are listed in the row titles at left. Although the column headings are omitted, the

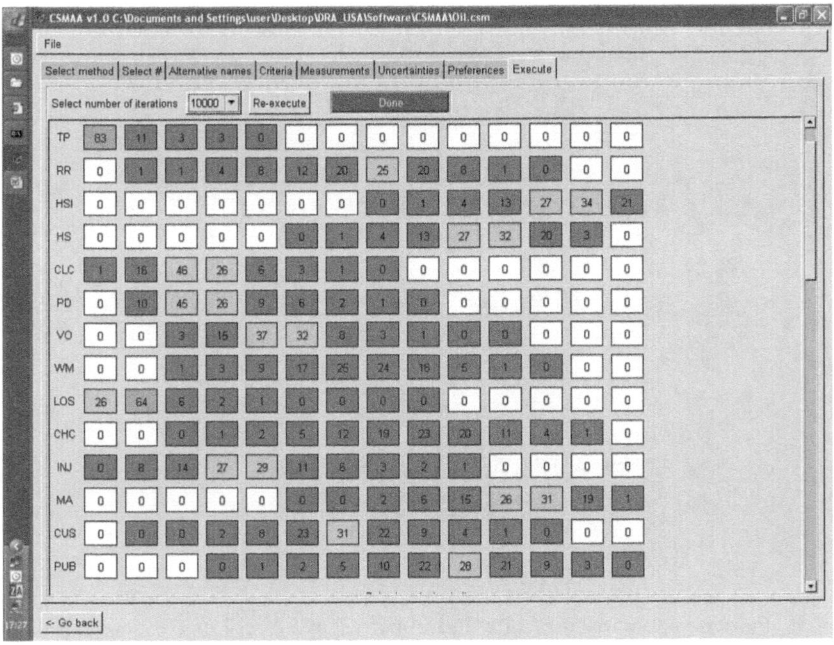

Figure 4. Rank acceptability analysis results show the likelihood of each alternative metric being ranked by decision makers as most preferred (left-hand box) or least preferred (right-hand box).

order of the alternatives is exactly the same. To determine the likelihood that a decision maker might prefer TP to LOS, it is necessary to follow the TP row to the right nine boxes, where we see the result is 74. (LOS is listed in the ninth row and therefore corresponds to the ninth column).

4.3. SENSITIVITY TO WEIGHTING BOUNDS

In some cases, decision makers will express a stronger preference for some assessment criteria than for others (compared with a randomly distributed, equally likely distribution of weights). In Figure 6 below we assign greater importance to cost by raising the lower bound on this weighting to 25% (without changing the upper bound of 49% or the bounds of any other criteria). For comparison, Figure 7 depicts the results of the rank acceptability analysis for over weighting of environmental considerations (Rel_Env) in the same way. It is interesting to note that in both cases, the results show that LOS is now most likely the preferred metric, compared with TP. While LOS is not the cheapest metric and it is not the most relevant metric to ecosystem health, its relative low cost and good performance on all criteria make it

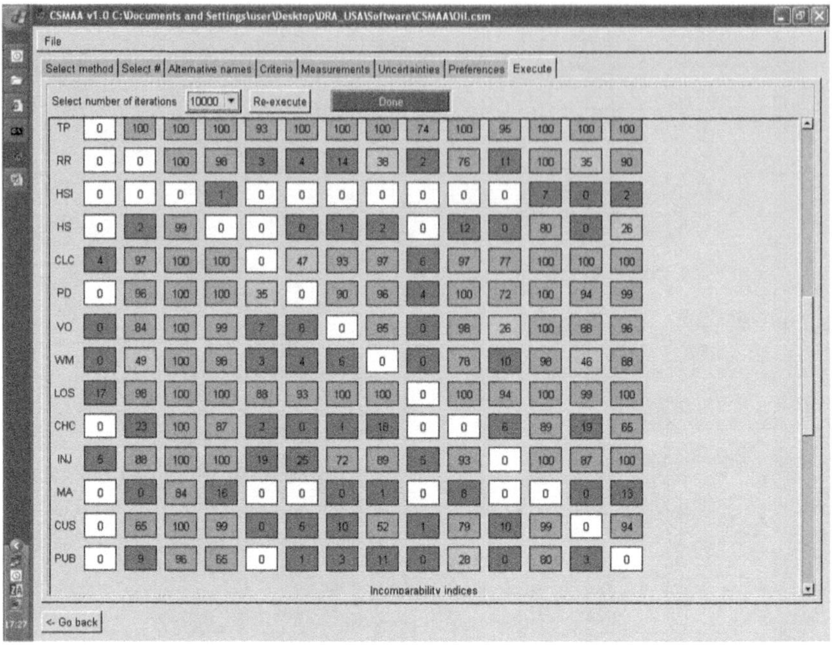

Figure 5. Pairwise comparison of alternative metrics in a 14x14 matrix showing the likelihood of the row metric dominating the column metric. Note that column metrics are presented in the same order as the rows, although the headings are omitted.

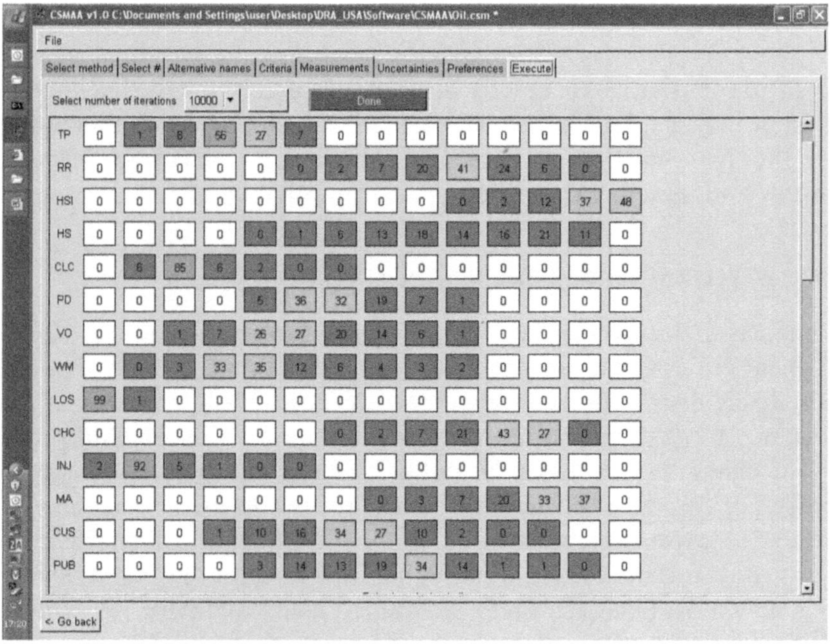

Figure 6. Rank acceptability analysis for potential overweighting of cost.

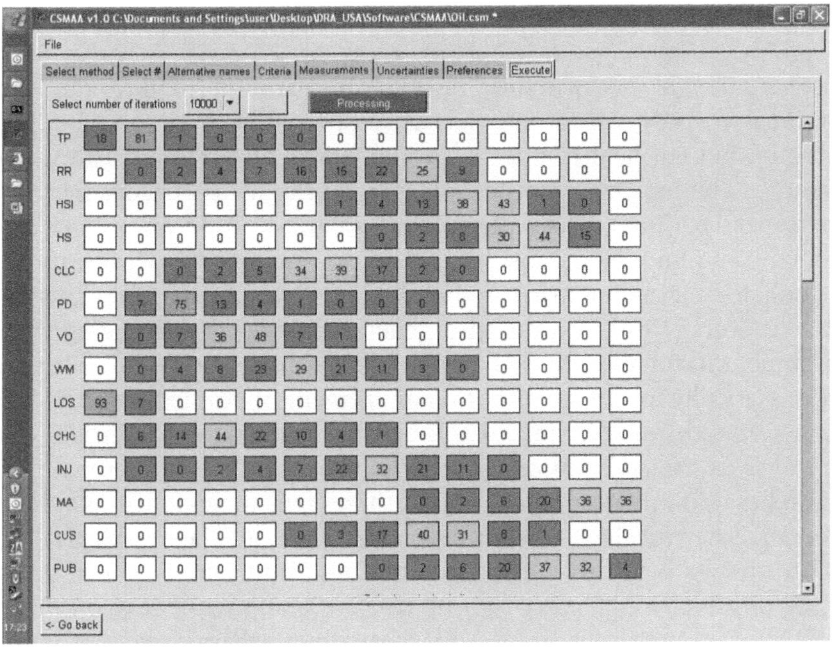

Figure 7. Rank acceptability analysis for potential overweighting of environmental relevance.

acceptable for both budget-conscious and environmentally-conscious stakeholder communities when these priorities are given greater weight.

5. Discussion

The example presented here is a simplified, hypothetical case to illustrate the utility of a SMAA-III approach to analysis of a suite of candidate metrics for gauging the success of oil spill response. Selection and ranking of performance metrics for specific region/situation requires a more detailed elicitation of candidate metrics, criteria, and assessments. Moreover, the number of performance metrics that may be relevant in an initial elicitation, such as the *Bouchard* and Chalk Point cases, is well above 100 (see Appendix). Prioritization and ranking of so many performance metrics may require an initial screening. Nevertheless, when populating the typology of metrics (e.g., Table 3), significant gaps may be found. The critical contribution of SMAA-III in this case is not that it facilitates gathering of information or generating the data required to meet the GPRA requirements, but that it provides a structured approach to *analysis* of the data.

6. Conclusion

Effective oil spill performance metric prioritization requires an explicit structure for jointly considering the environmental, ecological, technological, economic, human health, and sociopolitical factors relevant to evaluating alternatives and making a decision. Integrating this heterogeneous information with respect to human aspirations and technical applications demands a systematic and understandable framework to organize the people, processes, and tools for making a structured and defensible decision. Previous papers by the authors [14, 8] introduced a general framework for management of environmental challenges based on MCDA. We have also tailored this framework specifically for application to management of contaminated sediments [15]. In general, we recommend a four-stage process of *elicitation, characterization, assessment,* and analytic *deliberation*. The work presented in this paper relies upon the results of the first three steps to help facilitate the final, deliberative step in which a decision is actually made. The overall goal of any MCDA process is not to make a decision, but rather to *support* a decision-making process. In the case of group decision-making processes, one of the strengths of the approach presented is the ability to characterize the sensitivity of the result to different weightings, assessments, or perspectives—especially in recognition of the fact that these perspectives may change. As managers gain experience with new spills or policies evolve, it is possible to revisit the process by considering new metrics, revising assessments, or weighting boundaries and revisiting the suite of metrics considered most applicable to the next spill. An adaptive management process is called for. (See Linkov et al. [16] for an example charting adaptive MCDA management processes.)

The result of the entire process is a comprehensive, structured process for selecting the GPRA metrics for in any given situation, and vetting these metrics with stakeholder groups in a way that incorporates their value judgments as well as scientific modeling and risk analysis. This process could be of great benefit to management of oil spill response, where there is currently no structured approach for making justifiable and transparent determinations about the success of any particular response. In particular, an MCDA framework allows visualization and quantification of tradeoffs. The proposed framework can also be used to prioritize research and information gathering activities and thus can be useful for value of information analysis.

8. Acknowledgements

The authors express their gratitude to Dr. Seth Tuler and Rebecca Kay for comments and suggestions. This study was funded by NOAA through the

7. Appendix: Objectives and Metrics Elicited for *Bouchard* and Chalk Point Oil Spill Response

7.1. BOUCHARD-120

Type of Metric	End-Point	Process	Resource
Protect worker and public health and safety	Number of people killed or injured	NA	Number of IRAC team members
	Number of mishaps during hours worked		OSHA/HAZMAT trained
	No re-oiling or residual oil causing chronic toxicity to something		
	Number of estuaries "shut off" (i.e., areas prevented from oiling from via booming)		
	Number of inlets that were able to boom off		
	Number of miles of shoreline impacted	Did getting required permits delay response action?	
	Amount of oil or globules on shore	Booming deployed around critical habitat (Roseate tern nesting habitat, salt marshes)?	
	Number of areas where oil contamination was prevented	Attempt to direct oil to "sacrificial areas"?	Number of sandbags deployed?
	Oil contamination is on sandy beach (easier to clean up than rocky intertidal zone)	Good choices made about where to deploy boom?	Number of people on cleanup crews to deal with oiled beaches?
	Shellfish beds protected	Oil being captured in open water before it hits the beach?	Amount of boom deployed

(continued)

(continued)

Type of Metric	End-Point	Process	Resource
	Number of miles that have been cleaned to an acceptable level	Proper placement of skimmers? Resources placed in the proper locations	Number of floating resources to pick up oil in open water oil?
Protect environment/ ecological systems	How many pounds of oil contaminated debris has been recovered and sent off for disposal?	How fast (# of days/ # of hours) rehabilitation efforts brought in and set up operations	Number of skimming units available and operable
	How many gallons of oil have been recovered?	Is rehabilitation center handling all live birds coming in?	Number of volunteers
	Number of birds oiled?	Speed at which able to handle birds coming in for rehabilitation?	Number of bodies to manage different aspects of response, including organizing SCATs?
	Number of birds released and survived	Amount of area covered in search and recovery	Is there a "bird searcher" on each team?
	Mortality/survival of wildlife (birds)	Volunteers able to get plugged in well?	
	Number of roseate terns lost	Time it takes to implement tasks, such as boom deployment?	
	Number of dead birds per unit search area		
	Number of dead birds		
	Percent of birds rehabilitated		
	Number of birds rehabilitated		
	Have cleanup standards for shoreline contamination been met?		
		How quickly decisions made?	
		Chain of command clear?	
		Resources placed in the proper locations	
		Are decisions correct (in hindsight)?	

		Location of command center convenient and easily accessible?	
		Are crews visiting hard hit areas every day?	
		Shoreline assessment teams record information in a unified way 1) where is oil, 2) how much, and 3) what does shoreline look like; want minimal variability	
		Organized, systematic way of recording where oil has come ashore and impacted is used	
	Modelers able to get 24 hours ahead of spill with accurate projections	Systematic, "non-political" approach used to deploy cleanup crews	
	Are pre-identified areas potentially being affected?	Presence of watchdog to see what's going on? Is there conflict or chaos in command center?	
	Did we have information to keep governor's office and other state senators and reps abreast of what was going on?	Are players familiar with each other? and reliable communication technology available (e.g., cell phones working)?	
	Understanding of whether oil is still stored offshore re-contaminating cleaned up beaches?	System established to track progress? Experts consulted for input on response strategies? Command center set up and people mobilized?	Number of teams of trained observers walk coastline and make observations of extent and coverage area of oil Number crews trained
Establish a coordinated and effective response framework	Is oil coming on shore several days later?	Plans communicated day in advance?	Type of oil

(continued)

(continued)

Type of Metric	End-Point	Process	Resource
	Accurate accounting of volume oil spilled?	Ability to effectively and timely monitor what happening in field?	Phones available and working
	Accurate accounting of amount of oil on the shore?	Are there clear protocols?	Number of hours worked
	How many areas have been cleaned as of today?	Are there clear schedules?	Number supervisors out in field
	Number of miles of shoreline impacted right now?	Time it takes to implement tasks, such as boom deployment? Is all pertinent data gathered and recorded?	
	Bad feelings among local responders toward the unified command staff?	Local responders notified quickly?	
	Level of conflict about plans and activities?	Conflicts and heated emotions among responders addressed? All goals worked on together? Attend to short, medium, and long term needs simultaneously? Follow "best response" protocol and integrated command system? Revise objectives and activities based on monitoring effort? Coordination of volunteers performed? Volunteers integrated into cleanup appropriately? Volunteers integrated into cleanup quickly? Equipment and personnel demobilized when no longer needed?	

		Cleanup of impacted areas organized to be manageable and able to monitor?	
		Access established for recovery and cleanup crews through private property?	
		Plans communicated day in advance?	
		Location of command center convenient and easily accessible?	
	Volume spilled compared to what's removed from the shoreline	Ability to effectively and timely monitor what is happening in the field?	
	What it looks like a year later	Muster all forces in town?	
	Evidence of oil on shoreline. How much oil was recovered off the water?	Local officials set up task force for how to handle spill response?	Number of people in the field? A "bird searcher" on each SCAT?
	How much oil was recovered off the shore?	Clear and reliable communication technology available (e.g., cell phones working)?	Hours worked? Supervisors out in field?
	Endpoints in plan achieved?	Do efforts correspond to tides?	Number of monitors in field to give direction and warning to clean up crews?
Implement an effective and timely response	Are desired endpoints achieved in reasonable amount of time?	Are there clear protocols?	
	Number of days until endpoints achieved?	Are there clear schedules?	Resources adequate for planned tasks?
	Are response actions having the desired effect? Members of the public voice support?	Data sheets available and sufficient?	Amount of boom deployed?
		Are cleanup crews assigned effectively to do a good job?	How much money spent? Types of skills represented on team?

(continued)

(continued)

Type of Metric	End-Point	Process	Resource
	Number of public meetings organized?	Recovery or rehabilitation of wildlife conducted?	Number of volunteers?
	Good working relationships with all parties involved?	How soon have experts been called and set up triage and rehab centers?	Number of packets of baby oil for oil removal distributed?
	Command center and command personnel located in accessible location	Time it takes to implement tasks, such as boom deployment?	
		Is entire shoreline being inspected—both oiled and non-oiled?	
		How often is beach searched, all shoreline searched?	
		Is all pertinent data gathered and recorded?	
		Basic training for volunteers on bird collection and rehabilitation conducted?	
		Equipment and personnel demobilized when no longer needed?	
	Lost rental income?		
	Lost tourism dollars?		
	Loss of summer rentals?		
	When beaches put back in use?		
	Recreational opportunities that were lost that are now back to what they were?		
Mitigate economic impacts	Acres of shellfish areas closed?		
	How many acres of closures reopened?		

	Duration of shell fish bed closures?	
	Acres of shellfish beds lost?	
	Costs of laboratory work?	
	Money spent on response?	
	Achieve termination endpoints?	
	No oil should come off to the touch	
	Shoreline back to conditions prior to spill?	
Meet legal and regulatory requirements	Evaluate response with respect to endpoints achieved	Number of days until endpoints achieved
	For sandy beach no visible oil	
	For groin (jetties between properties), riprap no sheen or no oil available when touched	
	For marshes no sheen	
	No odor of oil	
Mitigate social nuisance impacts	Presence of stained rocks that lead people to say that it's not successful because you can still see evidence of that?	
	Complaints about stained rocks?	
	Oil on the beach?	
		A timeframe for ending the cleanup is established
		Ongoing monitoring and addressing of issues post-spill?
		People have a place or someone to go to with concerns?

(continued)

(continued)

Type of Metric	End-Point	Process	Resource
		Public receives assurances that beaches will be cleaned up to the level of their expectation?	Number of pamphlets distributed to inform public of hazards
	Level of staining and other aesthetic damage?	Information fairs and public meetings set up?	Number of stakeholders involved in setting cleanup standards
	Complaints about stained rocks?	A forum is provided to public so they can hear what's going on in response action and give their feedback	Number of dispatch teams arranged to reach out to various stake-holders
Address needs and concerns of the affected public/ communities	Residual oil on shore (tar balls)?		
	Number of phone calls to complain?	Number of public meetings	
	Spirit of state regulations for public involvement was met	Level of conflict/anger	Frequency of information postings on BBP website
		People are happy	
		Number of visuals for media	Number of flyers passed out about educating public on risks
		People given examples of what was impacted and what kinds of cleanup was going on	
		Public provided the kinds of information it wants?	
		Establish and keep up to date website for public information	
		A timeframe for ending the cleanup is established	
		Ongoing monitoring and addressing of issues post-spill?	

		People have a place or someone to go to with concerns?	
		Public receives assurances that beaches will be cleaned up to the level of their expectation?	
	Members of the public voice support?	Information fairs and public meetings set up?	
	Level of staining and other aesthetic damage?	Relationships and trust with local officials developed?	Number of pamphlets distributed to inform public of hazards
	Complaints about stained rocks?	A forum is provided to public so they can hear what's going on in response action and give their feedback	Number of stakeholders involved in setting cleanup standards
Gain public support for the response	Residual oil on shore (tar balls)?	Number of public meetings	Number of dispatch teams arranged to reach out to various stakeholders
	Number of phone calls to complain?	Level of conflict/anger	
	Was there trust from the public?	People are happy	Frequency of information postings on BBP website
	Public comments of support from critics and local residents (about response effort, not about outcomes)?	Number of visuals for media	
		People given examples of what was impacted and what kinds of cleanup was going on	Number of flyers passed out about educating public on risks
	Spirit of state regulations for public involvement was met	Public provided the kinds of information it wants?	
		Quality of questions from media	

(continued)

(continued)

Type of Metric	End-Point	Process	Resource
		Establish and keep up to date website for public information Able to "stay on message" during public meetings, press conferences, etc.? Unified Command accessible for public questions and comments?	

7.2. CHALK POINT

Type of Metric	End-Point	Process	Resource
Protect worker and public health and safety	Presence of PHCs in water samples? Concentrations of oil in fish samples Level of PHCs in clam/oyster tissue		
	Number of (human) life-threatening situations Toxins in smoke plume if do in situ burning Number of oiled swans Number of fish killed Damage from cleanup actions remediated? Number of animals saved "Appropriate" wildlife saved (e.g., not exotics or "varmints") Oiling of osprey eggs		

	Number of dead ducks		
	Number of injured birds around marsh area		
	Oiling on Diamondback terrapin eggs		
	Mortality of wildlife	Accurate cataloguing and enumeration of what was found	
	How far sheen at surface extended out (miles)?		
	How long oil stayed?	Monitoring stations established	
	Presence of odors of oil	Deploy booming and double-booming in sensitive areas	Presence of booms
Protect environment and mitigate environmental impacts	How much grass was destroyed?		
	How many acres of marsh were impacted?	Immediacy of rehabilitator organiza tion's response to call for assistance	Amount of oil contain ment boom deployed
	How many oiled birds were observed?		
	Dead and stressed organisms found (rather than estimated)?	Change of helicopter flight patterns in response to requests from biolo gists to not disturb nesting birds	
	How long it takes to reach background levels/ concentrations		
	Observe water blowing over booms		
	How long does it take for area to recover from cleanup related damage?		
	Degree of change to beaches and sandbars from cleanup actions		
	Types of animals and vegetation present after spill cleanup		
	Areas that were impacted cleaned up to conditions before the spill happened		
		Did trench digging affect sites?	

(continued)

(continued)

Type of Metric	End-Point	Process	Resource
	Number of critical sites protected	Experts contacted early for input about sites potentially at risk?	Number of GIS and hard maps
Protect cultural resources	Residual presence of oil on artifacts	Command responsive to requests for protection of sites (e.g., deployment of booms)	Number of laptops
	Smell of oil		Amount of boom deployed
			Accurate spatial informa tion about location of sites and artifacts
	Soil concentrations of oil	Were less destructive response actions chosen (e.g., sorbents and booms rather than burning)	
Meet legal and regulatory requirements		All procedures followed (e.g., NIMS)	
	Devaluation of property?		
	Number of dead fish		
	Number of dead ducks and geese		
	Lost wages to watermen		
	Number of lost fishing days		
	Dollar value for days of beach closures		
	Change in earnings relative to previous years		
Mitigate economic impacts	Cost of response actions compared to value of lost resources (e.g., artifacts)	Funds spent on important and appropriate projects	Amount of money spent on studies
	Duration of recreational fishing closures	RP cleaned boats that were soiled with oil from spill	Total cost of cleanup
	Duration of beach closures		
	Presence of PHCs in water samples		

	Increase in crime rate (in southern MD due to influx of people from cleanup crews)		
	Monetary compensations		
	PEPCO made restitution to counties for employee overtime and personal equipment used in supporting response efforts		
		Number of hours to set up incident command center	
		Number of daily meetings	
		Daily reports for morning meetings held by UC?	
		Informed of meetings in advance (i.e., lead time)	
		RP is responsive	
		Key people became involved early?	
		Is Incident Commander able to reach key people, does their phone number work, did they respond?	
		Clear communication protocols?	
		Stable command posts established?	
		Time taken to re-staffing response people after contractor fired	
		Chauvinistic behavior among staff in command center?	
Establish a coordi nated and effective response framework	Clear chain of command established?	Frequency of resource and personnel changes	Number of supervisors assigned per section

(continued)

(continued)

Type of Metric	End-Point	Process	Resource
	Incident Command System used?	All federal responders notified quickly?	Enough Spanish-speaking crew supervisors?
		Pick-up and shipping schedule for waste generated by cleanup organized?	Number of radios available
		Response organized by discrete (river) segments	Availability of GIS and computers
		Frequency of meetings	
		Accurate information obtained from the wildlife surveys and SCAT teams	
		Clear standards for signoff established	
		People in field used to verify accuracy of information	
		Did Unified Command resist information that did not conform to their expectations?	
		Arguments and lack of cooperation among Unified Command	
		Clarity to all parties about stages of response effort	
		Clear understanding of rights the state trustees have as a state agency	
		Digging holes to look for oil on shoreline?	
		Quarterly checks to see if oil is present or not on beaches	

	Number of feet of shoreline impacted	Breach of water over boom?	
	Presence/absence of sheen	Protection equipment put in the right place at the right time?	
	Number of gallons of oil recovered		
	Number of days company (PEPCO) was shut down	Number of times a single order was given but still not done	Cost of response
	No oil in water		
	How long takes to reach endpoints	Accurate reporting and counting of crews in field	Number of cleanup teams and people on crews
	Oil present in sediments?		
Implement an effective and timely response	Areas that were impacted to be cleaned up to the ecological state the environment was before the spill happened	Number of newsletters published by RP	Amount of money spent by RP
		Number of public meetings held by RP	Number of laptops
		Time it takes to get response in order	Number of GIS and hard maps
	Patchiness of oil	Number/frequency of flyovers for real time aerial photography	Amount of equipment brought in Pounds of sorbent material used
	Solid sheet of heavy oil		
	Presence/absence of tarballs	Time taken to set up boom	Amount of oil contain ment boom deployed
	Is oil being contained?	Clear chain of command established?	Number of people working at one time
	Number of areas to be signed off compared to number actually signed off	Lead resources mobilized?	Number of people per SCAT

(continued)

(continued)

Type of Metric	End-Point	Process	Resource
	Number of areas that were identified as sensitive areas ordered to be boomed off but order was not followed through and area was impacted	Communication funneled out to the appropriate people?	
		Quality of contractor work (e.g., perform required tasks?	
		How quickly SCATs out in field?	
		Time it takes to resolve bad cell phone coverage, radio issues?	
		Public reimbursements for private property losses	
Address needs and concerns of the affected public/ communities	Number calls from public	Immediacy of public meetings (number days after spill occurs)	Number of hours agencies spent on public outreach (meetings)
	Number of fliers and informational packets delivered door to door	Incorrect information disclosed?	
	Amount of oil removed manually from shoreline		
	Public reimbursements for private property losses		
Gain support of public	Number calls from public	Immediacy of public meetings (number days after spill occurs)	
	Number of fliers and informational packets delivered door to door	Incorrect information disclosed?	Number of hours agencies spent on public outreach (meetings)

Coastal Response Research Center at the University of New Hampshire (NOAA Grant NA04NOS4190063, Project 05–983).

9. References

1. Belton V, Stewart T. Multiple criteria decision analysis: an integrated approach. Boston: Kluwer Academic Publishers; 2002.
2. Chess C, Johnson BB, Gibson G. Communicating about environmental indicators. J Risk Research. 2005; 8:63-75.
3. Figueira J, Greco S, Ehrgott M, editors. Multiple criteria decision analysis: state of the art surveys. New York: Springer; 2005.
4. Gore A. From red tape to results: creating a government that works better & costs less - report of the National Performance Review. Washington, DC: U.S. Government Printing Office; 1993.
5. Grabowski MR, Harrald JH, Roberts KH. Decision support and organizational forms in a high velocity environment: responses to catastrophic oil spills. In: Grabowski MR, Wallace WA, editors. Advances in expert systems for management, vol. 2. Greenwich: JAI Press; 1997.
6. Harrald JR. Preparing for success: a systems approach to pollution response. Paper presented to the Five Years After the Exxon Valdez Oil Spill Conference, Anchorage Alaska. 1994.
7. Harry MJ. Six Sigma: a breakthrough strategy for profitability - making strides along the road to defect-free work. Quality Progress. 1994; 31(5):60-66.
8. Kiker G, Bridges T, Varghese AS, Seager TP, Linkov I. Application of multi-criteria decision analysis in environmental management. Integr. Environ. Assess. Manage. 2005; 1:49-58.
9. Klein G. Sources of power. Cambridge: MIT Press; 1999.
10. Kravchuck R, Schack R. Designing effective performance-measurement systems under the Government Performance and Results Act of 1993. Public Administration Review. 1996; 56(4):348-358.
11. Lahdelma R, Salminen P. SMAA-2: stochastic multicriteria acceptability analysis for group decision making. Operations Research. 2001; 49(3):444-454.
12. Lahdelma R, Salminen P. Pseudo-criteria versus linear utility function in stochastic multicriteria acceptability analysis. European Journal of Operational Research. 2002; 141(2):454-469.
13. Lahdelma R, Salminen P, Hokkanen J. Using multicriteria methods in environmental planning and management. Environmental Management. 2000; 26(6):595-605.
14. Linkov I, Varghese A, Jamil S, Seager TP, Kiker G, Bridges T. Multi-criteria decision analysis: framework for applications in remedial planning for contaminated sites. In: Linkov I, Ramadan A, editors. Comparative risk assessment and environmental decision making. Amsterdam: Kluwer; 2004.
15. Linkov I, Satterstrom FK, Seager TP, Kiker G, Bridges T, Belluck D, Meyer A. Multi-criteria decision analysis: comprehensive decision analysis tool for risk management of contaminated sediments. Risk Analysis. 2006; 26(1):61-78.
16. Linkov I, Satterstrom FK, Kiker G, Bridges T, Benjamin S, Belluck D. From optimization to adaptation: shifting paradigms in environmental management and their application to remedial decisions. Integrated Environmental Assessment and Management. 2006; 2(1):92-98.
17. NOAA. Characteristics of response strategies: a guide for spill response planning in marine environments. Washington DC; National Oceanic and Atmospheric Association; 2001. Available at: http://response.restoration.noaa.gov/book_shelf/910_response.pdf.
18. Seager TP. Understanding industrial ecology and the multiple dimensions of sustainability. In O'Brien and Gere Engineers, editors. Strategic environmental management. New York: John Wiley & Sons; 2004.

19. Tervonen T, Figueira JR, Lahdelma R, Salminen P. Towards robust ELECTRE III with simulation: theory and software of SMAA-III. Decision Support Systems. Under review. 2007.
20. Tuler SP, Seager TP, Kay R, Linkov I. Objectives and performance metrics in oil spill response: the Bouchard-120 and Chalk Point spill responses. In Linkov I, Kiker G, Wenning R, editors. Environmental security in harbors and coastal areas. Amsterdam: Springer; 2007. In press.
21. Tuler SP. Individual, group and organizational decision-making in technological emergencies. Indust. Crisis Quarterly. 1988; 2:109-138.
22. Yoe C. Tradeoff analysis planning and procedures guidebook. Prepared for Institute for Water Resources, U.S. Army Corps of Engineers. April 2002.

CHAPTER 15

RISK-COST-BENEFIT ANALYSIS FOR PORT ENVIRONMENTAL SECURITY INVESTMENTS

J.H. LAMBERT, P.E., PH.D., D. WRE

University of Virginia
Charlottesville, Virginia 22904 USA

Abstract: This paper reviews experience in implementing risk-based multiobjective tradeoff analysis and decision making for water resources, multimodal and maritime transportation, and related topics. The paper addresses principally situations in which the decision space of risk management is large and combinatorial, engaging the decision maker in the allocation of resources to the security, robustness, and resilience of critical infrastructures. The applications are hurricane preparedness and resilience of multimodal transportation systems; optimal disaster protection of interdependent floodplain industry sectors; redundancy, robustness, and resilience of critical water resources infrastructure; priority setting of transportation systems that are vulnerable to extreme events; use of uncertain benefit-to-cost ratios to support tradeoff analysis in resource allocation to infrastructure improvements; requirements identification for major navigation structures that are vulnerable to accidents and natural hazards; and addressing the criteria of multiple stakeholders in management of the risk of extreme events for navigation structures. We address the relevance, lessons, and implications of the above for sustaining and improving the security and environmental security of ports and related critical infrastructures.

1. Introduction

According to the United States Government Accountability Office [8]:

> The U.S. Coast Guard is the lead federal agency responsible for protecting domestic ports. In this role, the Coast Guard must identify, evaluate, and mitigate many kinds of security challenges. Ports are often sprawling enterprises that contain key infrastructures besides docks, piers, ships, barges, and warehouses. Many ports are also home to power plants, chemical factories, bridges and tunnels, and a variety of other assets of critical importance to the nation's economy and its defense.

GAO [8] further reports that the Coast Guard is implementing a risk-based approach to port security that will ultimately implement risk assessment,

I. Linkov et al. (eds.), Managing Critical Infrastructure Risks, 299–307.
© 2007 *Springer*.

alternatives evaluation, management selection among alternatives, and implementation and monitoring.

Morel and Linkov [40] suggest extending the classical risk-based approaches to infrastructure protection to encompass *environmental security*. They report that environmental security can relate to any of the following:

1. Security issues that arise from environmental constraints or needs.
2. A sustainable relationship of regions or states to natural and environmental resources.
3. The fact or urgency of a cooperation of regions and states to address their joint environmental problems.

An aim of this paper is the opportunity to address the protection of ports and their related infrastructure in a paradigm that joins environmental security with risk assessment and management. The paper describes several experiences in risk assessment and risk management that benefit from theory and methodology that has been developing in the water resources engineering profession across many years. The cases recall the use through centuries of a holism of approaches that includes risk identification, risk assessment, risk management, environmental management, and policy analysis for national and international security; as well as several other disciplines that are closely allied to risk analysis: namely, systems engineering and multiple-criteria decision making.

2. Three Rs of Port Security: Redundancy, Robustness, Resilience

The numerous distributed activities of modern ports will assure that redundancy, robustness, and resilience (the "three Rs") are emergent properties—observed but not necessarily designed—of these complex systems. The three Rs will be observed to greater or lesser degree across and within ports. These fundamental emergent properties must be addressed in risk assessment and management. For example, Haimes, Matalas, Lambert, et al. [13] describe redundancy, robustness, and resilience as key features of water resources systems, in the context of protecting these systems from willful attack, as follows.

Redundancy implies additional capacity or components beyond what is needed for normal operation. *Robustness* implies the strength of the system to withstand an attack. *Resilience* implies the ability of the system to rebound from an attack at a reasonable schedule and cost.

2.1. PORTS AND INFRASTRUCTURE INTERDEPENDENCIES

The interdependence of ports with other critical infrastructures assures that there will be cascading impacts of security breaches. For example, Olsen et al. [43,42] address the economic interdependencies of water resources systems subject to extreme events, studying the portfolio of types and levels of protection across the distributed systems. Haimes et al. [10, 11] also address the cascading impacts of security incidents with large-scale systems.

2.2. PORT NETWORK RELIABILITY AND RISK ANALYSES

The security of the logistics networks of ports, including the delivery of fuel, water, and chemicals, are amenable to network reliability models; e.g., as described by Park et al. [44], who develop a specialized metric of reliability related to the uncertain capacities and demands associated with the network components. Schneiter et al. [47] also address this topic for water distribution systems.

2.3. PORT RECONFIGURATION FOLLOWING FROM SECURITY INCIDENTS

The reconfiguration of port systems in response to security incidents will benefit from a risk-based multiple-objective approach. Pinto and Lambert [45] develop tools for optimal configuration of such large-scale technology systems that are vulnerable to extreme events. Horowitz and Lambert [15] demonstrate metrics of cost, schedule, and technical risk in configuration of large-scale systems.

2.4. RESOURCE ALLOCATION IN A PORT SECURITY AND SAFETY PROGRAM

The allocation of resources across the security program of a major port can be addressed by the approach of Lambert et al. [32], who consider multiple-objective analysis of the allocation of resources in a safety program for transportation infrastructure. Lambert and Turley [26] use risk and uncertainty analysis to prioritize locations for safety and security upgrades in a transportation network. Baker and Lambert [1] address the balance among risks, costs, and benefits in resource allocation to infrastructure projects. Frohwein and Lambert [5] and Frohwein, Haimes, and Lambert [7] address how to account for the impacts of current risk management decisions to future options.

2.5. USES OF QUANTITATIVE AND QUALITATIVE EVIDENCE IN PORT SECURITY

The multiple stakeholders and diverse uses of modern ports assure that security programs will evaluate alternatives according to both the quantitative and the qualitative evidence of risk reduction, performance gain, and cost. Lambert et al. [34] describe how to integrate the quantitative evidence with the qualitative evidence in allocation of funding to an infrastructure safety program. The quantitative metrics inevitably fall short of covering the broad goals of the organization; the qualitative classification of resources in the decision space helps to account for the diversification of the solution to cover the goals as extensively as possible.

3. Risk Analysis of Extreme Events Affecting Ports

Security management in ports must consider multiple scenarios of extreme events in planning. Lambert and Haimes [28] develop impact analysis for hurricane impacts to the region of a major port. The impact analysis shows the tradeoffs between preparedness and the cost and schedule of recovery for various classes of equipment that are vulnerable to hurricane winds. The analysis demonstrated that the tradeoffs are sensitive to what level of hurricane (Saffir-Simpson level 1, 2, 3, 4, 5) is considered as the scenario for planning purposes. Bier et al. [2] and Haimes, Lambert, and Li [12] address general risk assessment and risk management for critical infrastructures that are vulnerable to extreme events. Olsen, Lambert, and Haimes [41] address how the risk of extreme events to large-scale systems such as ports can change in time due to changing loads and demands, environment, and wear and tear on the infrastructure.

3.1. EXTREME EVENTS AFFECTING THE INDIVIDUAL NAVIGATION STRUCTURES

Furthermore, Lambert and Tsang [22] and Tsang et al. [49] demonstrate multiple-objective analysis for design of major hydraulic structures subject to various levels of severity of earthquake and barge collision. The analysis successfully integrates data from historical incidents with both severity- and likelihood-defined scenarios of earthquakes and collisions. The analysis considers multiple stakeholders (user, owner, operator) and their multiple performance measures (cost of recovery, time to recovery, cost of operations losses).

3.2. PLANNING THE PORT SCHEDULE OF RECOVERY FROM EXTREME EVENTS

The vulnerabilities of a port to major extreme events include the dependencies among activities and agencies that are engaged in a recovery that could take months or years. Lambert and Patterson [21] describe a risk-based prioritization of schedule dependencies in the recovery of a port region from a major hurricane that heavily damages the physical infrastructure.

4. Diversity/Plurality of Investments in Port Security

Diversity of resource allocation to the security program of a port may be of use when a multiple-objective analysis is thought to incompletely describe program requirements or benefits. Diversity can include the distribution of security resources across system components, geographic regions, temporal periods, and qualitative goals. Joshi and Lambert [16] report that metrics of diversification of investment in a resource allocation can be useful to distinguish among nearly-optimal combinatorial solutions to a multiple-objective optimization involving safety and security of a transportation infrastructure.

4.1. ALLOCATING PORT SECURITY SENSORS AND MONITORS

The allocation of security sensors and monitors around a port and the surrounding environment can be addressed from a risk-based perspective. Lambert and Farrington [23,24] and Lambert, Joshi, and Farrington [36] address how risk analysis, cost-benefit analysis, and multiple-objective analysis can contribute to improve the allocation of air-toxics sensors around a critical infrastructure system. The efforts describe building a multiple-objective sensor selection based on uncertain knowledge of sensor performance and the vulnerabilities of localities throughout an infrastructure network.

4.2. CLASSIFYING AND TRACKING PORT SECURITY INCIDENT REPORTS

The analysis of security logs and incident reports around the operations of a major port should be performed continuously to appropriately focus the attention of security managers to critical precursor events. Lambert and Sarda [29] develop a superposition of infrastructure networks to better identify threats to large-scale infrastructures. Lambert et al. [30] use the methodology of statistical quality control to look for anomalies in a weekly or monthly sequence of security-incident reports surrounding major infrastructure. Haimes [9], Lambert et al. [38], Haimes, Kaplan and Lambert [14], and Dombroski et al. [3] describe applications of hierarchical holographic modeling to identify, assess, and manage the diverse sources of risk to a major system according to relevant programs, time periods, and geographic regions.

4.3. RISK-BASED PRIORITY SETTING IN PORT SECURITY PROGRAMS

The security of a major port involves allocation protection to critical transportation links, including rail, highway, and tunnels. Leung, Lambert, and Mosenthal [39] develop a risk-based approach to allocating protection to the security of thousands of bridges across a region that includes a major port. Haimes, Kaplan, and Lambert [14] address generally the risk filtering, ranking, and management methodology for sources of risk to engineering systems.

4.4. SECURITY OF PORT INFORMATION INFRASTRUCTURE AND CONTROL SYSTEMS

The security vulnerabilities of the information infrastructure and automation systems of ports will continue to be a concern. Ezell, Haimes, and Lambert [4] address the modeling, risk assessment, and risk management of security vulnerabilities of the control systems of a water resources infrastructure. The effort describes the application of expert evidence, event and fault trees, and multiple-objective analysis to evaluate alternatives to improve the information security of a water resources infrastructure.

4.5. SOURCES OF RISK IN PORT SECURITY BUSINESS PROCESSES

The business processes associated with the security of major ports have themselves associated sources of risk. Security business processes include personnel hiring and recordkeeping, and policies and procedures. Lambert, Jennings, and Joshi [37] address how to account for identified sources of risk in business process models involving major infrastructure systems. Sources of risk are associated with activities, mechanisms, inputs, and controls according to the integrated definition (IDEF) modeling syntax.

5. Conclusions

This paper has reviewed some literature that is relevant to port security risk assessment and management in several areas: the three Rs, infrastructure interdependencies, network reliability and risk analysis, reconfiguration following from security incidents, resource allocation in a security and safety program, uses of quantitative and qualitative evidence, risk analysis of extreme events, extreme events affecting individual structures, schedule of recovery, diversity/plurality of security investments, allocating security sensors and monitors, classifying and tracking of incident reports, risk-based priority setting, security of information infrastructure, and sources of risk in the security business processes.

6. Acknowledgements

The efforts described in this paper were supported by the Virginia Governor's Office of Commonwealth Preparedness, the Virginia Department of Transportation, the Virginia Transportation Research Council, the National Science Foundation, the Multimodal Office of the Commonwealth of Virginia and VTrans 2025, the Defense Threat Reduction Agency, the President's Commission on Critical Infrastructure Protection, and the U.S. Commission on the Threat of Electromagnetic Pulse Attack.

7. References

1. Baker JA, Lambert JH. Information system for risks, costs, and benefits of infrastructure improvement projects. Public Works Management and Policy. 2001; 5(3):199-210.
2. Bier VM, Haimes YY, Matalas NC, Lambert JH, Zimmerman, R. A survey of approaches for assessing and managing the risk of extremes. Risk Analysis. 1999; 19(1):83-94.
3. Dombroski MJ, Haimes YY, Lambert JH, Schlussel K, Sulcoski M. Risk-based methodology for support of operations other than war. Military Operations Research. 2002; 7(1): 19-38.
4. Ezell BC, Haimes YY, Lambert JH. Risks of cyber attack to water utility supervisory control and data acquisition systems. Military Operations Research. 2001; 6(2):23-33.
5. Frohwein HI, Lambert JH. Risk of extreme events in a multiobjective decision tree: severe events. Risk Analysis. 2000; 20(1):113-123.
6. Frohwein HI, Lambert JH, Haimes YY, Schiff LA. A multicriteria framework to aid the comparison of roadway improvement projects. Journal of Transportation Engineering. 1999; 125(3):224-230.
7. Frohwein HI, Haimes YY, Lambert JH. Risk of extreme events in a multiobjective decision tree: rare events. Risk Analysis. 2000; 20(1):125-134.
8. GAO. Risk management: further refinements needed to assess risks and prioritize protective measures at ports and other critical infrastructure, report GAO-06-91. Washington, DC: United States Government Accountability Office; December 2005.
9. Haimes YY. Risk modeling, assessment, and management, second edition. New York: John Wiley and Sons; 2004.
10. Haimes YY, Horowitz BM, Lambert JH, Santos JR, Crowther KG, Lian C. Inoperability input-output model (IIM) for interdependent infrastructure sectors - I: theory and methodology. Journal of Infrastructure Systems. 2005; 11(2):67-79.
11. Haimes YY, Horowitz BM, Lambert JH, Santos JR, Crowther KG, Lian C. Inoperability input-output model (IIM) for interdependent infrastructure sectors - II: case study. Journal of Infrastructure Systems. 2005; 11(2):80-92.
12. Haimes YY, Lambert JH, Li D. Risk of extreme events in a multiobjective framework. Water Resources Bulletin. Journal of the American Water Resources Association. 1992; 28(1):201-209.
13. Haimes YY, Matalas NC, Lambert JH, Jackson BA, Fellows JFR. Reducing the vulnerability of water supply systems to attack. Journal of Infrastructure Systems. 1998; 4(4):164-177.
14. Haimes YY, Kaplan S, Lambert JH. Risk filtering, ranking, and management framework. Risk Analysis. 2002; 22(2):383-397.
15. Horowitz BM, Lambert JH. Assembling off-the-shelf components: learn as you go systems engineering. Transactions on Systems, Man, and Cybernetics Part A. 2006; 36(2):286-297.

16. Joshi N, Lambert JH. Equity metrics for the prioritization and selection of transportation projects. Transactions on Engineering Management. (In press.)
17. Lambert JH. Aviationextendedtool.xls, Railextendedtool.xls, Portsextendedtool.xls, Transitextendedtool.xls: four MSExcel workbooks implementing comparison of multimodal transportation projects. Available at: http:// www.virginia.edu/crmes/multimodal. 2006.
18. Lambert JH. Comparisontool.xls, Extendedcomparison.xls: two MSExcel workbooks implementing the comparison of highway infrastructure improvements Available at: http:// www.virginia.edu/crmes/VDOT and http://www.virginia.edu/crmes/priority. 2006.
19. Lambert JH. Hurricanepriorities.xls: MSExcel workbook implementing the screening of locations across a region vulnerable to hurricanes. Available at: http:// www.virginia.edu/crmes/2001hurricane. 2006.
20. Lambert JH. Multimodalpriorities.xls: MSExcel workbook implementing multiobjective comparison of multimodal transportation corridors. Available at: http://www.virginia.edu/crmes/multimodal. 2006.
21. Lambert JH, Patterson CE. Prioritization of schedule dependencies in hurricane recovery of a transportation agency. Journal of Infrastructure Systems. 2002; 8(3):103-111.
22. Lambert JH, Tsang JL. Riverlocks.xls: MSExcel workbook implementing the comparison of alternatives for a river lock/dam that is vulnerable to barge collisions and earthquakes. Available at: http://www.virginia.edu/crmes/corps_project. 2006.
23. Lambert JH, Farrington MW. Cost-benefit functions for the allocation of security sensors for air contaminants. Reliability Engineering and System Safety. 2007; 92:930-946.
24. Lambert JH, Farrington MW. Risk-based objectives for the allocation of chemical, biological, and radiological air emissions sensors. Risk Analysis (in press). 2007; 26(6):1659-1674.
25. Lambert JH, Turley T. Priority setting for the distribution of localized hazard protection. Risk Analysis. 2005; 25(3):745-752.
26. Lambert JH, Turley T. Screening methodology for needs of roadway lighting, report FHWA/VTRC 03-CR14. Charlottesville: Virginia Transportation Research Council; 2005.
27. Lambert JH, Turley T. Lightingscreening.xls: MSExcel workbook implementing the risk-based screening of roadway lighting needs. Available at: http://www.virginia.edu/crmes/lighting. 2006.
28. Lambert JH, Haimes YY. Hurricaneretrofit.xls, Hurricanespares.xls: two MSExcel workbooks implementing the comparison of alternatives for hurricane preparedness. Available at: http://www.virginia.edu/crmes/recovery. 2006.
29. Lambert JH, Sarda P. Risk identification by superposition of infrastructure and terrorist networks. Journal of Infrastructure Systems. 2005; 11(4):211-220.
30. Lambert JH, Schulte BL, Sarda P. Tracking the complexity of interactions between risk incidents and engineering systems. Systems Engineering. 2005; 8(3):262-277.
31. Lambert JH, Pinto CA, Peterson KD. Extended comparison tool for major highway projects, report VTRC 03-CR18. Charlottesville: Virginia Transportation Research Council; 2003.
32. Lambert JH, Baker JA, Peterson KD. Decision aid for allocation of transportation funds to guardrails. Accident Analysis and Prevention. 2003; 35(1):47-57.
33. Lambert JH, Baker JA, Peterson KD. Risk-based management of guardrails: site selection and upgrade, report VTRC 02-CR1. Charlottesville: Virginia Transportation Research Council; 2002.
34. Lambert JH, Peterson KD, Joshi NN. Synthesis of quantitative and qualitative evidence for risk-based analysis of highway projects. Accident Analysis and Prevention. 2006; 38:925-935.
35. Lambert JH, Peterson KD, Wadie SM, Farrington MW. Development of a methodology to coordinate and prioritize multimodal investment networks, report FHWA/VTRC 05-CR14. Charlottesville: Virginia Transportation Research Council; 2005.
36. Lambert JH, Joshi N, Farrington MW. Integration of multiobjective tradeoff analysis with cost-benefit analysis in the allocation of resources to system safety. Proceedings of

the Eighth International Conference on Probabilistic Safety Assessment and Management (IAPSAM), New Orleans, LA (in press).
37. Lambert JH, Jennings RA, Joshi NN. Integration of risk identification to business process models. Systems Engineering. 2006; 9(3):187-198.
38. Lambert JH, Haimes YY, Li D, Schooff R, Tulsiani V. Identification, ranking, and management of risks in a major system acquisition. Reliability Engineering and System Safety. 2001; 72(3):315-325.
39. Leung MF, Lambert JH, Mosenthal A. A risk-based approach to setting priorities in protecting bridges against terrorist attacks. Risk Analysis. 2004; 24(4):963-984.
40. Morel B, Linkov I, editors. Environmental security and environmental management: the role of risk assessment. Dordrecht: Springer; 2006.
41. Olsen JR, Lambert JH, Haimes YY. Risk of extreme events under nonstationary conditions. Risk Analysis. 1998; 18(4):497-510.
42. Olsen JR, Beling PA, Lambert JH. Dynamic models for floodplain management. Journal of Water Resources Planning and Management. 2000; 126(3):167-175.
43. Olsen JR, Beling PA, Lambert JH, Haimes YY. Input–output economic evaluation of a system of levees. Journal of Water Resources Planning and Management. 1998: 124(5): 237-245.
44. Park JI, Lambert JH, Haimes YY. Hydraulic power capacity of water distribution networks in uncertain conditions of deterioration. Water Resources Research. 1998; 34(12):3605-3614.
45. Pinto CA, Lambert JH. Risk of extreme events in the configuration of priority systems. Reliability Engineering and System Safety. 2002; 76:265-271.
46. Pinto CA, Lambert JH. Configuration of inter-office switch for extreme traffic with zone configuration evaluator diagram. Reliability Engineering and System Safety. 2003; 79(3):369-375.
47. Schneiter C, Haimes YY, Li D, Lambert JH. Capacity reliability of water distribution networks and optimum rehabilitation decision making. Water Resources Research. 1996; 32(7):2271-2278.
48. Tsang JL, Lambert JH, Patev RC. Extreme event scenarios for planning of infrastructure projects. Journal of Infrastructure Systems. 2002; 8(2):42-48.
49. Tsang JL, Lambert JH, Patev RC. multiple-criteria decision-making in the design of innovative lock walls for barge impact; phase 2, implementation methodologies. Vicksburg: ERDC Vicksburg Publications; 2002. Available at: http://libweb.wes.army.mil/uhtbin/hyperion/ITL-TR-02-5.pdf.

CHAPTER 16

APPLYING GAME THEORY TO BALANCE RISK AND COST FOR SECURITY INSPECTION SYSTEMS

J. BOON, B.M. GORMAN

Johns Hopkins University Applied Physics Laboratory
11100 Johns Hopkins Road
Laurel, MD 20723

Abstract: The emergence of asymmetric threats, whether terrorist or insurgent, suggests a need to reexamine analytic methods applied to the design of risk mitigation systems. Emerging threats have shown ability to adapt strategies rapidly, suggesting that threat characterization must account for the response of the threat to defensive measures. We examine policies of targeting and random selection within checkpoint inspection systems, such as those used for seagoing containers. Specifically, we account for the effects of the transparency of targeting rules on the threat's strategy preferences and ultimately on the requisite balance of targeted and random inspections. Static risk analysis fails in this case, because it does not take into account the effects of policy design on the threat's strategy. We analyze inspection policy and process from the perspective of mathematical game theory. From this perspective we gain important insights into the role of targeting and are able to identify design factors that critically impact the balance between sunk cost and risk. We further extend the analysis by addressing the allocation of inspection resources over multiple checkpoints.

1. Introduction

The Customs and Border Protection (CBP) Directorate at the U.S. Department of Homeland Security shares with the U.S. Coast Guard the burden of securing the nation's maritime transportation system. CBP has been the lead agency in addressing the security risk that containers holding illicit cargo pose to vulnerable assets, including U.S. seaports. The Homeland Security Presidential Directive/HSPD-11 specifically calls on the Secretary for Homeland Security to:

> ...outline a strategy to enhance the effectiveness of terrorist-related screening activities... by developing comprehensive, coordinated, systematic terrorist-related screening procedures and capabilities... [1]

that account for (among others) the needs to:

I. Linkov et al. (eds.), Managing Critical Infrastructure Risks, 309–326.
© 2007 *Springer*.

1. Maintain no less than current levels of security created by existing screening and protective measures.
2. Incorporate security features, including unpredictability, that resist circumvention to the greatest extent possible.
3. Facilitate legitimate trade and travel.
4. Limit delays caused by screening procedures that adversely impact foreign relations, or economic, commercial, or scientific interests of the U.S.

These points, in addition to the mitigation of the security risk, summarize the main objectives in developing a strategy for inspection of cargo containers.

1.1. CONTAINER SECURITY MEASURES

In reaction to the perceived risk, CBP has been building a system around specific initiatives to help ensure that high-risk containers are identified early and inspected before the cargo reaches its intended destination. Three major components of this system are the 24-Hour Advance Manifest Rule, the Container Security Initiative (CSI), and the Customs-Trade Partnership Against Terrorism (C-TPAT).

The 24-Hour Advance Manifest Rule [2], commonly called the 24-hour rule, requires submission of certain manifest information to CBP 24 hours in advance of the vessel cargo being laden at the foreign port. This rule is designed to give inspectors adequate information and time to perform a risk assessment on the cargo shipments and execute targeting procedures before shipments reach the border.

CSI specifies placement of customs officials at foreign ports to target containers for inspection before they reach U.S. seaports [3]. It also specifies prescreening of targeted containers before they are shipped, as well as the use of tamper-evident container seals to ensure that prescreened containers remain intact. Presently, 50 foreign ports participate in the CSI, and many more ports are planning to participate.

C-TPAT offers to companies and carriers a lower probability of their containers being inspected, provided they assess the security of their supply chains and implement measures to address any weaknesses discovered [4]. The examination of security practices extends not only to shipping companies, but also to the companies that provide them with services. Presently, there are nearly 5,800 participating businesses [5].

1.2. TARGETING AND ITS PROBLEMS

CBP targets containers for inspection using an Automated Targeting System (ATS) [6]. ATS was a joint effort by CBP Office of Field Operations and Office of Information and Technology, and is currently operated and maintained by the CBP's National Targeting Center. The system creates an integrated shipment record from a standardized electronic bill of lading. It then integrates information from government, commercial, and enforcement databases to evaluate the risk posed by each shipment. The system uses over 300 weighted rules derived from the targeting methods of Customs experts in order to generate a risk-related score [7].

Shipments that score in a high-risk category are targeted for secondary inspection. Every high-risk container is scanned using one or more non-invasive or minimally invasive procedures, including radiation screening, x-ray or gamma-ray scanning, or cursory physical inspection. If secondary scanning raises suspicion about a container, the contents are unpacked and physically examined.

The scoring rules and threshold are varied according to intelligence reports, but they are not subject to further adjustment based on limitations in inspection resources. In establishing a scoring threshold, however, there is no associated threshold of expected damage loss or other quantitative risk measure. The difficulty in establishing a quantitative risk measure is due in part to the difficulty in assessing expected damage loss associated with an inspection policy.

A targeting system has great apparent value in reducing cost associated with inspecting containers that are unlikely to pose a threat. However, targeting also increases predictability of the inspector's actions, prompting changes to the adversary's strategy in order to evade targeting and thereby reduce the inspection policy's deterrent value. This effect can be mitigated by introducing random inspections of containers that are not targeted. CBP does have a random inspection program called Supply Chain Stratified Examination [6].

The proper conduct of random inspections ensures that no terrorist action guarantees that the transported cargo will enter the U.S. undetected. For random inspections to be an effective deterrent, however, adversaries must perceive the expected likelihood of interdiction and severity of retaliation as sufficiently high that they would not determine seagoing containers to be a preferable means of conveyance. Deciding on the number of random inspections needed depends on a quantitative risk assessment that includes a quantitative assessment of the mutually dependent strategies of both the inspector and the adversary. The focus of the present paper is on the factors that relate to the effectiveness of targeting and random inspection of cargo containers.

1.3. SCOPE OF THIS WORK

This paper summarizes findings we have made using an approach based in game theory to assess cost and risk associated with designs of inspection systems with focus on the container inspection problem. We use the approach to determine the appropriate balance of targeting and random selection in order to balance the costs associated with security inspections with the risks that the inspections are supposed to mitigate. Our inspection model and approach differ from earlier work [8] primarily in that we use the extensive-form model for the game to better account for the adversary's accumulation of information on the inspector's policies. This extensive-form model includes a continuum of branches to represent the inspector's continuum of options [9].

We demonstrate our approach in the context of the problem of inspecting seagoing containers entering U.S. ports. In generalizing the approach to encompass the range of possible preferences for both inspector and adversary, we also demonstrate a means to determine appropriate requirements for policies, technologies, and systems in order to drive adversary preferences in the inspector's favor.

2. Game Theory Approach

The analysis process we use to assess interactions among strategies is based in mathematical game theory. Game theory was devised for the purpose of accounting for interactions among strategies of rational decision makers, and it is essential for determining a preferred strategy where such interactions are in play. Game theory casts the decision makers, or players, as rational in the sense that their preferences are self-consistent and their behavior is consistent with their preferences. Game theory is thus heavily based in utility theory and requires similar information—capabilities, scenarios, environmental parameters, and utility criteria—in order to determine recommended and anticipated strategies [10]. Game theory can also capture uncertainty and learning as situations unfold [11,12]. Mathematical games can be structured serially, in parallel, or may be nested [13]. Game theory also frames the decision-making problem in a way that clearly identifies what information must be obtained.

2.1. ANALYSIS PROCESS

The basic process is summarized in the block diagram of Figure 1. We begin the analysis by identifying the players, characterizing their desired end states, and describing both their capabilities to bring about those end states and limitations to using those capabilities. From this information, we postulate the options open to each player. We then form admissible combinations of those options in scenario models to estimate outcomes of the interactions. We then evaluate each potential outcome from the perspective of each player, bringing to bear all relevant tactical metrics, operational measures of effectiveness, and models of derived strategic value. Finally, we use the core

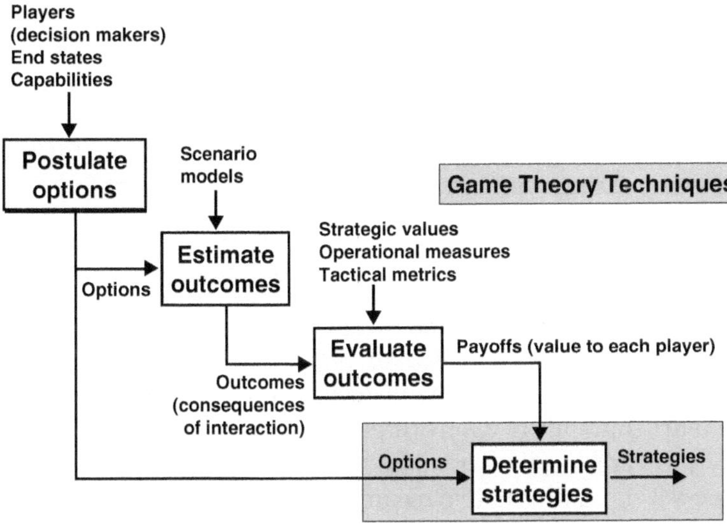

Figure 1. Game theory analysis process.

techniques of game theory to determine the combination of strategies, or plans of action, that can be expected to be chosen by each player.

The value of using this approach is manifold, including:

1. Providing a conceptual framework to articulate the dynamics and interdependencies of rational decision making in conflicts.
2. Enabling prescription and anticipation of strategies.
3. Helping to quantify motivations behind others' strategies.
4. Prescribing strategies that can influence others' strategies or preferences.

2.2. INSPECTION AS A GAME

For the purpose of this analysis, we define an inspection as any formal or official examination whereby an inspecting office is charged to enforce rules regarding the disposition of some object, the owner of which might have reason not to comply with the rules. Here we are careful to distinguish the goal of an individual inspection from that of an inspection system. The goal of an inspection is to detect noncompliance with the established rules. In contrast, the goal of an inspection system is to deter or dissuade noncompliance as a strategy. This distinction is critical in discussing inspection as a game.

It is natural to identify inspection as a mathematical game. Both the inspector and the owner of the inspected object act intelligently and interdependently. Also, the owner of object in question has some motive not to comply with rules. Examples of motive include cost savings by shortcutting safety or security measures, direct economic payoff by selling smuggled items, and political influence in the case of terrorist acts. The inspector balances sunk cost with expected losses because inspecting increases sunk cost but also mitigates risks that noncompliance poses. In the end, both parties choose a strategy that maximizes payoff or minimizes loss.

2.3. ROLE OF TARGETING

At this point it is prudent to discuss what an effective targeting system does from a game perspective. The assessment of a targeting system's effectiveness has been an ongoing issue with regard to container security [6,14,15]. It is particularly difficult to assess targeting performance against threats that are simply too rare to allow a sufficient accumulation of sample statistics. There is, however, an alternative means to assess targeting effectiveness that is motivated by game theory.

An object is targeted if available information suggests that it poses a high risk, specifically if a risk-based score S exceeds some threshold σ. The score should therefore be defined in some way that is related to the relative risk. However, risk actually depends on strategic choices, which in turn depend on an analysis of future outcomes. This type of information is rarely, if ever, available for scoring.

Another way to score the object is to consider, as a surrogate measure, what cost B_E the owner incurs to ensure that the object will score below the targeting threshold and thereby evade the targeting system. We call the quantity the evasion cost. The lower a desired target score, the greater the evasion cost, and the more the owner can expect to pay to achieve it, regardless of strategy.

As targeting relies on indirect information related to the actual evasion cost, the relationship between the score and the cost is not exact, but statistical. However, we expect that if the targeting system is effective, the two quantities are highly correlated and the evasion cost is sensitive to the target score.

The ellipse in Figure 2 illustrates a notional distribution of scores and the costs to achieve them. The down-sloping line shows the major axis of this distribution, which approximately relates any score to an expected cost to achieve it. If the inspector increases targeting by lowering the scoring threshold, say from σ to σ', then the cost to the owner to lower the object's score in order to evade targeting should increase. A limit on inspection resources imposes a lower limit on the threshold.

Here we are careful to make the distinction between the owner in a security compliance inspection and the owner in a security threat scenario, such as a weapon of mass destruction hidden in a container. In the threat

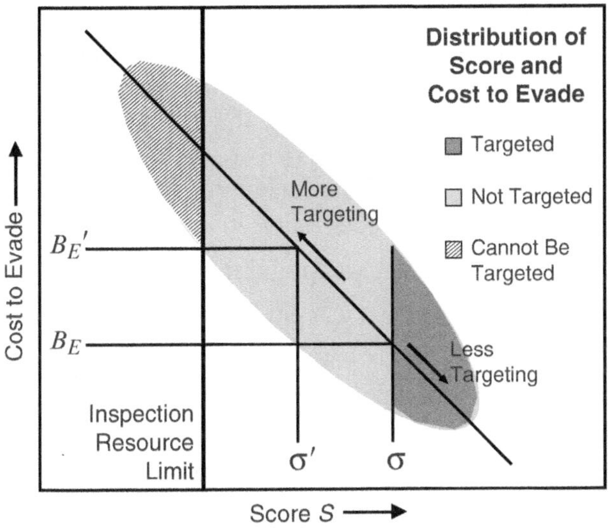

Figure 2. Notional evasion cost distribution.

scenario, the *owner* is the owner of the weapon; i.e., the entity who makes the decision to pose a threat.

3. Container Inspection Example

We suppose an inspecting authority for a U.S. port is given an annual budget B to inspect arriving containers. The inspecting authority, or inspector, must implement an inspection system that minimizes the expected loss to U.S. interests due to the smuggling of weapons (or bomb-making materials or combatants) into the port.

Given the budget B, the inspector can afford to inspect some proportion R_I of the incoming containers. Each container may be selected for inspection in one of two ways:

1. It is targeted for inspection based on a risk score.
2. It is selected randomly.

The inspector wishes to choose an appropriate balance between targeted and randomized inspection that minimizes expected loss.

3.1. INSPECTION MODEL

If a container containing a weapon is selected for inspection, the inspection system has some probability $1 - \beta$ $(0 \leq \beta \leq 1)$ of detecting it. The inspection system also alerts when there is no weapon with some false alarm probability α ($0 \leq \alpha \leq 1$), causing further delay and additional costs for a more thorough inspection. However, we shall not specifically address false alarm issues in this example. We assume that costs due to false alarms are embedded in the relationship between R_I and B.

Given the distribution of inspection resources between targeted and randomized inspection, a single adversary infers the targeting rules and level of randomization and then chooses between one of three options:

1. Do not attempt to smuggle a weapon (comply).
2. Make an attempt to smuggle a weapon in a container that will be targeted (confront).
3. Make an attempt to smuggle a weapon in a container that will not be targeted (evade).

The adversary chooses to comply, confront, or evade based on the expected values of utility functions that take into account the payoff associated with a successful smuggling attempt, the penalty associated with an unsuccessful smuggling attempt, the cost of acquiring the weapon, the cost to evade targeting, and the known effectiveness of the inspection system (i.e., the probability of detection: $1 - \beta$).

The inspection process is represented in Figure 3. Starting from the left of the figure, two types of containers are queued into the system:

1. Those containing a weapon, represented throughout the figure by black connectors.
2. Legitimate containers, represented throughout the figure by gray connectors.

Containers targeted for inspection at Node T are sent into a queue to be scanned at Node S. Note that threat containers move from Node T to Node S only if the adversary chooses the confronting strategy. Containers not targeted for inspection may still be sent to be scanned at Node S if they are

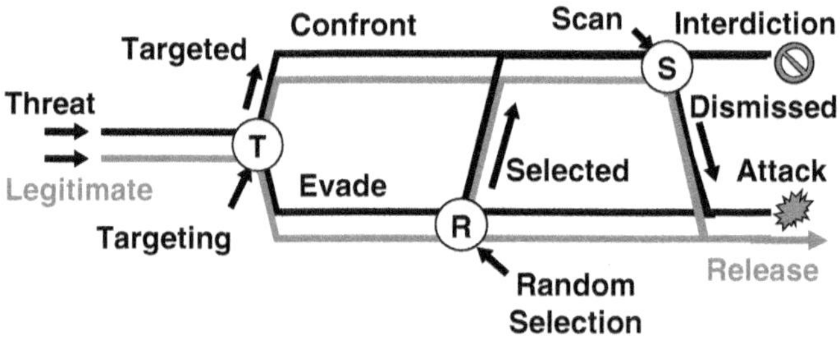

Figure 3. Nodal representation of the inspection process.

selected randomly at Node R. Note, however, that if the adversary chooses to evade targeting, random selection is the only way for the container to reach Node S. From Node S, containers are eventually dismissed or interdicted, depending on the cargo. Threat containers are dismissed with probability β and are interdicted with probability $1 - \beta$, while legitimate containers are all eventually released.

3.2. GAME TREE ANALYSIS

Because the choices of the inspector and of the adversary are interdependent, the optimal strategies can be identified only using game-theoretical techniques. Before these techniques can be applied to the inspection problem, we must express the problem mathematically. Let θ represent the proportion of inspections being done through random selection, so that $1 - \theta$ is the proportion of inspections being done through targeting. The proportion of containers inspected randomly is thus $R_I\theta$, and the proportion of containers targeted for inspection is $R_I(1 - \theta)$. The inspector must choose the value for θ that, anticipating the choice of the adversary, minimizes the expected loss due to damages resulting from a successful attempt. Given this choice by the inspector, the adversary chooses to comply, confront, or evade. For each of these three options, there is an expected payoff both to the inspector and to the adversary. The interplay between these choices is captured in Figure 4.

If the adversary chooses to comply, the payoff to the inspector is simply the sunk cost ($-B$) of the inspection system. For the adversary, the choice to comply involves neither a loss nor a gain.

If the adversary chooses the confronting strategy, the inspector's expected utility includes the term ($-D\beta$), which is minus a damage loss of magnitude D for a successful smuggling attempt times the probability β of the threat container being released after scanning at Node S (recall Figure 3). For the adversary, the expected utility has three terms:

1. $A\beta$, which is the payoff A to the adversary for a successful smuggling attempt.
2. $(-I)(1 - \beta)$, which is a penalty of magnitude I imposed after interdiction times the probability $1 - \beta$ of being interdicted.
3. $(-B_0)$, for the cost of acquiring the weapon and container.

Finally, if the adversary chooses to evade targeting, the inspector's expected utility is again the sunk cost ($-B$) of the inspection system and the expected cost of a successful smuggling attempt. In this case, though, the probability of a threat container being released is one minus the

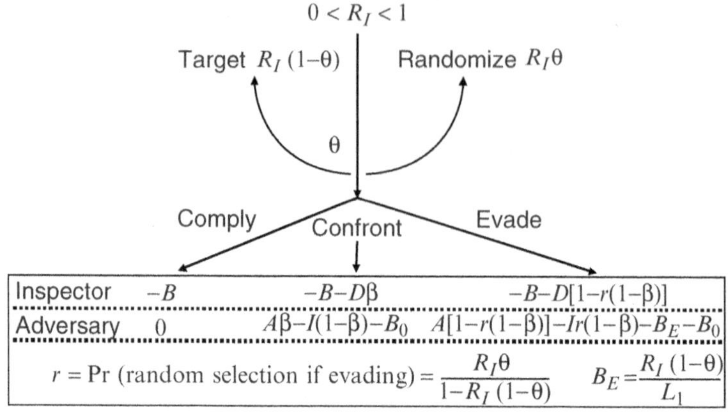

Figure 4. Game tree representation of the inspection problem.

probability r of being sent to Node S based on random selection times the probability $1-\beta$ of being interdicted at that Node. For the adversary, the expected utility has four terms:

1. $A[1-r(1-\beta)]$, which is the payoff A to the adversary for a successful smuggling attempt times the probability of the threat container being released.
2. $(-Ir)(1-\beta)$, which is the penalty I for being interdicted times the probably of being interdicted.
3. $(-B_E)$, for the cost to evade targeting.
4. $(-B_0)$, for the base cost of acquiring the weapon and container.

As the cost to evade targeting should depend on the targeting rate $R_I(1-\theta)$, we employ a simple model of direct proportionality: $B_E = R_I(1-\theta)/L_1$, where L_1 is a parameter representing the targeting rate for which the cost to evade equals the attack payoff A. In the results to follow we take $A = 1$ without loss of generality.

The inspector's expected utilities (as defined by the expressions in Figure 4) depend on five parameters: B, D, β, R_I, and θ. As defined, this game has the desirable property that, for every combination of these five parameters, the inspector's expected utility is such that he always prefers compliance over confrontation and confrontation over evasion. In other words, from the inspector's perspective, the expected utilities associated with the adversary's three strategies are ranked, regardless of the values of the parameters that influence his expected utility functions. This ranking is shown graphically in Figure 5.

Because of this preference structure, maximizing the inspector's expected utility almost always involves choosing θ that induces the adversary to

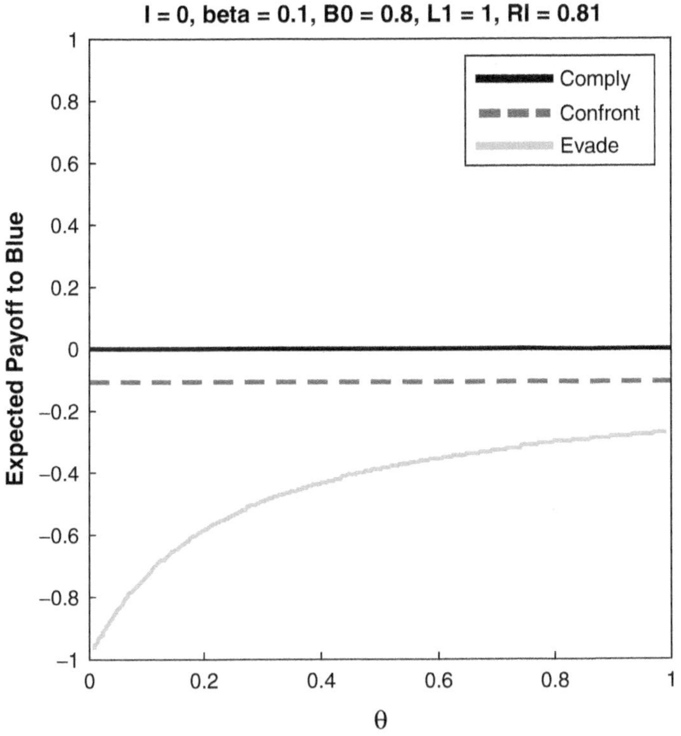

Figure 5. Representative example of inspector's utility as a function of θ.

choose the most favorable strategy from the inspector's perspective. For example, setting all parameters other than θ constant, suppose that for $\theta < 0.3$ the adversary will choose to comply, while for $\theta \geq 0.3$ the adversary will choose to evade. Given the ranking of the inspector's preferences, it is in his interest to choose $\theta < 0.3$. The one exception to this rule is the following: of the three strategies available to the adversary, the inspector's expected utility function associated with evasion is the only one that depends on θ. Thus, if the parameters are so unfavorable to the inspector that the adversary will evade for all possible values of θ, the inspector will set θ equal to one (making the inspection system one based solely on random inspections) in order to minimize his expected loss. The payoff to the adversary in this scenario is shown in Figure 6.

After considering all possible combinations of the nine parameters driving this problem, we have identified seven domains of the adversary's response to changes in θ. One is shown in Figure 6 and the remaining six are shown in Figure 7. One obvious characteristic of these plots is that the adversary's expected utility for confronting is constant with respect to θ. This follows as the payoff for the confronting strategy depends only

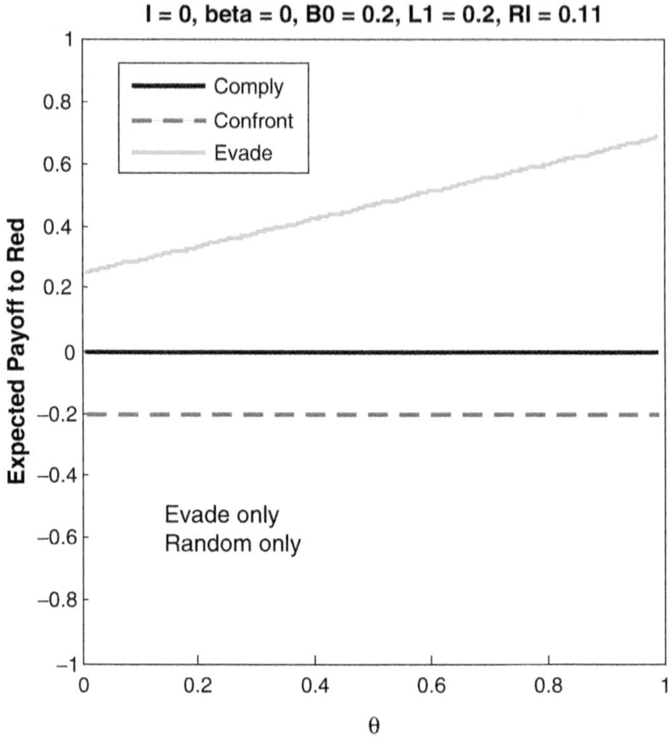

Figure 6. Payoff for evade-only adversary as a function of θ.

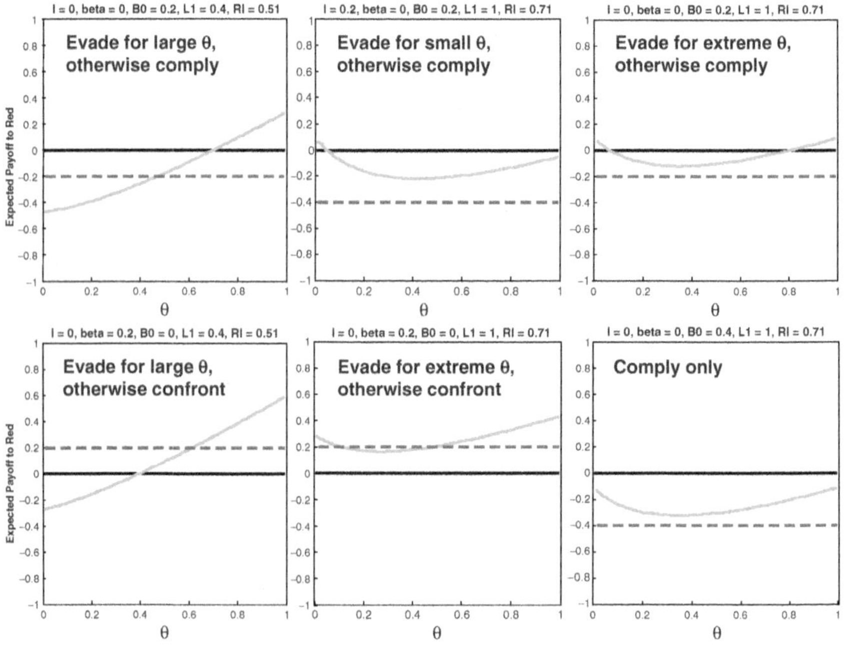

Figure 7. Examples of adversary's (Red's) expected utility as a function of θ.

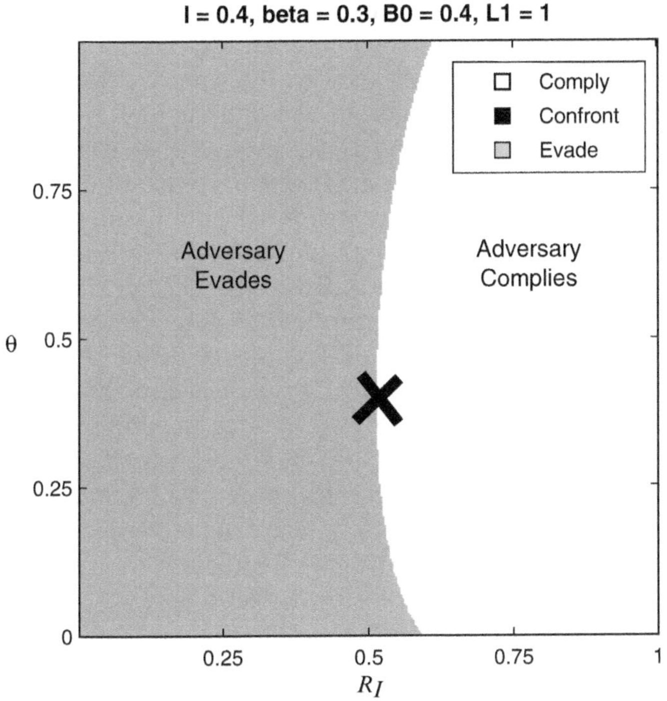

Figure 8. Example of minimum inspection rate needed to induce compliance.

on the parameters A, β, I, and B_0. Note also that the expected payoff for the confronting strategy is also not a function of the overall inspection rate R_I.

3.3. RECOMMENDED STRATEGIES

With this approach, we can recommend a range of θ values that induces the adversary to choose according to the preference structure of the inspector. It may be the case, though, that the inspector's choice of θ makes no difference to the adversary choice of strategy, as in the example shown in Figure 6. Such situations are typically due to a small inspection rate R_I resulting from a small inspection budget B. By considering combinations of R_I and θ, one can identify the minimum value of R_I and corresponding θ that induce a compliant or confronting strategy in the adversary. Figure 8 shows an example of this for one combination of parameters. In this example, an inspection rate $R_I \cong 0.5$ together with $\theta \cong 0.4$ induces compliance at minimal cost. Note that any inspection rate less than 0.5 would encourage the adversary to evade targeting.

To help set performance requirements for inspection systems, we can further explore the response of the minimum inspection rate to induce compliance (or confronting, as applicable) with technical design parameters for scanning or targeting performance. By varying the performance parameters for scanning and targeting, β and L_1, respectively, we can investigate what performance is required to deter an attempt or to deter the evasive strategy.

The surface displayed in Figure 9 shows the minimum rate of inspection R_I to deter evasion in favor of compliance or confronting for a particular choice of parameters, $I = 0.5$ and $B_0 = 0.6$. The surface is a boundary below which inspection rates are too low to deter the adversary from an evasive strategy. Thus, for rates of inspection below this boundary the adversary is expected to be evasive, and the inspector should conduct only random inspections to minimize risk. Above the surface the inspection rate is sufficient to prevent evasion, but the optimal strategies depend on the performance parameters for scanning and targeting. Each shaded domain on the surface is labeled to indicate the recommended strategy for the inspector and the expected strategy for the adversary. The inspector's options are labeled as "Target" for targeting only, "Randomize" for

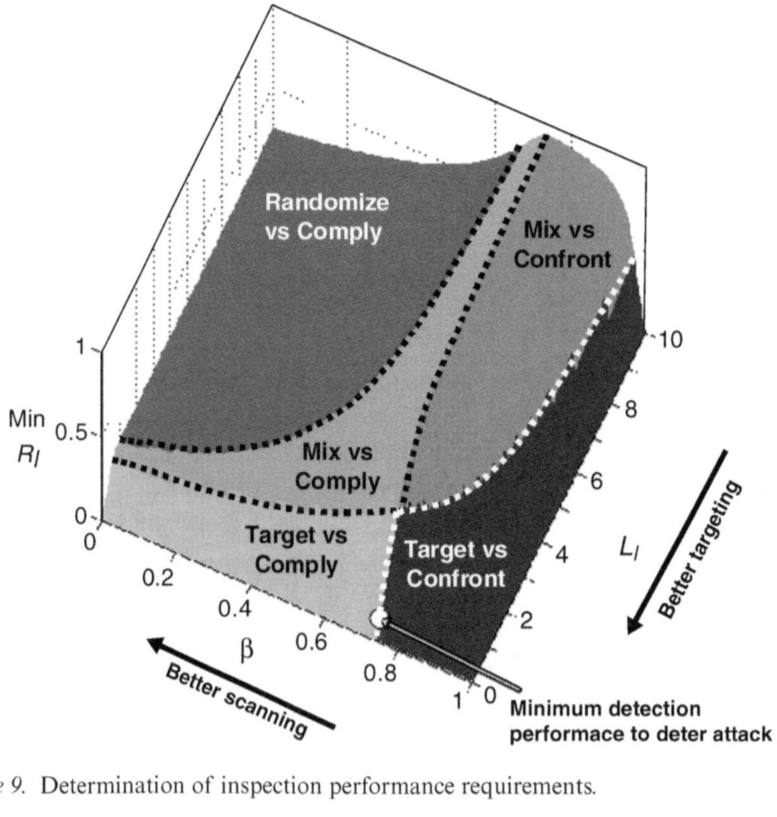

Figure 9. Determination of inspection performance requirements.

random selection only, and "Mix" for an optimized mix of targeting and random selection.

Note that in the two regions to the right there is no possibility to deter an attempt. To deter an attempt, a minimum detection performance is required, shown in the figure at $\beta \cong 0.75$. Also, for a set value of "good" scanning performance, say $\beta = 0.4$, there is a minimum targeting performance, near $L_1 \cong 3$ in this case, at which the inspector's optimal strategy begins to involve targeting as part of a mixed strategy.

4. Extension to Multiple Ports

The modeling and analysis applied to the single-port inspection problem can be easily extended to handle multiple ports. Instead of the entire inspection budget B being allocated to a single port, suppose that the budget must be shared between two port sites, Site 1 and Site 2. The highest level decision making body must set an allocation parameter γ such that Site 1 receives a budget $B(1 - \gamma)$ and Site 2 receives a budget $B\gamma$ to conduct inspections.

If B is large enough, it is possible for one or both sites to be given more funds than are necessary to inspect all of the inbound containers. For this reason, we expect γ to be set such that $(B - B_{1\max}) / B < \gamma < B_{2\max} / B$, where $B_{1\max}$ and $B_{2\max}$ represent the minimal funding necessary to inspect every container at Site 1 and Site 2, respectively. This condition guarantees that neither site receives more than is necessary to inspect every container. Also, let $B_{1\max} + B_{2\max} = 1$, so that every container coming through either site is inspected when $B = 1$. These modeling assumptions are captured in the game tree shown in Figure 10.

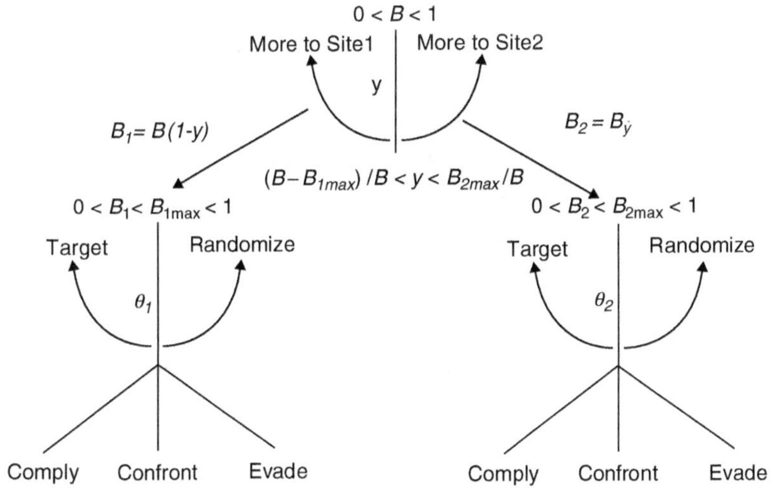

Figure 10. Game tree representation of the two-port inspection problem.

Although the expected utility functions are not shown in Figure 10, they are identical to those shown in Figure 4, with each site having its own A, D, I, β, B_0, L_1, and B_E parameters. Based on the expected utilities associated with the six leaf nodes of the game tree in Figure 4, a single adversary chooses one among five strategies:

1. Confront at Site 1.
2. Evade at Site 1.
3. Confront at Site 2.
4. Evade at Site 2.
5. Comply (i.e., do not attempt to smuggle a weapon through either site).

As with the single-site case, it is possible to infer the most attractive strategy available to the adversary given each possible combination of budget B and distribution parameter γ. An example of this mapping is shown in Figure 11. Three regions on the surface are labeled according to the adversary's anticipated strategy. The four unlabeled regions in the foreground indicate budget allocations that are not in the inspector's interest as they exceed the budget required to inspect all containers at one of the sites.

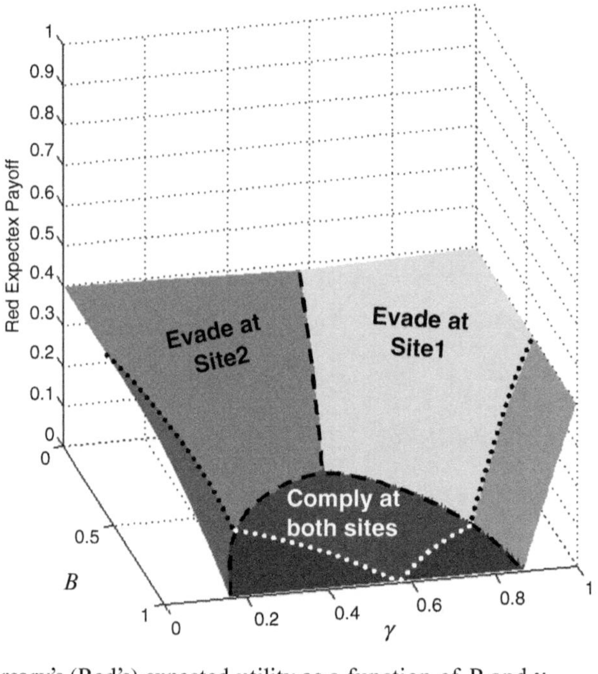

Figure 11. Adversary's (Red's) expected utility as a function of B and γ.

For the set of parameters represented in the figure, we see that the minimal budget B necessary to induce compliance at both sites is about 0.35 (i.e., about 35% of the total number of containers entering Site 1 and Site 2 combined) with $\gamma \cong 0.58$, meaning that Site 2 receives 58% of the budget. The asymmetry with respect to sites is a result of the chosen parameters ascribing somewhat better performance to Site 1.

By approaching the inspection problem from a game-theoretical perspective, we are able to make recommendations regarding high-level budgets, the distribution of inspection resources to sites of varying vulnerability and inspection capability, and the balance between random and targeted inspections at each site while simultaneously taking into account the adversary's most attractive course of action.

5. Discussion

Simple exercises like the container inspection problem drive home a number of important points when making decisions about the design of a security inspection system, or, for that matter, any inspection system. First, traditional risk analysis does not have a mechanism to account for the effects of common knowledge about the inspection rules. Such knowledge can be expected to influence strategy on both sides. Others will exploit known targeting rules to their advantage, and the resulting costs and risks may depend heavily on the strategic choices subsequently made. Second, game theory combined with traditional risk analysis provides a framework in which to quantify the costs and probabilistic factors that drive strategy. The result is a more accurate accounting of the cost-risk tradeoffs that drive the design of an inspection system. Third, game theory can uncover policy decisions (and associated technological needs) that drive an adversary's strategy in a desired way.

6. Acknowledgements

We gratefully acknowledge support from the Johns Hopkins University Applied Physics Laboratory. We also wish to thank Duane Boniface for his support of this work and Igor Linkov for his generous invitation to contribute to these proceedings.

7. References

1. Bush GW. Comprehensive terrorist-related screening procedures. Homeland Security Presidential Directive/HSPD-11. 2004.
2. U.S. Department of Homeland Security. Secure seas, open ports: keeping our waters safe, secure, and open for business. Press release. 2004.

3. U.S. Customs and Border Protection. Container security initiative fact sheet. Press release. 2006.
4. U.S. Customs and Border Protection. Securing the global supply chain: Customs-Trade Partnership Against Terrorism (C-TPAT) strategic plan. Press release. 2004.
5. U.S. Department of Homeland Security. Fact sheet: securing U.S. ports. Press release. 2006.
6. U.S. Government Accountability Office. Homeland security: preliminary observations on efforts to target security inspections of cargo containers. Report GAO-04-325T. 2004.
7. U.S. Office of Management and Budget. Program assessment: customs and border protection border security inspections and trade facilitation. 2006.
8. Gorman BM, Boniface DE. Analytic methods for enhancing security inspection systems. In: Working together: R&D partnerships in homeland security, paper no. 1428. Boston, MA. 2005.
9. Rosendorff BP, Sandler T. Too much of a good thing? The proactive response dilemma. Journal of Conflict Resolution. 2004; 48:657-671.
10. Luce RD, Raiffa H. Games and decisions. New York: Wiley; 1957.
11. Harsanyi JC. Games with incomplete information played by "Bayesian" players. Management Science. 1967-8; 14:159-182, 320-332, 486-502.
12. Bennett PG. Hypergames: developing a model of conflict. Futures. 1980; 12:489-507.
13. Selten R. Spieltheoretische behandlung eines oligopolmodells mit nachfragetragheit. Zeitschrift für die gesamte Staatrwissenschaft. 1965; 121:301-324, 667-689.
14. U.S. Government Accountability Office. Homeland security: summary of challenges faced in targeting oceangoing cargo containers for inspection: report GAO-04-557T. 2004.
15. U.S. Government Accountability Office. Cargo container inspections: preliminary observations on the status of efforts to improve the Automated Targeting System: report GAO-06-591T. 2006.

CHAPTER 17

COMPLEX SECURE SOLUTIONS FOR SUSTAINABLE PRACTICES IN ARMENIA

V. SARGSYAN

Yerevan State Institute of Economy
Nalbandyan 128
Yerevan 375025 Armenia

Abstract: Transitional and development processes in Armenia and other post-Soviet states are accompanied by technological incursion, which is inevitable. In order to establish effective sustainable policies, these countries must be ready for notable technological changes. Each step in economic development and technological change must be balanced by effective environmental solutions. It is important to work out a methodical basis and common strategies for the development of secure solutions for sustainable practices and risk management. This paper presents an integrated approach to the problem of sustainable and secure development in Armenia. Under economic, social, and political systems, the concept of complex "triangle" analysis for sustainable policymaking in the region—incorporating framework, tools, and strategies—is considered. The model employs spiral dynamics. The framework of problems, tools, and outputs for sustainable development in Armenia, which must be analyzed in one system, is addressed.

1. Background

Environmental problems are serious and pervasive in today's world. Countries in transition have inherited a badly damaged environment; energy and raw material-intensive economies; and obsolete, polluting technologies. Furthermore, they are facing new constraints associated with the shift from a centrally planned to a market economy, such as mounting unemployment, high inflation, and public deficits. It is, however, widely understood that the aim of transition is not only to improve economic performance in the short and the long term but also to move towards secure and sustainable development. In post-Soviet countries, particularly in Armenia, environmental problems, such as deforestation, soil erosion and the expansion of desert areas, and degradation of water resources are aggravated [1,4]. These problems have severely threatened the sustainable development of these countries

I. Linkov et al. (eds.), Managing Critical Infrastructure Risks, 327–338.
© 2007 *Springer*.

and have caused great concern at all levels, from the general public to national governments and international agencies.

The efforts of these countries to strengthen the institutional capabilities for environmental policymaking, planning, and management are being undermined by numerous weaknesses, deficiencies, and difficulties, such as weak or inadequate legislation, the lack of political clout of environmental authorities, low public awareness, lack of well-established target groups which represent specific interests, lack of managerial skills, inadequate information base, and general lack of financial resources.

2. Principles for Integrated Approach

The problem is focused on the establishment of foundations for sustainable policies in Armenia. Environmental security and sustainability are major issues for development of post-Soviet areas. Under transitional and development processes, these countries must get ready for notable technological changes. Economic development and technological change must be balanced with effective environmental solutions. First of all it is important to develop a general approach to the problem of secure and sustainable policymaking, on which further investigations will be based. This approach must fully consider the specific character and gaps in post-Soviet development and democratization processes.

Solutions for secure policymaking must be obtained under one "economic-social-political" system. A keystone for success is the development of paths for effective management of environmental resources, which is based on the establishment of economic principles in natural resource management. In order to balance economic development and technological changes with environmental sustainability, it is necessary to realize multifaceted analysis. Sustainability concerns not only economic, but also social and political systems. Social and political sustainability are important conditions of secure development. Setting up the paths of sustainable development has its special aspects under transitional economies in post-Soviet countries, where democratization processes are in the establishment stage. It is extremely important not to separate democratization processes from economic development. On the other hand, secure solutions often require political influence. Integration of economic, social, and political values in environmental decision making is the main condition.

Problems concerning sustainable development in the region must be analyzed in one framework. Numerous weighable investigations around sustainable development in Eastern Europe and post-Soviet states were done and are in the process of development. Each of these investigations touched on separate concepts of sustainable development and often had no common solutions; they even sometimes conflicted with each other. Separate solutions of individual problems can cause other problems. In order to achieve security and environmental quality, it is important to find an appropriate balance between interrelated regional problems with common roots.

According to the above-mentioned principles, the concept of complex "triangle" analysis for sustainable policymaking in the region—framework, tools, and strategies—(Figure 1) is considered. Per triangle analysis, different sustainable development problems must be examined on an integrative basis in one (regional) economic, social, and political system.

Such an analysis offers a new quality, facilitates the definition of new approaches, and serves as a condition for formulating valuable recommendations. After definition of the framework of regional problems connected with secure development, appropriate tools and methods must be worked out. Application of defined tools could generate strategies for solutions of the problem's framework with the features of regional development. Progress is spiral. In the dynamic contest, after each round, the framework of the problems changes, which could characterize a new set of tools and accordingly new strategies.

The proposed model is an attempt to present principles of complex and versatile analysis of sustainable policymaking for the region, particularly in Armenia, with the goal of setting up strategies with minimum possible economic, social, and political losses.

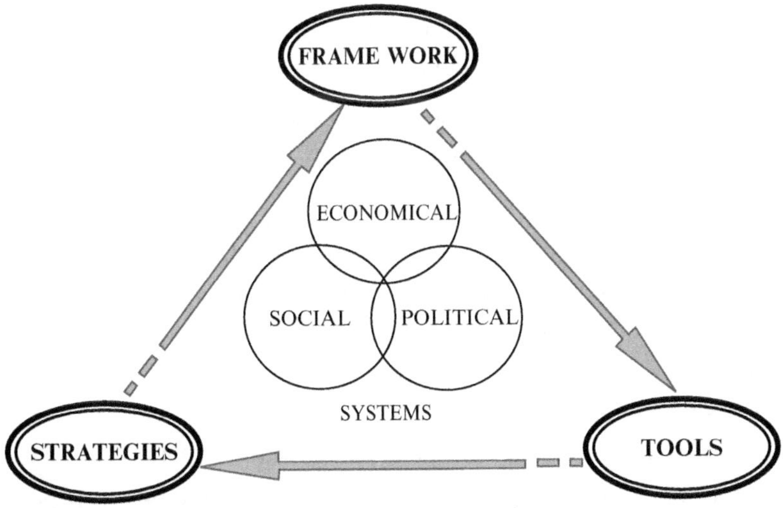

Figure 1. Triangle analysis for sustainable policymaking.

3. Framework of Sustainable Development Problems in Armenia

Secure policymaking is based on set of topics that touch different sides of environmental activities—balancing economic development and environmental management, saving nature, characterizing economic tools for environmental policy, defining alternative methods of environmental management, and analyzing obstacles.

Despite the fact that sustainable practices address common problems, each region or state can be characterized with specific descriptions of problems and objectives within the above-mentioned topics. This depends on the character of economic development, markets, social and political systems, and national strategies. In order to cover the range of problems concerning secure development in Armenia, the following main objectives were defined:

1. Definition of value of returns to land and water and costs of degradation and development of datasets that relate to the economic (agricultural) aspects of natural resource management in Armenia.
2. Analysis of protected area management issues, in order not to lose outstanding natural landscapes and strategic natural resources for the region.
3. Analysis of natural resource pricing as an effective management tool.
4. Investigation of opportunities for voluntary environmental management arrangements as a new type of activity in the region.
5. Definition of impediments to effective policy implementation for identifying potential gaps in the institutional framework.

3.1. DEFINITION OF VALUE OF RETURNS TO LAND AND WATER AND COSTS OF DEGRADATION

This problem presents the development of datasets that relate to the economic aspects of natural resource management in Armenia. There is a focus on resources used to support agriculture and resources impacted by agriculture. The agricultural sector in Armenia was regarded as the main sector of economy, taking into account its infrastructure, the general socioeconomic condition of the republic, and the problem of food security.

The following main activities must be realized:

- An overview of the economic returns from the nation's land and water resources used in agriculture.
- Analysis of economic aspects of the degradation of natural resources.
- Developing information on willingness to pay to slow rural population decline and improve environmental attributes that are not part of the market for agricultural products.

Eventually, realization of this objective will result in recommendations to different groups of farmers and extension services on optimal and sustainable uses of their resources from an ecological and economic point of view. Also, the insight gained from farmers' decision making regarding soil use management may become relevant for policy makers. It allows for predictions of possible reactions to new technologies, incentives, or changes in the macroeconomic environment by different classes of farmers. As such, the methodology may also help to bring decision makers and farmers closer together.

3.1.1. Challenge to Sustainable Agricultural Development in Armenia

Agriculture is the main "working" branch in Armenia. During the first years of independence, reforms began in agriculture, aiming to transform this sector into a stable and productive one, based on private property and market relations. Armenia was first among the post-Soviet states to initiate land privatization. Rural households cultivate or use most available land resources through a combination of private ownership and leasing of state lands. Reform processes generated fresh problems and several constraints appeared in the way of sustainable development. Obviously, hurried privatization and land ownership has resulted in new and unfamiliar legal and social conditions relating to land and water use in agriculture. It is not clear how this will affect biodiversity. Difficult economic conditions and a lack of resources for farming have reduced the effectiveness of agriculture. However, the absence of effective regulations for the use of private lands could result in even greater environmental impacts in the future.

Because of the weak management of agriculture and incorrect financial policy, the agricultural system is undergoing structural degradation. Many land issues arise from inefficient land administration, poorly defined institutional responsibilities, and limited enforcement. Little scientific basis exists for current land use planning. Current legislation is in a state of transition and has not caught up with changing political and economic conditions. The legal framework defining the rights and obligations of land holders and regulating the use of land resources and ensuring their protection is unclear and incomplete. There are serious difficulties in enforcing laws and regulations due to limited public awareness. Awareness of the rights and obligations of land users is limited. Under present conditions in Armenia, farmers are free to determine which crops to sow and where to sell them. However, their options are limited by numerous constraints. For appropriate solutions, it is important to consider the special characteristics of agriculture in Armenia:

- Arable land is limited and the land provision in Armenia is very low (0.13 ha of arable land per capita). The country is mountainous, with only 28% of land below an altitude of 1500 m. Only 20% of the land is suitable for crops (other than pasture). Large numbers of farmers cultivate highly eroded land on steep slopes.

- The bulk of agricultural land has been privatized but there is little support for small holdings. After the agrarian reform, 25% of arable lands are not in use because farmers do not have the means to cultivate it due to their difficult socioeconomic situation. The continuous deterioration of land productivity is directly due to a lack of land improvement. Access to formal credit for many small farmers remains difficult. As a result, access to quality inputs and machinery is difficult.
- Lack of basic knowledge of farming and management skills is a major bottleneck. The proper use of fertilizers and pesticides is critical. Violation of agro-technical rules, due to lack of knowledge, could cause numerous environmental problems. Agricultural research and the limited extension services have not been functioning properly for many years. For instance, the Agricultural Research Institute has suffered significant financial cuts since 1991 and has managed to produce only four varieties of wheat in the past 20 years.
- Marketing constraints are possibly the chief obstacle to increasing farm income and the most difficult for farmers to overcome. Geopolitical constraints make export of fresh produce virtually impossible, except by air, and therefore the products become expensive and cannot find their place in external markets. Despite progress in re-establishing some agro-processing capacity in the last few years, a large number of food and agro-processing enterprises (including some large wheat mills) are not working or are working at very low capacity, reducing the competitiveness of output. Home markets are not developed. The bulk of the population has very limited purchasing power and auto-consumption and barter probably account for up to two-thirds of consumption of all domestic produce. Appropriate marketing and transport infrastructure for small farmers is lacking. Roads are poor and transport is expensive. Moreover, there is no market information system, with the result that many farmers in an area tend to produce the same crops if they were previously profitable, causing a glut.
- Institutional shortcomings exacerbate the above constraints. The government of Armenia has drawn up a strategy for the development of an enabling environment for agriculture. Though the government is trying to stop the process—passing resolutions and adopting laws—they are not sufficient and often inapplicable due to the absence of mechanisms for their implementation.

The problem is to guide farmers through broad-range analysis in improving their sustainable practices and developing their knowledge about organizing sustainable production. Sustainable agricultural development in Armenia is based on several tasks, such as analysis of the diversity of farmers' strategies and development of regional indicators and measures of sustainability; modeling and estimating the process of land degradation and pollution on the basis of consideration of local principles and features; and economic and cost-benefit analysis of alternative forms of farming. Each of these tasks is presented as an important analysis to be conducted on an integrative basis to create weighable factors for the establishment of valuable recommendations.

3.2. ANALYSIS OF PROTECTED AREA MANAGEMENT ISSUES

The natural and near-natural features of protected areas offer attractions which in many countries have become the cornerstone of tourism and recreation. However, promoting tourism for the economy is not the primary role of most protected areas. Their primary role is the conservation of biodiversity, and the provision of a rich natural resource that permits scientists, educators and the community to meet their various needs. Generally speaking, however, the market alone does not support a system of protected areas—hence society, through its various levels of government, must provide environmental management and protection as a public service in the same manner that it provides health, education, defense, and legal systems. Failure to provide these public services impoverishes the quality of life for individuals and indeed for entire nations.

The debate over protected area management is often about the balance between leaving areas in their natural or near-natural state, and developing and exploiting them. This choice is fraught with tension—for example, should a forest be left uncleared or logged and converted to agriculture? Should wetlands and mangroves be left in their natural state or cleared and developed? This objective is intended to help answer such questions. The goal is to develop an overview of how the economic and social values of protected areas can be assessed, provide new insights, and inform the debate. It is important to identify those sites where protecting the environment has made a significant contribution to the economy—increasing national wealth, national incomes, and levels of national economic output.

Despite the fact that Armenia is small country with total area of 29.74 thousand sq. km, it contains one national park, five state reserves, and 22 protected areas. Significant areas include:

- Sevan National Park (largest lake of Transcaucasus Region and one of the largest freshwater high-mountain lakes of Eurasia).

- Khosrov State Forest Reserve (contains 1,686 species of flora and 213 species of fauna, 47% of which are registered in the “red book”).
- Dilijan State Forest Reserve (812 species of flora and 115 species of fauna, 49 of which are rare).
- Shikahog State Reserve (unique natural landscapes; species are not yet registered).
- Erebuni State Reserve (290 species of flora, 48 of which are rare).
- Sev Litch State Reserve (160 species of flora, 7 of which are rare).

Reserved areas form almost 11% of the total area of the republic and contain more than 60% of Armenia’s flora and fauna species [1]. Therefore, analysis of protected area management is an important issue for the region and there is a strong need to set up effective management standards in order to preserve outstanding natural landscapes and strategic natural resources. Management of protected areas related to agricultural and industrial development; that is why it is included in the presented problem. The debate over environmental protection is often about the balance between leaving areas in their natural or near-natural state, and developing and exploiting them. Besides versatile investigation of the problem, research includes developing strategies for protected-area managers. The recommendations must first of all consider regional characteristics and include methods of implementing economic principles in the process of managing protected areas.

3.3. ANALYSIS OF PRICING NATURAL RESOURCES AS A TOOL FOR MANAGEMENT

Pricing of land, water, and other environmental resources is an important tool for natural resource management. A key conclusion of this proposed activity is that there are significant opportunities to improve resource use through pricing policy changes. Many of the price policy opportunities outlined are challenging, because they require action from state government and local agencies and cannot be implemented by farmers and businesses. Decision makers are influenced by good information and persuasive arguments. This objective targets the creation of strategic information, for presentation to pertinent government bodies.

A great deal of research is available on the valuation of resources. Contemporary Environmental Economics has a large toolkit for valuing non-market goods [2, 5], but the application’s pricing methods are heavily reliant on the availability of corresponding data and organizational processes. The substitute service, productivity, and hedonic approaches are more likely to be applied as they are closely tied to market relations. Another objective considers the pricing of non-market goods, which is more difficult to implement.

3.4. INVESTIGATION OF OPPORTUNITIES OF VOLUNTARY ENVIRONMENTAL MANAGEMENT ARRANGEMENTS IN THE REGION

Applications of command methods for environmental management were sufficiently investigated and implemented in the post-Soviet era and it is now important to establish new types of activity—voluntary arrangements for environmental management—in order to fill the gap left by command and control strategies. This involves taking and implementing the experience of developed countries, where such solutions are part of nature protection programs. This is a new type of environmental managerial strategy for Armenia. In this period of transition and development in Armenia there is an emerging interest in voluntary environmental management arrangements to encourage businesses to prove and improve their environmental management credentials. Some arrangements could involve labeling or certification to allow businesses to demonstrate adherence to environmental standards. There are at least three good reasons to believe that voluntary programs can yield significant improvements in resource management sustainability that other approaches might not.

- Information/education-based voluntary programs show businesses and households ways to make "eco-efficiency" management changes—management changes that both enhance environmental performance and save costs. Farmers can learn how to do business by preserving the environment. The households or businesses that participate and realize eco-efficiency gains generally would not in the absence of a voluntary program, because they would not make the necessary information/education investment on their own.
- Many of the businesses and households who participate in voluntary programs are first movers who are willing to undertake environmental action beyond regulatory requirements when the information that is prerequisite for such actions is offered through voluntary programs.
- Voluntary programs can be a good way of inducing environmental performance improvements in the small-business sector which is often not subject to the same stringent environmental regulations as big industry.

The purpose of this objective is to disseminate the principles of voluntary environmental management among farmers, small businesses, and households. Recommendations for best practices must be developed and distributed to achieve voluntary environmental arrangements.

3.5. DEFINITION OF IMPEDIMENTS TO EFFECTIVE POLICY IMPLEMENTATION

The problem also requires highlighting impediments to best practices from the perspective of economic efficiency, fairness and the impact on the stock and quality of natural resources in Armenia. The institutional framework and government policies within the region will be considered. The purpose is to:

- Identify potential impediments to best practices in the management of the resource.
- Identify potential gaps in the institutional framework as it exists.
- Ensure that our understanding of the framework is sufficient before moving forward to consider various incentive mechanisms for resource management.

4. Data Provision: Problems and Solutions

Obtaining accurate data is one of the main problems for environmental analysis in Armenia. For environmental systems, the problem of accuracy of data processing is common. Any analysis of the links between economic change and the environment in transition economies are connected with the problems posed by the very uneven data quality.

Armenia's environmental information network is already established, and is in the development stage. State ecological monitoring is an important source of environmental information. The Ministry of Environmental and Nature Protection is responsible for environmental monitoring and ecological statistical information collection. Enterprises submit their respective reports to the State Statistics Committee and the water resources copy is sent to the Ministry of Agriculture. The Ministry of Agriculture generalizes all the water consumers' reports annually. Nongovernmental organizations do not collect ecological information; mainly they use official statistics in their activities. Despite the fact that there are still serious problems in this field, positive dynamics are evident.

5. Tools and Methods

The research presented here is based on a variety of tools and techniques. The investigation must be realized by analyzing and modeling the conditions of sustainable development. The basic principle is to integrate research and practice by using existing knowledge and implementing a participatory approach. The analysis will be based on the data and facts directly collected from the investigated field. Realization of surveys and interview procedures here are main strategies. Mathematical and statistical modeling tools are used for analysis and quantitative estimations. Up-to-date computer and information technologies are the tools for natural resource dataset creation, versatile analysis, and online presentations.

5.1. SURVEY PROCEDURE

Surveys comprise one of the key methods of data collection, defining indicators and measures of sustainability. Different types of surveys could be implemented with varied topics, regions, and groups of respondents. A socioeconomic survey (eliciting opinions concerning alternative practices and approaches, willingness to pay, or pricing) could engage farmers' participation in developing solutions and policies. Such surveys must be organized in primary elected communities with different conditions of development. Different social groups could be examined.

5.2. ECONOMIC MODELS

It is natural to analyze the problem of sustainability from the point of view of economic models, which may be helpful in at least identifying the main factors behind such preoccupations and perhaps in suggesting policy measures [6]. Indeed the models are useful to discuss some relevant factors. As for environmental systems, the problem of data accuracy is common; here probabilistic methods and models considering situations with incomplete information could be more acceptable.

6. Proposed Output

The research presented here addressed answers to some of the following questions, which presented scientific and practical interest:

- How have transition, development, and technological change affected sustainability options?
- How should post-Soviet states' characteristics and differences be considered in sustainable policy making?
- How can economic, social, and political problems be balanced in secure policymaking?
- Which is the best arsenal of tools and methods for the implementation of best practices for effective management of natural resources in the post-Soviet era?

The investigation of the above-listed questions is proposed to analyze important and specific ideas relating to sustainable policymaking. This investigation is multidisciplinary and includes legal, economic, biophysical, political, and social approaches.

7. Concluding Remarks

The proposed research addresses a regional problem as it is more acceptable to speak about sustainable solutions not only for one country but for the region as well. Yes, there are also local characteristics of development for each country, but the conditions and constraints of economic development and use of natural resources are similar and solutions will have common characteristics. In the developing world, the environment mostly suffers because of economic needs; economic policy constraints cause numerous environmental problems.

The analysis of presented problems is considered to be a notable step. The results could lead to the implementation of new research goals around secure development topics. This will bring opportunities for the activation of relations and collaborations around current and potential environmental problems in Armenia, between scientists, policymakers, and businesses.

8. References

1. Armenia National Program of Protection of Environment. Main report. The World Bank; 1999.
2. Berck P, Roberts M. Natural resource prices: will they ever turn up? Journal of Environmental Economics and Management. 1996; 31:65-78.
3. World Bank. Economic reform and environmental performance in transition economies; technical paper no 446. 1999.
4. UNEP, UNDP, OSCE. Environment and security: transformation risks into partnership, on the example of South Caucasus. 2004.
5. Munasinghe M, editor. Environmental economics and natural resource management in developing countries. The World Bank; 1992.
6. Hafkamp W. Economic-environmental modeling in a national-regional system. Amsterdam; 1984.

CHAPTER 18

GLOBAL CHANGES, NEW RISKS, AND NOVEL METHODS AND TOOLS OF THEIR ASSESSMENT

Modeling and Management of Environmental Security in Ukraine

K. ATOYEV

Glushkov Institute of Cybernetics
Ukraine

A. TOMIN, T. AKSIONOVA

National Technical University of Ukraine
Kyiv Polytechnic Institute

Abstract: The main goal of this work is to strengthen the capacity in such fields as impact of uncertainty and unpredictability caused by global change (GC) on transformation of security and sustainability for which standard modeling methods and tools are inadequate. Advanced modeling method and tool for system risk analysis has been elaborated. It is based on catastrophes bifurcation, and chaos theories. A "fractal pyramid" of parametrical bifurcation sets, characterizing transition of security from one its sublevel to another, is constructed. On the basis of this pyramid, a hierarchical risk matrix of system transformation can be determined. The following questions will be examined:

1. How do new risks caused by GC escalate the extreme manmade and natural disasters?
2. Why do rare events take place with increasing frequency?
3. How do new challenges deform security area in the Black Sea region?

1. Global Change and Escalation of Extreme Manmade and Natural Disasters

During the past decade we have witnessed novel forms of global challenges, resulting from a wide range of asymmetric threats, strengthening of the destructive power of rare events, and increased possibilities that minor disturbances can cascade into disasters with unpredictable ecological, economic, and political consequences. A short list of generators of new risks includes:

I. Linkov et al. (eds.), Managing Critical Infrastructure Risks, 339–351.
© 2007 *Springer*.

1. Increasing role of social amplification of threats.
2. Existence of longer-term links and causal chains of great extent.
3. Increasing role of system risk.
4. Increasing geographic concentration of wealth, human capital, knowledge, and communication links.

Progress in communications and information processing associated with a networked society also generates new risks. Uncertainty and unpredictability principles have become keywords characterizing various aspects of GC. That is why our times were named "epoch of bifurcation" [1]. The deformation of global security causes crises of control systems, leads to sharp increases in environmental and manmade disasters and financial and economic catastrophes and decreases the horizon of forecasting for extreme events, especially when the probability density function of destructive power is described by power low. Its curve decreases more slowly than Gaussian probability low (see Figure 1).

High susceptibility to initial conditions is a common feature of nonlinear systems such as the real world and makes it impossible to predict the development of GC. Traditional methods to forecast system development are not efficient enough for such events.

GC (per F. Capra [2]), can be considered a giant web of fractal networks, where each network contains another. The quickly changing world with its instability and jumps to new forms of organization is a good example of

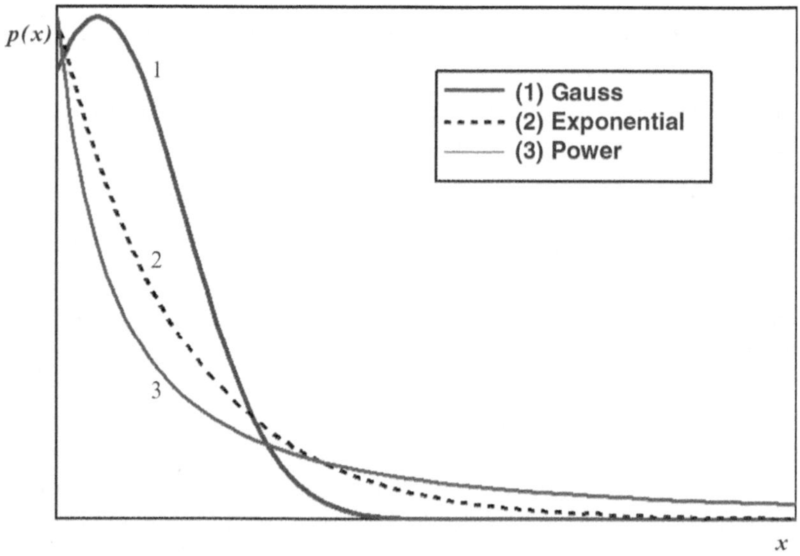

Figure 1. The typical view of Gauss (1), exponential (2), and power (3) lows.

Prigogine's theory of dissipative structures [3]. These instabilities appear to be a result of fluctuations amplified by positive long-term feedback loops. The amplifications of GC play an important role in the appearance of the supported chaotic environmental and social regimes of global flows.

GC plays an important role the existence of new risk. One typology of new risks has been proposed by O. Renn [4]. This typology classifies risks into five main categories:

1. Psychosomatic risks.
2. System breakdown risks.
3. Systemic risks.
4. Knowledge management risks.
5. GC risks.

Other new taxonomies of risks also include risks resulting from:

1. Homogenization of infectious diseases.
2. Occurrence of new pathogens (AIDS, Severe Acute Respiratory Syndrome (SARS), Bovine Spongiform Encephalopathy (BSE) or mad cow disease).
3. Over prescription of antibiotics.
4. Advances in functional genomics, DNA technologies, genetic and protein engineering, and genetically modified food.
5. Insidious risks, which emerge slowly but surely over time [5, p. 386].
6. Terrorist threat.
7. Mental disadaptation of personality connected with atomization of society and increased social volatility.

Hacker-type-risks and transboundary risks also are included in new taxonomies of risks.

Climate change is another enormous global threat. Various scenarios of the Intergovernmental Panel of Climate Change indicate some possibility of irreversible catastrophic effects that could wipe out a great part of the world's population due to global warming. Moreover, climate change causes disturbances to traditional biogeocenosis, changing the migration routes of birds of passage. One possible consequence of these processes is the expansion of avian influenza (bird flu).

The next group of new risks is subjective risk, which is associated with prospect theory. One very important result of Kahneman and Tversky's work is the finding that people's attitudes toward risks concerning gains may be quite different from their attitudes toward risks concerning losses [6]. A. Boholm has proposed separating this phenomenon as the subjective component of risk:

Subjective risk acknowledges that people's beliefs and opinions tend to deviate from such scientific assessments. People understand and judge risks in terms of locally defined values and concerns [5].

Subjective risk can be examined as an additional powerful source of sharp escalations in manmade mistakes and behavioral unpredictability that decrease security.

So, the transformation of security is carried out in space with its principal new features caused by globalization and increasing uncertainty. The changing of fundamental time-space continuum characteristics generates the fields of new risks, leads to a more unpredictable world, and increases the degree of global mega threats (atomic, genetic, chemical, mental, environmental) and their damage. Moreover, the revolution in the quality of communication links, the necessity of fast solutions, and the increasing role of psychological factors in decision making in the face of uncertainty also transform the area of security.

New risks demand new assessment methods. That is why theories of defense planning, psychological and synergetic economics, and the theory of catastrophes have taken central places in risk analysis in this era of uncertainty [7, 8]. Adaptiveness describes the ability of a system to transform its internal structure in response to environmental changes. It has become one of the most significant characteristics of sustainable development for systems. The objective of this article is to develop an advanced modeling method and tool for the investigation of adaptiveness to GC impacts and risk assessment of irreversible changes causing cardinal shifts in the area of security.

2. The Advanced Modeling Method for Management of Environmental Risk

The advanced modeling method for extreme manmade and natural hazards assessment has been elaborated A global society is a complex interrelationship in a "square of security" (SS), which is determined by the risks of manmade and natural disasters and economic and social crises. The disturbance of balance within the SS is a most formidable threat, as it leads to a transition from norm to disaster. GC increases the possibility of some fluctuation destroying an intricate balance within the SS, cascading into a disaster with unpredictable consequences.

Catastrophe theory (CT) is a mathematical formalism for modeling nonlinear systems whose discontinuous behavior is determined by smooth changes in a small number of driving parameters [9]. CT has been applied to a wide range of different systems from physics, engineering, biology, psychology and sociology. A small subset of specific phenomena that were analyzed and modeled using CT includes the formation of caustics in ray optics, the stability of ships, the stability of black holes, the size of bee societies, morphogenesis or cell differentiation in embryology, bistability of perception, and sudden transitions in attitudes [10]. In all these applications, the behavior of the system under study shows sudden, discontinuous changes or phase transitions as a result of small, continuous changes in variables that influence the system (e.g., the freezing of water when temperature is gradually decreased or the collapse of a bridge under slowly mounting pressure). In general, CT applies to systems that may respond to continuous changes in control variables by a discontinuous change from one equilibrium state to another.

Let us propose that global society satisfies all the requirements of a potential system and can be described by some potential function $U(X)$ of hazard - X. This function has some steady states; some of them are stable and others are unstable. The transformations of a system from one steady state to another, or transformation of the character of a steady state (for instance, from stable to unstable) are functions of control parameters A. These parameters control both the movement of the system's current position upon surface U, and transformations of the surface itself. To make it simple, it is possible to put forward the following suppositions. Potential function $U(X,A)$ has three stable steady states. The first one characterizes normal conditions (low level of X). The second characterizes a state with a medium level of X. The third characterizes crisis conditions (high level of X^x).

Society is changing in many ways. In [11] four key GC drivers were identified:

- Increasing global connectivity.
- Atomization of society.
- Rapidly advancing technology.
- Growing ability to measure and monitor.

To examine a GC impact problem we also have chosen a four-dimensional typology. The transformations of security are defined by one of the universal deformations of CT: a butterfly:

$$-\partial U(X, A)/\partial X = X^5 + A_1X^3 + A_2X^2 + A_3X + A_4 \tag{1}$$

where A_i (I = 1,4) are control parameters, characterizing various aspects of the GC impacts on economic, social, environmental, and technogenic dimensions of sustainable development:

- i = 1 – index of global connectivity (economic index), determined by the powerful forces of economic globalization, market liberalization, and economic possibilities to prevent and minimize consequences of disaster.
- i = 2 – index of atomization of society (social index), determined by factors that provoke the rise of violence, mental disadaptation of personality, and

probability of epidemics; and reinforce the notions of individual ability, identity, freedom, and responsibility.

- i=3 – index of ability to measure and monitor (index of natural disasters), determined by ecological and climatic factors that impact the probability of natural disaster and man's growing ability to observe and analyze his environment.
- i=4 – index of advancing technologies (index of man-made disasters), determined by factors that impact the development of technologies, reducing their lifespan and increasing their disasters.

The risk estimation of sudden and unexpected transformation from one steady state to another may be performed using the theory of catastrophes allowing the calculation of bifurcation values, curves, and surfaces of control parameters. The risk is estimated as an extent of a control parameter approaching these bifurcation values characterizing a system's transition from one steady state (norm) to another (catastrophe) [4]. The main advantage of this approach is a determination of risk as a function of dynamic variables of the investigated system. It also allows identification of the weakest link of the system under examination and the areas in need of improvement.

The following algorithm of risk assessment can be proposed:

1. The collection of information characterizing the above-mentioned indices input from contemporary databases.
2. Determination of indices characterizing appropriate groups of parameters and calculation by means of developed mathematical models using input data.
3. Calculation of the bifurcation parameter values at which the number of system states changes.
4. Estimation of restoration possibilities of each considered system by the remoteness of the parameter characterizing the appropriate index from its bifurcation value.

Under more detailed investigation, parameters A_i also may be examined as *n*-dimensional functions $A_i\ (F_j)$ (j = 1,n), that are arguments of some potential functions U_{Ai} [12]. Variables F_j describe each index in more detail. Such iterative process leads to the next sublevel of bifurcation sets, determined by:

$$dA_i / dt = -\partial U_{Ai} / \partial A_i,\ \ A_i^M + B_{il}\, A_i^{M-2} + \cdots + B_{i(M-3)}\, A_i^2 + B_{i(M-2)} A_i + B_{iM-1} = 0 \quad (2)$$

So, a peculiar fractal pyramid of bifurcation sets can be constructed; by using this pyramid, a hierarchical risk matrix of system transformation from one sublevel of security to another can be determined.

For instance, the economic component of threat (A_1) can be determined on the basis of integrated indices of political engagement (B_{11}), technological development (B_{12}), personal contacts (B_{13}), and economic integration (B_{14}). The level of political engagement (B_{11}) can be determined on the basis of number of memberships in international organizations (C_{111}), U.N. Security Council missions in which each country participates (C_{112}), and foreign embassies that each country hosts (C_{113}). The level of technological development (B_{12}) can be determined on the basis of number of Internet users (C_{121}), Internet hosts (C_{122}), and secure servers (C_{123}). The level of personal contacts (B_{13}) can be determined on the basis of intensities of international travel and tourism (C_{131}), international telephone traffic (C_{132}), and cross-border transfers (C_{133}). The level of economic integration (B_{14}) can be determined on the basis of parameters characterizing trade (C_{141}), foreign direct investment and portfolio capital flows (C_{142}), and income payments and receipts (C_{143}), as has been proposed [13].

Relationship shifts within SS may cause transfers from norm to pre-crisis or to crisis. Bifurcation values of these parameters can be calculated by means of the proposed mathematical method. Reaching such critical values abruptly increases the probability of a transfer from one functional state to another. Thus, for a given state of a global system, it is possible to determine ranges of functional parameter values corresponding to the states of norm, pre-crisis, and crisis. The main advantage of this approach is the determination of risk dynamics as the function of dynamic variables connected with SS. It also allows identification of the weakest link of the system under examination and the areas in need of improvement.

Fitting a catastrophe model to noisy data constitutes a serious challenge, however, because CT was formulated specifically for deterministic systems. Cobb [14] has developed a stochastic version of CT (SCT) based on the Ito stochastic differential equations. In SCT, the stable and unstable equilibrium states of the system correspond to the modes and the antimodes of the empirical probability density function, respectively. Unfortunately, SCT is not invariant under smooth and invertible transformations of variables—this was an important limitation, since invariance to diffeomorphic transformations is essential in deterministic CT.

Wagenmakers et al. [15] have generalized the Cobb's method by taking into account the Ito transformation rule. In contrast to the Cobb's method, this generalized SCT is unaffected by smooth and invertible transformations. It was shown that an invariant function is obtained by multiplying the probability density function with the diffusion function of the stochastic process. This invariant function can be estimated by a straightforward time series analysis based on level crossings. So there are no serious restrictions for application of this theory to noisy data. In future we are planning to adopt the results obtained in [15,16] to our risk assessment tool. It essentially increases the quality of systemic risk analysis and management of sustainable development.

3. The Decision Support System for Environmental Security Assessment

The decision support system for integrated security assessment and management has been elaborated on the basis of the fractal pyramid approach. The control board of a computer system is presented in Figure 2.

The distinctive feature of the elaborated computer system is joining technological gear software that provides forecasting and optimization task solutions that minimize the level of environmental risk. It calculates optimal controls and minimizes the negative impacts of GC on sustainable development. The impacts of manmade and natural disasters are investigated with the help of various mathematical models.

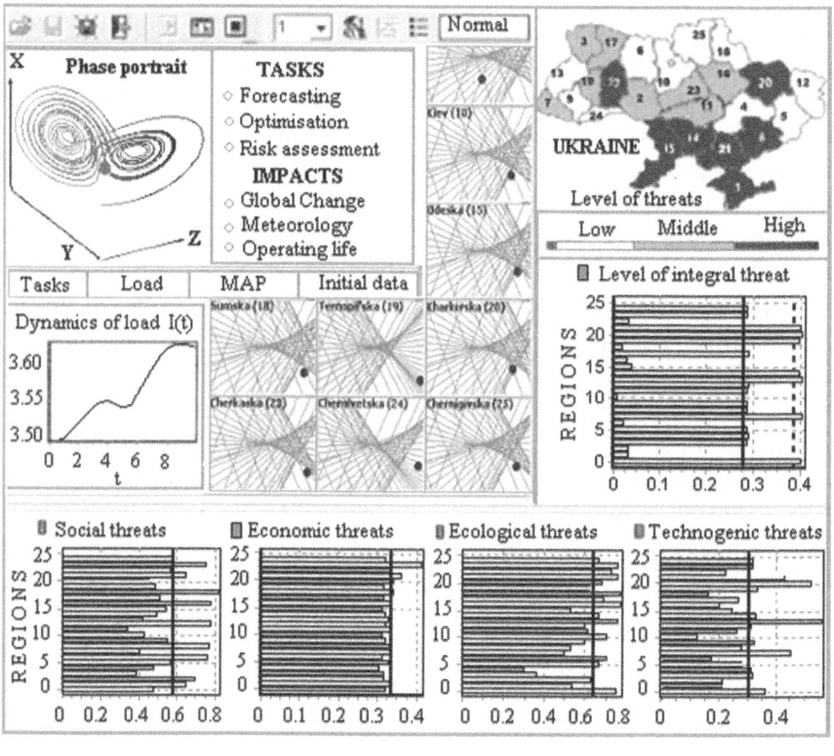

Figure 2. The control board of a computer system.

Ukraine's 25 regions were ranked by degree of environmental vulnerability (EV) and technogenic, ecological, social, and economic components. The projections of catastrophes on the surfaces of technogenic and ecological security are shown in the middle of Figure 2. The changing phase portraits that correspond to the dynamics of chosen environmental impacts, also as changing operating life at chosen load, are presented at the left of Figure 2. The ranking of regions on EV and its components that were calculated with the help of the examined method are presented at the bottom. A solid vertical for components corresponds to the median value of threat indices for Ukraine. A solid vertical on the figure for integral threats separates regions with low and middle levels of threat; a dashed vertical separates regions with middle and high EV.

4. The Challenges to Security in the Black Sea Region

Let us examine the main challenges to security in the Black Sea region caused by the risk of manmade and natural disasters, and economic and social crises.

The first environmental threat to the Black Sea is due to its status as the "kidney" of Eastern Europe. The Black Sea collects waters from an enormous drainage basin which includes a large part of Turkey and Central and Eastern Europe. The Mediterranean and the Black Sea form an integrated oceanographic system. Due to pollution (brought mainly by rivers), the Black Sea is dying.

The second environmental threat is oil pollution that can cause substantial damage to the environment and traditional economic activities in the Black Sea (tourism and fishing). Since 1960, there have been about 2000 oil spills. 24.8% of oil spills from vessels are due to groundings. 21% of oil spills from vessels are due to collisions. Additional pressure on the environment of the Black Sea area may be due to:

1. Increased tanker traffic transporting oil from Azerbaijan, Kazakhstan, the Russian Federation, and Turkmenistan.
2. The Blue Stream Pipeline, designed to run 400 km under the Black Sea to Samsum and further to Ankara (it will pump 16 billion cubic meters of gas per year to Turkey).
3. The elaboration of possible gas fields on Crimea's shelf and near Zmeinyi Island.

The third threat has an epidemiologic genesis and connected with the migration of birds of passage through this region. Turkey, Romania, Ukraine, and some Mediterranean countries may become centers of bird epidemics.

Social and economic threats are interrelated. The Black Sea area now presents one of the bifurcation zones bridging European security, especially if we take into account its location near the arc of geopolitical conflicts and the escalation of terrorism threats. Without establishing security and stability in this region, threats to oil/gas transport flows to Europe will increase. The main economic threat is connected with energy supply. The diversification of energy sources, energy traffic, and securing energy supply are important necessities for sustainable development. One of the ways to solve this problem is to route energy flows from Middle Asia, over Georgia, the Black Sea, and Ukraine to Western Europe (Pipeline Odessa-Brody-Gdansk). This pipeline will strengthen economic stability in the Black Sea area and will reduce the environmental threat to Istanbul and its surrounding area due to decreased tanker traffic though a narrow bottleneck (Bosporus Strait).

The impact of abovementioned factors on security dynamics in the Black Sea region was investigated with the help of the elaborated risk assessment tool. The dynamics of control parameters A were investigated on the basis of the Lorenz model of deterministic nonperiodic flow [16]. There are positive and negative feedback loops between levels of GC and EV. The formalization of these interrelations permits the investigation of possible mechanisms of chaotic behavior that cause instability in the security area. It was shown that relationships between levels of GC, energy resources, and threats may be described by the Lorenz model of Metastable Chaos [17]. A transition from stability (low level of vulnerability) to instability (high level of vulnerability) could be analogous to the transition from a laminar to turbulent flow.

The preliminary ranking of states by degree of security and economic, social, ecological, and technogenic components was carried out. In Figure 3, projections of surfaces describing the catastrophe on the surfaces of economic threats (ET) and social threats (ST) for some states of Black Sea area are presented.

The deterministic nonperiodic dynamics of threat and the projection of catastrophe on the surface of technogenic and ecological security for selected region of Ukraine (Crimea) are presented in Figure 4.

On the basis of obtained modeling results, some countries of the Black Sea region were ranked for vulnerability to ET and ST. Georgia faces the highest level of threats. The point characterizing the current state of this country corresponds to pre-crisis area parameters. Ukraine and Romania

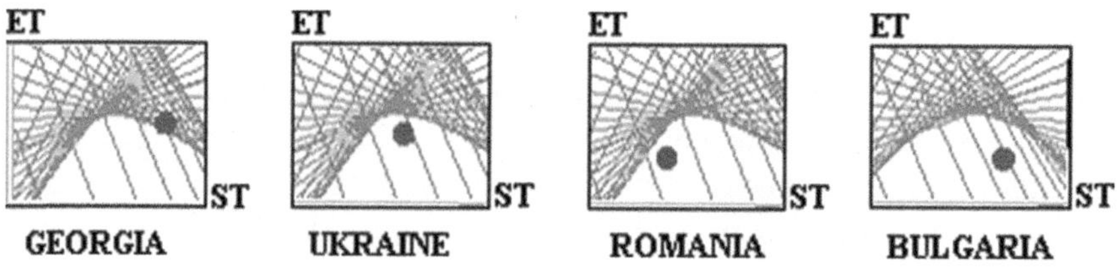

Figure 3. Projections of catastrophe on the surfaces of ET and ST.

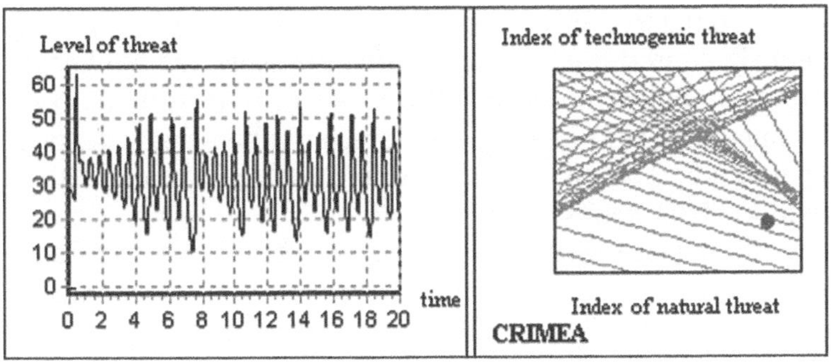

Figure 4. Dynamics of threat and phase portrait of security area for Crimea.

have the same remoteness from bifurcation lines. Bulgaria faces the lowest level of threats. The majority of the new democracies of the Black Sea area have a similar past (with roots in Byzantine civilization), are transforming in similar directions as they move toward the North-Atlantic community, and face similar problems and threats.

So, the intensification of regional cooperation in the fields of energetic, environmental, and political security is a major agenda item for new democracies of the Black Sea area. The optimal trajectory of their sustainable development demands the synergetic and coordinative efforts of these neighbors.

5. Conclusion

The threats that will be most challenging in the future are connected with synergistic interactions between risks of various origins. There are constellations of interconnected risks within a community described as a "risk envelope" [5]. That is why our regulatory focus moves from managing point sources and risk in isolation to one of managing and trading risks from multiple sources and interdependencies of ecological, economic, social, and technogenic risks.

The cumulative effects of various stressors emanating from multiple sources become important dangerous generators of uncertainty and instability in modern society. Indeed, interdependencies of various risks increase the difficulty of identifying the principal cause of injurious effects and make systems already weakened by stress more susceptible to the effects of additional stressors.

One of the most significant indicators of system security is adaptiveness to various impacts: the ability of a system to change its internal structure in accordance with environmental changes due to mutation or activation of internal systemic transformations. The main forces driving these transformations are dependent on energy support. The stress response is also associated with the mobilization of energy to enable the system to deal with threat or danger. However, energy cannot be diverted in all directions simultaneously, so in the case of cumulative stressors, the nonlinear increase in system vulnerability and decrease in its adaptive possibilities takes place. These nonlinearities lead to system parameter values that often are characterized by unpredictable behaviors, as when small impacts cause the essential transformation of system security.

The traditional methods of risk analysis often are not efficient enough for such events when a system is very susceptible to initial conditions and its behavior shows sudden, discontinuous changes or phase transitions as a result of small, continuous changes in variables that influence the system. Indeed:

> …risk management tools designed on the assumption that historical variability and existing trends are relevant to future context are likely to be increasingly challenged by growing interdependencies of environmental, technological, social and economic hazards [5].

In this paper, a decision support strategy is proposed for the investigation of interconnected risks with the help of CT, chaos theory, and bifurcation. It increases our possibilities in the field of systemic risk analysis and management. The main advantage of this approach is a determination of risk as a function of the dynamic variables of the investigated system.

The developed approach may be applied to a wide range of different complex systems from ecology (assessment of EV, GC impact, reducing water stress, and optimal water resource management), epidemiology (estimation of HIV/AIDS, sharp escalation for management of epidemics, and risk ranking) and biology (healthcare, cellular energetic trigger, metabolic and hormonal regulation), to psychology (formation of attitude, mutual penetration of western and eastern worldview matrices, and terrorism) and sociology (deformation of social risk fields).

6. Acknowledgements

I would like to thank all organizers of the workshop and especially Dr. Igor Linkov, Dr. Abou Bakr Ramadan, and Dr. Stefano Della Sala for the high level of scientific programming and their cordial attention to all participants of workshop. Travel to present this paper and stay during workshop was funded by NATO ARW.

7. References

1. Laszlo F. The creative cosmos. Edinburgh: Floris Books; 1993.
2. Capra F. The hidden connections: a science for sustainable living. New York: Random House; 2002.

3. Prigogine I, Stengers I. Order out of chaos. New York: Bantam; 1984.
4. Renn O. Typology of risks. Journal of Risk Research. 2003; 6(4-6):519-524.
5. Wilkinson A, Elani S, Eidinow E. Background and dynamics of the scenarios. Journal of Risk Research. 2003; 6(4-6):365-401.
6. Kahnemann D, Tversky A. Subjective probability: a judgement of representativeness. Cognitive Psychology. 1972; 3:430-454.
7. Lofstedt RE, Ikeda S, Thompson K. Risk management across the globe: insights from a comparative look at Sweden, Japan and the United States. Risk Analysis. 2000; 20(2):157-161.
8. Klinke A, Renn O. A new approach to risk evaluation and management: risk-based precaution-based and discourse based strategies. Risk Analysis. 2002; 22(6):1071-1094.
9. Thom R. Structural stability and morphogenesis. New York: Benjamin-Addison Wesley; 1975.
10. Poston T, Stewart I. Catastrophe theory and its applications. New York: Dover; 1978.
11. Wilkinson A, Elani S, Eidinow E. Riskworld scenarios. Journal of Risk Research. 2003; 6(4-6):297-334.
12. Atoyev KL. Mathematical modeling of metabolic and hormonal regulation: risk assessment of environmental and radiation influence on various links of endocrine system. HIT Journal of Science and Engineering B. 2005; 2(1-2):31-53.
13. Kearney AT. Measuring globalization: who's up, who's down. Foreign Policy. 2003; 1:60-72.
14. Cobb L, Zacks S. Applications of catastrophe theory for statistical modeling in the biosciences. J. Am. Stat. Assoc. 1985; 80(1985):793-802.
15. Wagenmakers EJ, Molenaar PCM, Grasman RPPP, Hartelman PAI, van der Maas HLJ. Transformation invariant stochastic catastrophe theory. Physica D. 2005; 211(3-4):263-276.
16. Lorenz E. Deterministic nonperiodic flow. Journal of Atoms.Sci. 1963; 20:130-141.
17. Atoyev KL. The challenges to safety in east Mediterranean: mathematical modeling and risk management of marine ecosystem. In Levner E, Linkov I, Proth JM, editors. Strategic management of marine ecosystems. Dordrecht: Springer; 2005.

CHAPTER 19

COLLABORATIVE PUBLIC-PRIVATE RISK ASSESSMENT IN VESSEL TRAFFIC SAFETY[1]

Two Case Studies

R.G. ROSS

U.S. Department of Homeland Security
Office of Comparative Studies

Abstract: This paper presents collaborative approaches as a strategy for addressing "wicked" problems in the area of vessel traffic safety and risk assessment. Two U.S. case studies demonstrate the effectiveness of collaborative decision making. Strategies for productive collaboration are provided and special considerations related to terrorism risks are discussed.

1. Decision Making in "Wicked" Environments

In their seminal initial statement of the "wicked problem," Horst Rittel and Melvin Webber, professors of Design and City Planning, respectively, noted:

> The search for scientific bases for confronting problems of social policy is bound to fail, because of the nature of these problems. They are "wicked" problems, whereas science has developed to deal with "tame" problems [1].

Note that the authors are using the term *wicked* in the sense of tricky or vicious rather than evil. They go on:

> ...as distinguished from problems in the natural sciences, which are definable and separable and may have solutions that are findable, the problems of governmental planning—and especially those of social or policy planning—are ill-defined; and they rely upon elusive political judgment for resolution. Social problems are never solved. At best they are only re-solved—over and over again.

Rittel and Webber identified 10 characteristics of wicked problems. Seven of these have been restated in Table 1 below and an eighth characteristic has been added to reflect the Law of Unintended Consequences.

An additional characteristic of wicked problems is that there are numerous stakeholders whose interests may be equally valid from a societal perspective

---

[1] The views expressed in this paper are those of the author and do not necessarily reflect the official positions of the U.S. Coast Guard, the Department of Homeland Security, or any other agency of government.

I. Linkov et al. (eds.), Managing Critical Infrastructure Risks, 353–367.
© 2007 *Springer.*

TABLE 1. Characteristics of wicked problems [2]

1. You don't understand the problem until you have developed a solution.
2. Wicked problems have no stopping rule.
3. Solutions to wicked problems are not right-or-wrong, but good-or-bad, better-or-worse.
4. Every wicked problem is essentially unique and novel.
5. Every solution to a wicked problem is a "one-shot" operation.
6. Wicked problems have no given alternative solution.
7. Every wicked problem can be considered to be a symptom of another problem.
8. Every solution to a wicked problem will generate more problems.

but which are also irreconcilably opposed. This aspect of a wicked problem is perhaps best illustrated with an example. Consider this question facing a city planner: "Should the highway be built through the city or around it?" Clearly, there is only one shot at building the highway. Once built, it won't be torn down and rebuilt. Equally clear are that the problem is unique (every city and highway is different) and that there will be numerous stakeholders with competing interests, including property owners, business owners, and residents of neighborhoods that would be displaced. This multiplicity of stakeholders and competing interests creates a highly complex social context within which the decision must be made. Finally, after all the studies, models and projections are done, the final decision will invariably create winners, losers, and unanticipated consequences (i.e., new problems). For these reasons, public policy decisions on how to deal with wicked problems necessarily call for judgment based on political and other relevant factors. Problem-solving and risk assessment approaches suited to tame problems simply will not work when faced with a wicked problem.

In the wake of the Exxon Valdez oil spill, the political environment surrounding oil spills and the related issue of preventing collisions and groundings in restricted waters were such that managing vessel traffic risks in ports and other restricted waterways was a truly wicked problem. Even in open waters off the California coast, the environmental sensitivities were such that battle lines were drawn and the arguments were, to say the least, heated. The two case studies which follow offer vivid proof that, beyond being merely beneficial in addressing wicked problems, collaborative approaches are absolutely essential for politically and operationally effective public policy decision making.

2. Case Study 1: The Port and Waterway Safety Assessment Process

2.1. INTRODUCTION

The Port and Waterway Safety Assessment (PAWSA) process was developed by the U.S. Coast Guard to satisfy objections to its previous approach for determining if a Vessel Traffic Service (VTS) would be an appropriate risk management tool in U.S. ports. The PAWSA process features significant stakeholder involvement, use of expert opinion and historical information, consideration of multiple alternative approaches to mitigating vessel traffic risk and a blend of several different analytic techniques. This case study provides both a description of the process and an examination of the factors that lie behind its design. PAWSA is also considered in light of both long-established and more recent observations on the politics of risk-informed public policy decision making, which provides useful insights into the whys and wherefores of the PAWSA process, and why it was so well accepted.

2.2. BACKGROUND

A VTS is a shore-based service which provides information, navigational assistance, and/or traffic organization to transiting vessels. VTSs are established in ports and other waterways based on need. The VTS concept is most easily understood by thinking of the approach and ground control elements of air traffic control. Approach control deals with aircraft coming into the traffic pattern around an airport as the aircraft is preparing to land. Ground control directs the movement of aircraft once they are on the taxiways between the runway and terminal. Of course, just as not every airport or airfield has a full suite of traffic control capability, not every port needs a VTS. The air traffic control/VTS analogy is not perfect (e.g., an air traffic controller can't tell a plane in the air to stop and hold its position while other traffic proceeds) but it is close enough for purposes of this paper.

The March 1989 EXXON VALDEZ oil spill was the largest such event in U.S. history and it sparked both public and congressional outrage. This outrage[2] resulted in the Oil Pollution Act of 1990 (OPA '90) (P.L. 101–380) which required a host of improvements in spill prevention and response preparedness. One requirement was that the Coast Guard perform a Port Needs Study for major oil ports to determine if a VTS would be an appropriate risk reduction measure for any of them. The task of performing the Port Needs Studies was given to the Volpe Transportation Systems Center, a federally funded research and development center located in Cambridge, MA.

[2] The anger over the spill was so intense that OPA '90 passed both houses unanimously and was signed with great fanfare by a former Texas oilman, President George H.W. Bush. To gauge the depth of feeling behind OPA '90, it is worth noting that the vote for the declaration of war against Japan after the sneak attack on Pearl Harbor was not unanimous.

The studies conducted by the Volpe Center used traditional operations research methods including development of models and use of statistical analyses based on the Coast Guard's accident database. The study showed a positive net VTS benefit in seven ports, a negative net benefit in eight and sensitive (i.e., indeterminate) net benefit in eight others. Existing VTSs were distributed among all three categories. For a variety of reasons to be discussed later, the conclusions reached in these studies raised strong, and sometimes contradictory, objections in many sectors of the maritime shipping community and sympathetic members of Congress. As a result of these objections, the Coast Guard was directed to reconsider the issue and provide answers that had better analytic support and political acceptability.

Objections to the original studies were not entirely fair but were nonetheless based on a number of very real concerns. At the time (1996), the Clinton administration was proposing user fees for a variety of navigational services. These would be imposed on vessels calling in U.S. ports. The potential for such fees raised concerns in the shipping community, both on the vessel side and on the port side. Vessel operators were opposed to any fee for services that had previously been paid for out of general revenues and which arguably benefited a wider pool than those who would have to pay. Port authorities were opposed to the fees on the grounds that different fee structures, based on different levels of service in each port, would alter the existing competitive balance between ports[3]. A number of port authorities opposed having a VTS due to the potential for a higher user fee. Interestingly, some port authorities were upset by findings that their port did not merit a VTS! These port authorities probably discounted the likelihood of user fees and were more concerned about environmental issues and an adverse impact on competitiveness, perhaps by creating a perception of second-tier status or of being less safe than a VTS port. Port authorities are creatures of local and/or state governments and understandably did not want federal user fees or the establishment of a Coast Guard-operated VTS to tilt their playing field—especially when they did not know in advance which way it would tilt.

[3] The collection of user fees to fund navigation (ATON) services poses several problems, not the least of which is how to calculate and assess the value of the services received by a given beneficiary. One of the principles governing the establishment of user fees is that any fees established should be based on the costs of delivering a specific service to each beneficiary or identifiable class of beneficiary. If adhered to for navigational services, a port with a VTS or other high ATON costs would be adversely impacted relative to a competitor with lower ATON costs. Other possible collection schemes posed their own problems based on restrictions in international trade agreements or, for one approach, the U.S. Constitution.

Further complicating the picture, the Coast Guard was then planning a major recapitalization of the existing VTSs (which would have resulted in yet higher user fees) and new navigation technologies were in the offing with the potential to reduce the future need for a VTS in some number of ports. A final objection, which was only partially fair, was that local stakeholders were not consulted as a part of the Port Needs Study process. This was unfair in that local stakeholders had been included, but it had some basis as local stakeholder involvement was limited and thus not widely known or appreciated. Moreover, in some cases those who were involved may not have been the best or most appropriate stakeholder representatives available.

A final reservation about the initial Port Needs Studies stems from the old saying "If every tool in your tool chest is a hammer, every problem looks like a nail." The Port Needs Studies carried out by the Volpe Center may have been impacted by an assumption, more likely implicit than explicit, that if unacceptable risks were found then a VTS would be the best available risk mitigation measure, or perhaps even the only one possible. A more holistic view of the vessel traffic management tool kit recognizes that there are a number of different tools available, each with its own strengths and weaknesses.

2.3. THE PAWSA PROCESS

2.3.1. Design Goals and Risk Focus

The PAWSA process was designed to meet a number of specific criteria set out in Table 2. These criteria were based on Congressional direction, our own internal examination of "What went wrong?" and information generated through the National Dialogue Group (a stakeholder focus group) conducted for the Coast Guard by the Transportation Research Board of the National Science Foundation.

TABLE 2. Ports and Waterways Safety Assessment design goals

1. Focused on Vessel Traffic Risk Drivers – Probability and Consequences
2. National Approach – Broadly Applicable – Not Tailored to a Specific Port
3. Structured Analytical Process – Documented – Repeatable
4. Low Cost – Widely Deployable
5. Locally Focused
6. Expert and Stakeholder Involvement
7. Select Locally Appropriate Risk Mitigation Measures

2.3.2. Participants and Expert Opinion Solicitation

The PAWSA evaluation process takes place at the local level and includes up to 30 representatives of various stakeholder groups, which are identified by the Coast Guard Captain of the Port. Stakeholders will typically include the Coast Guard Captain of the Port, Coast Guard Aids to Navigation service providers responsible for the port, Army Corps of Engineers personnel, the local Pilots Association(s), towing vessel operators, port authorities, locally active environmental groups, designated Natural Resource Trustees (e.g., National Park Service or Fish and Wildlife), and others as deemed appropriate based on local concerns. PAWSA evaluations are typically facilitated by outside contractors who take care of logistics, report preparation, data capture, and so forth. Outside contractors are used for three primary reasons—to minimize the burden on already stressed Coast Guard operating units, to avoid the impact of repetitive trips up the learning curve and, most importantly, to ensure that the process will be seen as impartial and not driven or steered by uniformed members of the Coast Guard. However, while outside facilitation is strongly preferred, full documentation and the appropriate software to execute the process are also available to every Captain of the Port for self-help efforts when funds are not available for facilitation services.

The PAWSA process (shown in Figure 1 in simplified form) takes place over a two-day period, during which participants are taken through a structured agenda in which concepts are introduced and the next step in the process is described before the group is asked to discuss issues or perform any evaluations. An important aspect of PAWSA is that the various experts are actually brought together physically and encouraged to exchange views on how the various factors apply in their port. This exchange is vital because each party will be basing his or her opinion on only a partial understanding of a more complicated whole. For example, Pilots/Docking Masters and tugboat operators are both involved in the final berthing of a ship, but their perspectives are very different. Similarly, operators of deep-draft and shallow-draft vessels are impacted by factors such as channel configuration in very different ways. Multiple views are necessary for a more complete understanding and the frank exchange of ideas occasionally reveals issues of which no stakeholder was aware. Additional side benefits of the process are the improved understanding and lines of communication that can be opened between historically antagonistic parties.

The other reason for this live forum is related to a deficiency in the original Port Needs Studies. Those studies were based on data in the Coast Guard's accident investigation files. This information, however, only represents the tip of the iceberg. There are many more near-misses than there are actual accidents. Near-misses are not generally reported and even if

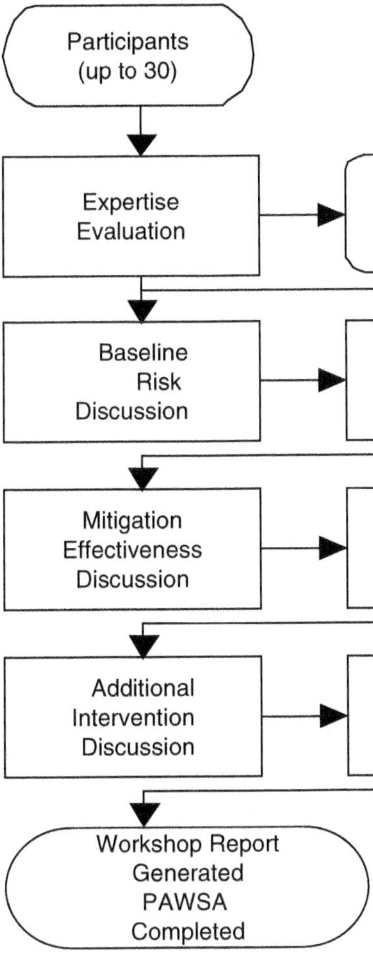

Figure 1. PAWSA process flowchart.

they were, would never make it into an accident database. Participants have actually been more open about such incidents in not-for-attribution group sessions than in one-on-one expert solicitations.

2.3.3. Rating and Ranking Port Risks

The PAWSA process starts by focusing on those factors that create risks associated with vessel traffic. These factors, called risk drivers, are associated with both the probability of an accident and the consequences that could result from such an accident. These risk drivers can be grouped into categories such as traffic conditions (i.e., the number and type of vessels present

Vessel Conditions	Traffic Conditions	Navigational Conditions	Waterway Conditions	Immediate Consequences	Subsequent Consequences
Vessel Quality	Volume of Commercial Traffic	Winds	Visibility Impediments	Personal Injuries	Health and Safety
Deep Draft Mariner Proficiency	Volume of Small Craft Traffic	Currents	Dimensions	Petroleum Discharge	Environmental
Shallow Draft Mariner Proficiency	Traffic Mix	Visibility Restrictions	Bottom Type	Hazardous Materials Release	Aquatic Resources
Small Craft Operator Proficiency	Congestion	Obstructions	Configuration	Mobility	Economic

Figure 2. Risk categories and factors.

at a given time) and waterway conditions (i.e., the nature of the waterway through which vessels will be traveling in a given port). Figure 2 presents the categories of risk drivers and the factors related to each category[4].

The PAWSA process is based on a justifiable belief that the risk drivers are the same for all ports and waterways but it also recognizes that the impact of each risk driver will be determined by conditions that will be specific to each port and waterway. In order to gauge the degree to which each factor affects the risk situation in a given port, an anchored rating scale was developed for each factor. The initial rating scales were based on "Port Heaven" and "Port Hell," which were qualitatively described in detail at each end of a scale. Intermediate stages between these two extremes were also described in sufficient detail to make navigationally meaningful distinctions between levels. Quantitative values were used where appropriate to further illustrate and differentiate the qualitative descriptions. The current process has moved beyond the Port Heaven and Port Hell construct [3] but the essential characteristics are retained—anchored scales with a mix of qualitative and quantitative descriptions of discernibly different levels for each risk driver.

After the risk driving factors have been discussed, participants individually rate their port against each of the risk factors shown in Figure 2. These

[4] This is an example of morphological analysis. The PAWSA matrices provide a morphological framework which, when augmented by user/stakeholder provided insights, yields a morphological model of vessel traffic risk in the port/waterway represented by the user/stakeholder group. The Swedish Morphological Society has found morphological analysis to be very useful in dealing with wicked public policy issues, including homeland security and national defense issues. See http://www.swemorph.com/ for details.

results are then consolidated into a group score. Significant disparities will be noted when appropriate. Once the consolidated port factor rating has been produced, the risk factors are then ranked against each other using Analytic Hierarchy Process. At the end of the Analytic Hierarchy exercise, stakeholders will have produced both an absolute assessment of where the port rates against the known risk drivers as well as a priority ranking of the risk drivers based on their relative importance in creating risk in that local port.

2.3.4. Mitigating Identified Risks

The next step in the process is a tutorial on the variety of tools potentially available to mitigate risks in a port or waterway. Table 3 lists most of the available tools. Each of these tools has its specific utility and carries with it its own set of costs and other drawbacks (but discussing these in any detail is beyond the scope of this paper). The focus in this stage of the PAWSA process is on matching identified and prioritized risk drivers with appropriate risk mitigation measures. The overall objective is to develop the least costly mix of risk mitigation measures that delivers the level of risk mitigation deemed necessary by users and stakeholders for the port or waterway in question. Measures that are already in place, such as existing navigational aids and traffic control measures, will be considered for their effectiveness in mitigating identified risks. Existing measures are also considered for potential improvements or possibly even termination if that is judged appropriate. New traffic safety tools are added as necessary to drive the residual risk level down to a level deemed acceptable by waterway users and other stakeholders. Again, the objective is to identify and justify the most cost-effective set of vessel traffic and navigational safety tools for the assessed port or waterway.

The important points about this stage of the process are that (1) those who will have to live with the solution, and possibly have to pay for it as

TABLE 3. Vessel traffic safety management tools

1. Rules of the Road (International Rules for the Prevention of Collisions at Sea)
2. Vessel Equipment Requirements (radar, navigation systems, radio, etc.)
3. Aids to Navigation (ATON)
4. Regulated Navigation Area (RNA)
5. Traffic Separation Scheme (TSS)
6. Pilotage Practices and Requirements
7. Vessel Traffic Service (VTS) (includes several different sub-types/variations)
8. Dredging and Channel Configuration
9. Anchorage Regulations
10. Port Facility/Terminal Site Approval
11. Non-regulatory Vessel Movement Coordination (e.g., by a port authority)

well, have a major role in deciding what the solution should be and (2) they are prepared for making the necessary judgments by the discussions and educational aspects of the preceding steps.

2.4. OUTCOME

PAWSAs provide a structured, collaborative, well-founded, simple, and clearly risk-focused process for assessing vessel traffic and navigation risks and identifying cost-effective risk mitigation measures to manage that risk. By bringing stakeholders into the process, PAWSA takes maximum advantage of their knowledge and local expertise, while also avoiding potentially sticky political wickets.

PAWSAs have now been performed in 34 U.S. ports with better acceptance of the results and at far less cost than was incurred in the earlier Port Needs Studies. These PAWSAs validated the need for a VTS in every U.S. port that already had one. Two additional VTSs have also been established based on PAWSA results. VTSs have been determined unnecessary in a number of ports that had previously been considered good candidates. Other, less costly navigational improvements have also been identified in many places.

The PAWSA process was very well received by both stakeholders at the local level and by key members of Congress and members of the relevant committee staffs. It was also well received by the Coast Guard's senior leadership and decision-makers in the Department of Transportation. The PAWSA process has been used by Australia for the Great Barrier Reef and by Denmark for the Oresund area, including the approaches to Copenhagen and Malmo harbors. PAWSA was well received and beneficial vessel traffic management changes were identified for both areas under consideration.

Finally, in recognition of PAWSA's practical benefits as a powerful diagnostic tool for assessing vessel traffic risk in ports and restricted waterways, it has been adopted by the International Association of Marine Aids to Navigation and Lighthouse Authorities (the international governing and coordinating body for national authorities responsible for marine navigation aids and vessel traffic management practices) as the first phase of a two-phase process. Given the conditions that gave rise to the PAWSA process, that is not a bad outcome.

3. Case Study 2: California Coastal Vessel Routing Schemes

3.1. INTRODUCTION

While the PAWSA process does provide a standard assessment process that can be used essentially everywhere, it is not a universal panacea and it does not provide necessarily complete answers. For example, in the IALA recommended practice for vessel traffic risk assessment, the PAWSA process is only Phase 1 of a two-phase process. Phase 2 in that process is intended to provide more probabilistic and econometric information where such information is required by decision makers. Similarly, where PAWSA might recommend that a Traffic Separation Scheme (TSS) or Aids to Navigation in a given port or waterway be modified, the PAWSA process is not designed to provide full details of what those changes should be. Other analytic techniques and decision-support processes are needed to develop those kinds of details. The California experience which follows did not grow out of a PAWSA evaluation (n.b., the PAWSA process was still in development as the following story was unfolding), but it does show the necessity of collaborative processes when dealing with "wicked problems" while also illustrating the kind of analyses that might flow from the PAWSA process.

3.2. BACKGROUND

Due to memories of the Santa Barbara oil spill and concerns heightened by subsequent spills, including that of the EXXON VALDEZ, environmentalists in California are extremely sensitive to the possibility of a ship or tug and barge accident causing an oil or hazardous substance spill on the California coast. These concerns are particularly high with respect to the Monterey Bay National Marine Sanctuary and the adjacent coastline. The issues in question revolved around the entry of ships into the boundaries of the Marine Sanctuary and whether or not minimum offshore distance requirements should be established for ships moving along the California coast. Some environmental advocates were calling for such requirements. In a few cases, the minimum distances offshore called for were on the order of hundreds of miles. Not surprisingly, the shipping industry resisted such recommendations as an unnecessary expense to counter what they saw as an imaginary threat. The two sides of this issue were clearly miles apart, both literally and figuratively, and the differences appeared to be irreconcilable.

The initial effort to resolve the issues in question took the form of a traditional Coast Guard Port Access Route Study. Unfortunately, that process was inadequate for the strong emotions involved and, to be blunt, this first effort blew up. Both sides were left more convinced than ever about the unreasonableness of the other side and the rightness of their own position. The primary failing in this first effort was that the process was not sufficiently collaborative. The Port Access Route Study process does allow for public input. In this case, however, deeply concerned stakeholders did not feel that their concerns had been given adequate attention.

3.3. THE PROCESS

In the face of continuing pressure from environmentalist stakeholders that something be done, a second attempt was made to bring the sides together. This time, a different approach was tried. Rather than rely on the simple public input process used in less contentious Port Access Route Studies, this new effort sought a mutually satisfactory solution through the involvement of all stakeholders in a very open and collaborative process. Participants included the Coast Guard, maritime interests, staff from the Monterey Bay National Marine Sanctuary, the State of California, and environmental advocacy groups. This effort also started with all options on the table, including the possibility that no action was necessary. Ground rules were laid to keep things on track. Rule 1 was that the participants had to respect each other (i.e., there was to be no name calling and no interruptions while another was speaking). Another important rule was that any decisions reached had to be based on fact, rather than on emotions.

With this new approach in mind, representatives of the various stakeholders were brought together under the joint sponsorship of the Coast Guard, the National Oceanic and Atmospheric Administration, and California State agencies. The ground rules were explained and commitments to abide by them were extracted. The next step was to get all participants to agree on a "Win-Win" end state. All of the participants were able to agree that there should be no oil spills from stranded vessels on the California coast and that the shipping industry should not be forced to incur significant expense to mitigate highly unlikely risks. From there, the discussions moved to clear identification of the concerns and then—and this is the most important part—to a scientific and engineering examination of their validity.

Chief among the concerns was a perceived risk that a disabled ship could drift onshore before help arrived. The scientific and engineering facts considered in examining this risk included historical data on coastal winds and currents, including historic worst-case winds, known drift behavior of disabled ships, and the distribution of towing vessels along the coast and their reasonable emergency response times. With facts in hand, much more realistic assessments of the feasibility of the various nightmare scenarios envisioned by the environmentalists were developed. In the main, the scenarios were not that realistic. In some particulars, however, the risks were much greater than the shipping industry had wanted to believe. Both sides converged to an understanding located between the extremes of their respective entering positions.

3.4. OUTCOME

The end result of this process was a recommendation to Coast Guard Headquarters, agreed to by all parties, that there should be a system of Recommended Routes along the California coast that would tie the southern legs of the TSS at the entrance to San Francisco together with the TSS around Point Conception and ultimately leading into the northern TSS leg for Los Angeles/Long Beach. In point of fact, the Recommended Routes did result in shipping moving a bit further offshore, but not so far as the environmental advocates had first thought necessary and not so far that the shipping industry was unwilling to incur the fuel and time costs involved. A "win-win" outcome was achieved and the Coast Guard accepted the recommendations for action.

All three TSSs had to be revised as part of the process and all of it—the TSS amendments and the recommended routes—had to be approved both domestically and by the Navigation Safety subcommittee of the International Maritime Organization in London. The process by which these measures were developed was an important factor in winning their easy approval.[5]

4. Conclusion

A number of important lessons can be learned from these two case studies that should make risk assessment and decision making for "wicked problems" far less difficult. These lessons learned are captured in Table 4 below.

TABLE 4. Tips for successful collaborative risk management

1. Include Stakeholders
2. Respect Legitimacy of Stakeholder Concerns
3. Provide a Structured and Transparent Process
4. Arrange for Neutral Third-Party Facilitation and a Neutral Venue
5. Work Backwards from an Outcome All Parties Agree To
6. Identify Risk Drivers and Appropriate Interventions
7. Work from Facts, not Emotion
8. Experts add Immeasurably (but Get the *Right* Experts)

[5] The author was the Program Manager for vessel traffic safety issues in Coast Guard Headquarters and headed the U.S. delegation to the IMO negotiations that ultimately approved these measures for adoption.

Of course, even if successfully applied, the tips above will not guarantee an easy process or a successful outcome. But ignoring these lessons will make things far more difficult and may well result in an outright breakdown in the process and failure in public policy decision making.

5. Afterword

It will come as no surprise to readers of this volume that there are fundamental differences between terrorism-driven port security risks, of the kind considered elsewhere in these proceedings, and safety risks of the kinds discussed above. Among these differences are the levels of uncertainty inherent in the two risks. Closely related to this are differences in the natures of the drivers creating these very different risks.

With vessel traffic, the *sine qua non* for risk is the presence of one or more vessels. The uncertainties relate to questions of where, when and how an accident might occur, which vessel or vessels might be involved and how bad the consequences could be. However, in the absence of any vessel, a condition which is readily discernable as vessels don't generally hide, there is no risk of a vessel-involved incident.

With terrorism, the *sine qua non* for terrorism risk is the terrorist and terrorists *do* hide. If there is no terrorist visible, does that mean there is no terrorist there—or does that mean he is adept at hiding? How can we tell?

Finally, there are essential differences in the natures of these two categories of risk. Safety risks arise from the facts of normal daily life. Terrorism risks are driven by the actions of an adaptive, reactive and strategically driven adversary. Further, information sharing may be problematic when the issue in question is safety and there may be a threat of civil lawsuits lurking in the background. However, when sharing information may threaten the future viability of valuable intelligence methods and sources, the obstacles to sharing information are far higher.

These differences will drive changes in the specific analytic techniques used when assessing and addressing terrorism-related port security risks, and will certainly change the information requirements and the limitations on its distribution. However, the unifying effect and political value of maintaining as open and collaborative a process as is possible should be obtainable, even when assessing and managing the risks of terrorism to seaports. Given the number of legitimate stakeholders in the typical port community, doing otherwise might guarantee failure.

6. References

1. Rittel HWJ, Weber MM. Dilemmas in a general theory of planning. Policy Sciences. 1973; 4:155-169.
2. Conklin J. Wicked problems and social complexity. Napa: CogNexus Institute; 2001. Available at: http://cognexus.org/wpf/wickedproblems.pdf [6].
3. U.S. Coast Guard Navigation Center. PAWSA background. Available at: http://www.navcen.uscg.gov/mwv/projects/pawsa/PAWSA_Back.htm
4. Swedish Morphological Society. Available at: http://www.swemorph.com

[6] Table 1 and the discussion of "wicked problems" in this paper rely heavily on this page.

CHAPTER 20

DECISION EVALUATION FOR COMPLEX RISK NETWORKED SYSTEMS DEVELOPMENT PROGRESS

T. SULLIVAN

Brookhaven National Laboratory
Upton, NY USA

A. GREBENKOV

Joint Institute for Power and Nuclear Research
Sosny, Minsk Belarus

B. YATSALO

Obninsk State Technical University of Nuclear Power Engineering
Obninsk Russia

I. LINKOV

Intertox, Inc.
Brookline, MA USA

G. KIKER

University of Florida
Gainesville, FL USA

L. KAPUSTKA

Golder Associates
Calgary Canada

I. Linkov et al. (eds.), Managing Critical Infrastructure Risks, 369–391.
© 2007 *Springer.*

Abstract: Environmental management requires decision support tools that are able to integrate heterogeneous technical information with decision maker values and judgment. Decision Evaluation in Complex Risk Networked Systems (DECERNS) is a software system under development that has the objective of providing a methodology, computer models, and software tools to facilitate decision making in the field of sustainable land use and related areas. DECERNS will integrate risk assessment and decision analysis tools from multiple disciplines (e.g., geographical information systems tools for mapping and data analysis, ecological and human health risk models, economic analysis tools for evaluating costs, and tools for incorporating social choices) into a single, user-friendly software package. This paper provides an overview of the approach used in DECERNS followed by a brief overview of multi-criteria decision analysis methods and tools used in the project. Future plans for development and testing of the software tool will be discussed.

1. Introduction

Environmental problems such as those involved in protection and security of our ports and harbors are complex and involve multiple interconnected and often competing objectives. Many countries face the challenge of managing ports and harbors to balance local ecological, economic, and social processes while assuring port and harbor security and sustainability. Decision making in such an environment is a complex, multifactor task. This entails an integrated site management approach supporting economic development and parallel goals of natural resource conservation, restoration of habitat in surrounding ecosystems, and biodiversity, as well as reduction of present and future pollution. These complex problems generally have several potential solutions. In order to evaluate these alternatives against the competing goals of environmental protection and economic development, decision makers could benefit from a framework that lead them through systematic processes for choosing priorities, conducting strategic planning, and performing comprehensive assessments, while explicitly integrating human health risks, ecological risks, and socioeconomic measures.

Decision Evaluation in Complex Risk Networked Systems (DECERNS) is a computer system now being developed with the objective of providing a methodology, computer models, and software tools to facilitate decision making in the field of sustainable land use. The new system integrates tools from several different disciplines on a single platform. These tools include geographic information systems (GIS) tools for mapping and data analysis, ecological and human health risk models, economic analysis tools for

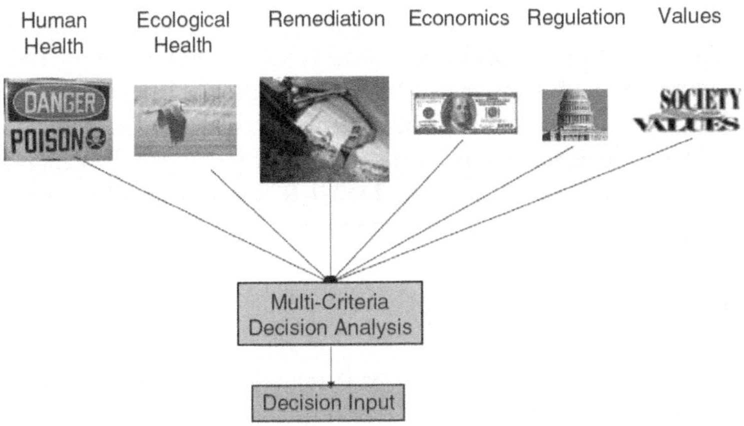

Figure 1. DECERNS framework for sustainable land use.

evaluating costs, and tools for incorporating social choices into the decision process (Figure 1).

DECERNS will be built based on the prior experience of team members, including models they have developed, and expanded with new models as necessary. Tools specifically under consideration include:

- PRANA (J-Prana) [18] can be used as a foundation for geospatial analysis.
- RiskTrace [9] can provide simulation algorithms for spatially explicit ecological risk assessment.
- RiskTrace Habitat Suitability Index (HSI) methodology [10] can be used to quantify habitat quality and usage by ecological receptors.
- Questions and Decisions™ QnD Model [6, 7] can provide a framework for incorporating scenario planning and gaming in model results visualization.
- MCDA methods and tools developed by the authors of this paper [9, 10, 11, 20, 6, 7] will be used for integrating heterogeneous information.

The main advantage of DECERNS in comparison to other decision support systems is its ability to integrate different models and measures as well as decision-maker values into a common framework using Multi-Criteria Decision Analysis (MCDA) tools. The MCDA tools will be incorporated into a systematic and modular framework that allows risk managers and stakeholders to appraise the relative importance of each factor to determine which alternative is most appropriate for a specific site. Sensitivity and uncertainty analysis will be incorporated within the framework to address the robustness of the decision to change the importance and value of each decision criterion.

This paper provides an overview of the approach used in DECERNS followed by a brief overview of MCDA methods and tools. The information collected in the reviews will provide information to decide which methods and models could be adapted into DECERNS. Future plans for development and testing of the software tool will be discussed.

2. Decerns

DECERNS will be developed as a web-based application with a modular structure that allows the user to develop problem-specific applications. Five tasks comprise the underpinnings of the DECERNS program (Figure 2):

1. Framework Development
2. Model Selection
3. Software Development
4. Testing and Evaluation
5. Documentation, Training, and Marketing

Each task is briefly described below.

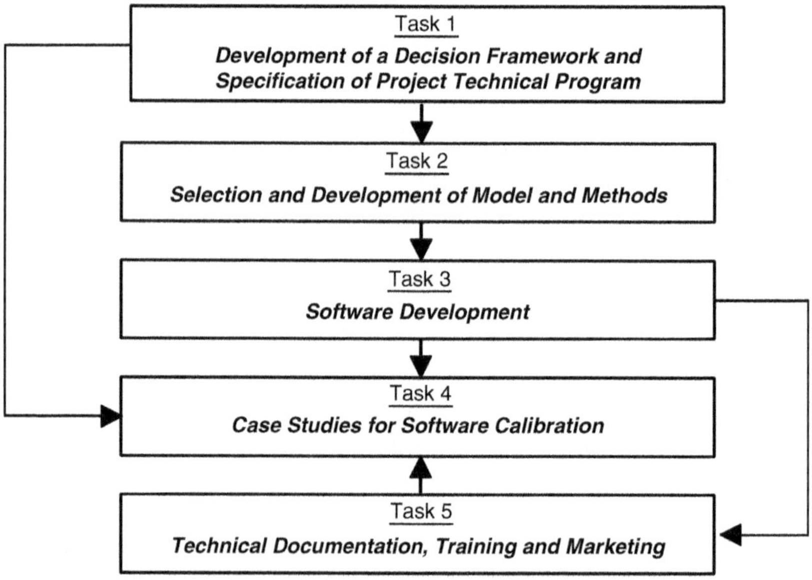

Figure 2. Project framework.

2.1. TASK 1: DEVELOPMENT OF A DECISION FRAMEWORK

The objective of this task is to develop a decision-making framework that integrates all of the decision endpoints (e.g., cost, risk) into the process. A web-based, modular system allows the user great flexibility in defining problems. DECERNS will integrate a number of developments and techniques currently in use to determine ecosystem state and the behavior and migration of populations in response to anthropogenic impact. DECERNS will also incorporate current techniques from the fields of comparative environmental risk analysis, environmental remediation, MCDA, GIS, information technology (IT), and computer science. In DECERNS, decision makers will be able to compare different options and select the most appropriate alternatives for each specific situation based on established criteria. The results of such complex comparative analysis will be the basis for ranking alternatives. Decision makers can tailor this framework to various sites; e.g., by weighting measures based on population, landscape, and environmental conditions. This framework will guide future development of software suitable for the goal of real-time presentation at stakeholder meetings. Upon presentation of alternative remedial scenarios, stakeholders may set up and use standard priorities (e.g., cost-effectiveness, general risk to human or ecological receptors, and more project- and stage-specific concerns) to individually develop weights for the various priorities, depending on particular concerns and issues. The proposed framework will support a value function that incorporates options and priorities being considered by stakeholders.

2.1.1. SDSS

A Spatial Decision Support System (SDSS) structure has been developed and preliminary evaluation has begun. The SDSS integrates GIS technology with models and tools for ecological assessment, human health risk assessment, cost, societal objectives, and stakeholder preference through MCDA methods.

2.1.2. Software Integration

Integration of the different software components into a single unit is a challenge. Existing software may be written in a variety of languages (e.g., C++, Visual Basic, FORTRAN). These programs are not automatically compatible with web-based programming languages such as JAVA. The bridging of software tools using different programming languages will be conducted using JAVA-Net tools. In addition, development of specifications for database management needed for the case studies is ongoing. Attention is being given to developing a project management system that will generate reports on the analysis.

2.2. TASK 2: SELECTION AND DEVELOPMENT OF MODEL AND METHODS

The objective of this task is to review and choose models and methods or identify where models need to be developed to support the program's decision framework.

The review and analysis of existing practices used in decision support systems (DSS) have been completed. In addition, models and techniques needed for the DECERNS program have been reviewed. Effective, advanced tools, involving several models and methods, will be integrated into the DECERNS project in a modular but heterogeneous structure (i.e., incorporating different software tools with different programming languages). The following methods have been reviewed and analyzed:

- MCDA methods
- Methods for incorporating GIS into DSS
- Methods of building SDSS
- Methods of building network-integrated DSS programs (Web-DSS)
- Methods and tools for developing network-integrated GIS applications (Web-GIS)
- Neural network methods applied to SDSS
- Methods for evaluating exposure and human health risk
- Methods of heterogeneous modeling on multiplatform systems
- Methods for evaluating environmental quality and habitat suitability
- Methods of modeling receptor behavior and stressor migration
- Environmental risk assessment methods, including comparative risk analysis
- Methods of evaluating the economic, ecological, and social effects of environmental protection measures

2.3. TASK 3: SOFTWARE DEVELOPMENT

The objective of this task is to develop the computer software integrated in a common interface with the models and methods received as a result of Task 2. In the beginning, the analysis of references to available software in this field will be carried out, and then the software beta version will be designed and tested. The compilation of necessary databases including the GIS applications database, HSI database, site remediation technologies database, and some additional electronic libraries, has been started. In general, discrete modules will be selected for performing certain tasks such as ecological risk assessment and very little development work will be required. The key component of software development in DECERNS will be the linking of the various models in a seamless, easy-to-use platform. The following tasks will be completed:

- Transformation of different models and design parameters to the unified interfaces of interaction with other software modules of the system as a whole.
- Interfacing with programming solutions, which are not present in the JAVA platform.
- Unification of models and organization of data processing from various sources.
- Simplification of interaction with GIS-DSS on a web server.

The software and developed models will be tested and calibrated based on hypothetical scenarios with known analytical solutions and will also be based on simplified case studies with collected available empirical data. For carrying out the calibration, separate modules and blocks of input data assignment, as well as the GIS module, will be adjusted to site-specific parameters.

2.4. TASK 4: CASE STUDIES FOR SOFTWARE CALIBRATION

The objective of this task is to conduct verification tests of the developed software product using two real cases. At this stage, the developed beta version containing formalized decision-making protocols and models based on estimations of risk and ecological-economic parameters for possible alternative strategies will be tested by applying this methodology to two test cases.

2.5. TASK 5: TECHNICAL DOCUMENTATION, TRAINING, AND MARKETING

The objective of this task is to finalize the project by releasing an operating version of the code along with all necessary infrastructure (e.g., user manuals, training guides). Several training seminars with potential software users and stakeholders, including representatives from environmental protection agencies and other institutions, will be held. For decision makers, all assumptions, advantages, and restrictions peculiar to the given product will be demonstrated. On the basis of feedback from potential users, regulating bodies, and environmental protection agencies—and with the help of industrial partners—a web-based marketing strategy will be developed.

3. Overview of Existing Software Tools

Environmental management is a large field and many tools support the evaluation of remediation alternatives based on their degree of human health risk, ecological risk, and environmental quality. As an overview, a brief description of four existing tools is provided to represent the state-of-the-art in this field.

Spatial Analysis Decision Assistance (SADA) is an environmental software product that incorporates tools from various fields—including visualization, geospatial analysis, statistical analysis, human health risk assessment, cost-benefit analysis, sampling design, and decision analysis—into a dynamic and interactive environment. Each of these modules can be used independently or in an integrated fashion to address site-specific concerns as part of characterization and remedial action design activities. SADA was designed to simplify and streamline several of the processes in environmental characterization and to bring the information together in a way that can help users make decisions about their particular site in a quick and cost-effective manner. SADA may be found useful by anyone who needs to look at data within a spatial context. Such users include statisticians, human health risk assessors, GIS/visualization users, project managers, and other stakeholders. SADA was developed by the Institute of Environmental Modeling at the University of Tennessee and is available at http://www.tiem.utk.edu/~sada/.

Adaptive Risk Assessment Modeling System (ARAMS™) integrates exposure and effects assessments to characterize risk. The U.S. Army Corps of Engineers developed this computer-based modeling- and database-driven analysis system for estimating human and ecological health impacts and risks associated with potential contaminants of concern. The U.S. Department of Defense and the Army conduct risk assessments to determine safe levels and cleanup target levels for military-relevant compounds and to evaluate remediation alternatives to provide the most cost-effective approach to reach target levels. ARAMS™ integrates multimedia and multi-pathway fate/transport, exposure, intake/uptake, and effects of relevant compounds to assess present and future human and ecological health impacts/risks associated with chronic exposure. ARAMS™ incorporates various existing databases and models for exposure, intake/update, and effects (health impacts) into an object-oriented, conceptual site modeling framework. With ARAMS™, the user has the flexibility to visually specify, through objects, multimedia pathways and risk scenarios. ARAMS™ is available at http://el.erdc.usace.army.mil/arams/.

Sustainable Management Approaches and Revitalization Tools. SMARTe is an open-source, web-based decision support system for developing and evaluating future reuse scenarios for potentially contaminated land. SMARTe contains guidance and analysis tools for many aspects of the revitalization process including planning, environmental (human health risk), economic, and social concerns. SMARTe is still under development, but many features are already available. Development will continue until 2010. Future enhancements will include MCDA tools to assist in the decision process. The expected release date for the MCDA tools is 2007. SMARTe is available at http://www.smarte.org/smarte/home/index.xml.

QnD [6, 7] is a screening model system created to provide an effective and efficient tool to incorporate ecosystem, management, economic, and sociopolitical issues into a user-friendly model framework. The QnD model utilizes a basic finite difference approach with simple Euler numerical integration of various rate transformation and mass-balance transfer equations as defined by the input files. The model is written in object-oriented Java and can be deployed as a standalone program or as a web-based (browser-accessed) applet. The QnD model links the spatial components within GIS files describing the abiotic and biotic interactions in an environmental system. The QnD model uses ArcView shape files for desired spatial inputs while utilizing a modular XML format for the input of other model-relevant information. The model can be constructed based on any combination of detailed technical data or estimated interactions of the ecological/management/social/economic forces influencing an ecosystem. Model development is iterative and can be initiated quickly through routine conversations with users or stakeholders, and model alterations and more detailed processes can be added throughout the model development process. QnD thus allows interesting concepts to be explored and subsequently expanded or discarded as further learning occurs.

These four examples suggest that well-developed tools already exist to address various functions needed for land management planning. A major difference between DECERNS and the existing tools is that DECERNS will emphasize decision support tools, particularly MCDA. Another difference is the narrower focus of existing tools. No single tool has all of the features contemplated for DECERNS. ARAMS™ and QnD are open source software products and some of their capabilities can be directly integrated into DECERNS in support of the new system's comprehensive functionality.

In addition to analyzing existing techniques, the DECERNS project will develop a new one, uniting a wide range of ecological impact and land reuse scenarios in an integrated tool that evaluates decision criteria for many purposes; e.g., minimum damage to the environment, minimum risk for population health, minimum cost, and optimum land reuse. The software will have the following major DSS elements:

1. An MCDA approach for defensible and sound decision making, including tools to incorporate value judgments in the decision process (e.g., prefer to minimize health risks as opposed to total costs).
2. GIS applications for effective criteria setting and presentation and comparison of available options.
3. Environmental risk assessment models for ecological and human health.
4. Cost-benefit or least-cost analysis tools.

According to Bardos et al. [1], decision support refers to:

> …the assistance for, and substantiation and corroboration of, an act or result of deciding; typically this deciding will be a determination of an optimal or best approach.

Environmental land management decision support integrates specific information about a site and general information such as legislation and guidelines to evaluate decision metrics. This should be accomplished such that the decision information is transparent, consistent, and reproducible. Although single decision metrics such as cost can be used, the complexity of environmental problems and the different viewpoints of the stakeholders generally require optimization among competing decision metrics (e.g., cost and human or ecological health). For this reason, there is a move to use MCDA methods in the process.

3.1. MCDA METHODS

Multiple choices of possible actions and multi-criteria approaches to their analysis require implementation of MCDA methods and techniques, which offer the possibility of comparing alternatives and taking into account different quantitative and qualitative criteria. The common purpose of MCDA methods is to evaluate and choose among alternatives based on multiple criteria using systematic analysis that overcomes the limitations of unstructured individual or group decision making [3,16]. The aim of MCDA in a broad sense is to facilitate decision makers' learning about and understanding the problem, about their own, other parties', and organizational preferences, values, and objectives; and exploring these in the context of the problem to identify a preferred course of action.

Three key phases of MCDA realization include [3]:

- Problem identification and structuring.
- Model development and implementation.
- Action plan development and decision making.

The following problem-solving steps are considered the basis of MCDA:

- *Sorting* alternative actions into classes/categories (e.g., "unacceptable," "possibly acceptable," "definitely acceptable").
- Screening alternatives: a process of eliminating those alternatives that do not appear to warrant further attention; i.e., selecting a smaller set of alternatives that (very likely) contains the best tradeoff alternative.
- Ranking alternatives (from best to worst according to a selected algorithm).
- *Selecting* the best alternative from a given set of alternatives.
- *Designing* (searching, identifying, creating) a new action/alternative to meet goals.

Three dichotomies within MCDA problems can be distinguished:

- Multi-attribute (MADM) versus multi-objective decision making (MODM)
- Individual versus group decision making
- Decisions under certainty versus decision under uncertainty

This classification is shown in Figure 3 [13]. Within MADM there are a finite number of explicitly given alternatives; within MODM we have infinite or a very great number of implicitly given alternatives.

Within this review we consider MADM approaches.

3.1.1. Cost-Benefit Analysis

One approach to MCDA is to convert all decision metrics into cost. Cost benefit analysis (CBA) and cost effectiveness analysis (CEA) aim to assess all the costs and benefits of alternative options, valuing all the main expected impacts of the options in monetary terms. When solving problems relating

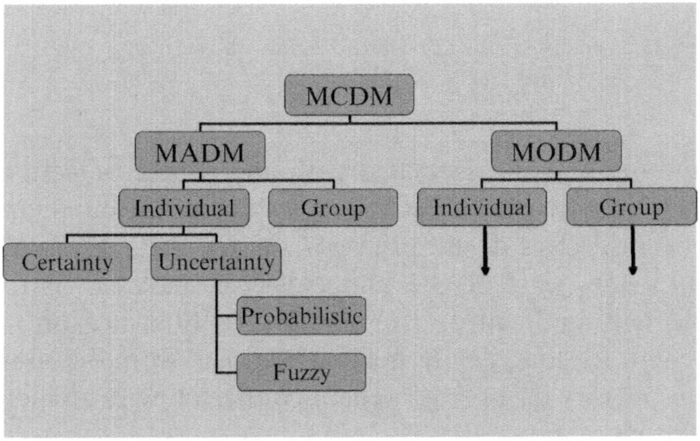

Figure 3. Classification of multi-criteria decision-making approaches.

to environmental and population protection or remediation of contaminated territories, many criteria are not expressed a priori in monetary terms. In this case, conversion factors that transform advantages (benefits) and disadvantages (damages/losses/risks) into monetary values are used. In general, the expression used within the CBA can be presented in the following way:

$$B(a) = \sum \alpha_i Y_i - \sum \beta_j X_j, \tag{1}$$

where B is a net benefit in monetary terms associated with an alternative ***a***, Y_i is as a benefit of *i*th type, and α_i is a corresponding conversion factor; X_j and β_j have the same meaning but for category damages/losses or risks.

CEA is used to compare different options to achieve the same objective and then to choose the option that achieves the objective at the lowest cost or with minimal cost associated with a unit of positive effect/benefit. The expression used within CEA can be presented as follows:

$$e = X_0 / Y, \tag{2}$$

where X_0 is the cost associated with a benefit/positive effect Y (e.g., X_0 is the cost of a protective measure—alternative ***a***, which averts collective dose $\Delta S = Y$).

It should be stressed that, in the general case, CBA and CEA are not equivalent; i.e., optimization tasks $B \rightarrow max$ and $e \rightarrow min$ are not equivalent.

CBA is effectively and formally used within radiation protection. Realization of the justification and optimization principles of radiation protection are based on implementation of CBA methods which take an objective view that alternative strategies should be selected according to a systematic comparison of the advantages (benefits) and disadvantages (costs) that result from the consequences of the choice.

Within the framework of radiation protection the following form of CBA is usually used:

$$B = \Delta Y - P_c + B_a = \alpha \, \Delta S - P_c + B_a \tag{3}$$

where B is the net benefit associated with any course of action (protective measure); ΔY is the avertable detriment, expressed in monetary terms; ΔS is the avertable collective dose to the population; $\Delta Y = \alpha \Delta S$, where α is the equivalent monetary cost of the unit collective dose; P_c is the monetary costs of implementing the protective measure (countermeasure); and B_a is an additional benefit (or loss; i.e., B_a may be negative) from the countermeasure expressed in monetary terms (e.g., benefit connected with an increase of crop productivity or its quality).

It should be pointed out that CEA cannot be used, in general, for analysis of justification of alternatives ($B > 0$). CEA may be considered a secondary (auxiliary) method; it is used if the full CBA cannot be correctly realized (e.g., when not all advantages/disadvantages can be effectively presented in monetary terms).

CBA has some drawbacks, including equity category realization on detriment compensation, lack of homogeneity of conversion factors within a territory or country, and impossibility of effective establishment of conversion factors for many criteria and/or their considerable uncertainty. CBA can be an effective method when conversion factors are determined and formalized at a departmental or national level (as for radiation protection). CBA has been implemented and [8, 2] used for analysis on radioactively contaminated territories.

Use of CBA/CEA in DECERNS may be implemented for the investigation of problems on land management within remediation/revitalization of contaminated sites/territories. New approaches for economic valuation of environmental measures (value of biodiversity including genetic biodiversity, value of life, and value of natural production) include another method that utilizes biotic and abiotic (non-living) indices and their monetary equivalents.

This approach introduces the cost of ecosystem services and biological resources. These two components of cost, which human society pays for sustainability of its habitat and living environment, have a long payback period and therefore cannot compete (from an economic point of view) with the short-term immediate benefits of industrial production. Therefore, this approach requires discounting of all costs and results of environmental protection measures.

3.1.2. MCDA

To remove the need to place a monetary value on all decision metrics, a number of other MCDA techniques exist. These are generally categorized into:

- Value function-based methods
- Outranking methods

These approaches are much more flexible than converting everything to cost and allow the analyst to better represent the decision process.

Value function-based methods. Approaches that use value functions comprise the multi-attribute value theory (MAVT) methods. Multi-attribute utility theory (MAUT) methods are also often used. While the methods are not always seen as fundamentally different [16], they are differentiated on the basis of certainty. A value function describes a person's preference regarding different levels of an attribute under certainty, whereas utility theory extends the method to use probabilities and expectations to deal with uncertainty [3,16].

The objective of MAVT is to model and represent the decision maker's preferential system into a value function $V(a)$,

$$V(\boldsymbol{a}) = F(V_1(a_1),\ldots,V_m(a_m)); \tag{4}$$

where alternative $\boldsymbol{a}$ is presented as a vector of the evaluation criteria $\boldsymbol{a} = (a_1, \ldots, a_m)$, a_i is the assessment of alternative $\boldsymbol{a}$ according to criterion i, and $V_i(a_i)$ is the value score of the alternative reflecting its performance on criterion i.

The goal of decision makers in this process is to identify the alternative $\boldsymbol{a}$ which maximizes the overall value of $V(\boldsymbol{a})$. The most widely used form of function $F(\ )$ is an additive model:

$$V(\boldsymbol{a}) = w_1\ V_1(a_1) + \ldots + w_m\ V_m(a_m), \tag{5}$$

$$w_i > 0\,,\ \Sigma w_i = 1, \tag{6}$$

where w_i, $i = 1, \ldots, n$, are the weights reflecting the relative importance of criteria or corresponding scaling factors [3,16, 5].

It should be stressed, however, that for a justified implementation of the additive model, some requirements/axioms of MAVT should be adhered to (e.g., preferential independence requirements [3,16, 5]). MAVT relies on the assumption that the decision maker is rational (preferring more utility to less utility, for example), that the decision maker has perfect knowledge, and that the decision maker is consistent in his judgments. Since poor scores on some criteria can be compensated for by high scores on other criteria, MAVT is part of a group of MCDA techniques known as "compensatory" methods.

Various sophisticated methods for defining partial value functions $V_i(x)$ and assessing weights (scaling factors) w_i have been developed both for quantitative and qualitative criteria.

Outranking methods. Outranking relation theory involves forming an ordered relation of a given set of alternatives. Outranking methods are based on a pairwise comparison of alternatives for each criterion under consideration with subsequent integration of obtained preferences according to a chosen algorithm. Among outranking approaches, the ELECTRE family of methods developed by Roy [14] and the PROMETHEE method developed by Brans and Vincke [4] are most used.

PROMETHEE is based on utilization of a performance matrix $\{z_i(\boldsymbol{a})\}$ (where $z_i(\boldsymbol{a})$ is an evaluation of alternative $\boldsymbol{a}$ against criterion i) and a chosen preference function $f_i(x)$, $0 \le f_i(x) \le 1$, with specified indifference and preference thresholds. It determines the intensity of preference for alternative ***a*** over alternative ***b***, $P_i(\boldsymbol{a},\boldsymbol{b}) = f_i(z_i(\boldsymbol{a}) - z_i(\boldsymbol{b}))$, and the preference index, $P(\boldsymbol{a},\boldsymbol{b})$,

$$P(\boldsymbol{a},\boldsymbol{b}) = \Sigma w_i \, P_i(\boldsymbol{a},\boldsymbol{b}) \tag{7}$$

where weights w_i reflect the relative importance of criteria and meet the requirements of Equation 6. According to the features of preference functions $f_i(x)$, if $P_i(\boldsymbol{a},\boldsymbol{b})>0$, then $P_i(\boldsymbol{b},\boldsymbol{a})=0$. Preference indices are used for determination of positive outranking flow $Q^+(\boldsymbol{a})$:

$$Q^+(\boldsymbol{a}) = \Sigma_b P(\boldsymbol{a},\boldsymbol{b}) \tag{8}$$

and negative outranking flow $Q^-(\boldsymbol{a})$:

$$Q^-(\boldsymbol{a}) = \Sigma_b P(\boldsymbol{b},\boldsymbol{a}), \tag{9}$$

summed over all alternatives $\boldsymbol{b} \neq \boldsymbol{a}$.

According to the PROMETHEE 1 method, a outranks b if $Q^+(\mathbf{a}) \ge Q^+(\mathbf{b})$ and $Q^-(\mathbf{a}) \le Q^-(b)$; ***a*** is indifferent to ***b*** if $Q^+(\mathbf{a}) = Q^+(\mathbf{b})$ and $Q^-(\mathbf{a}) = Q^-(\mathbf{b})$; ***a*** and ***b*** are incomparable if $Q^+(\mathbf{a}) > Q^+(\mathbf{b})$ and $Q^-(\mathbf{b}) < Q^-(\mathbf{a})$, or $Q^+(\mathbf{b}) > Q^+(\mathbf{a})$ and $Q^-(\mathbf{a}) < Q^-(\mathbf{b})$. Thus, PROMETHEE 1, like some other outranking methods, does not presuppose that a single best alternative can be identified, since some alternatives may be incomparable.

The PROMETHEE 2 method is based on the "net flow" criteria $Q(\boldsymbol{a})$:

$$Q(\boldsymbol{a}) = Q^+(\boldsymbol{a}) - Q^-(\boldsymbol{a}), \tag{10}$$

and it may be used for a complete ranking of alternatives (and alternative ***a*** outranks ***b*** if $Q(\mathbf{a}) > Q(\mathbf{b})$).

PROMETHEE, like other outranking methods, is considered an attractive and transparent method, although both positive and negative flows depend on the complete set of alternatives under consideration [3]. However, a drawback of outranking is that "indifference" and "preference" thresholds—though often based on expert knowledge—are essentially arbitrary, and the relationship representing preferred alternatives depends on selection of those thresholds. One way to analyze the robustness and check consistency between thresholds is to perform uncertainty analysis through manipulation of the thresholds.

Outranking techniques allow inferior performance against some criteria to be compensated for by superior performance against others. They do not necessarily, however, take into account the magnitude of relative underperformance against a criterion versus the magnitude of overperformance in another criterion. Therefore, outranking models are known as *partially compensatory*.

3.1.3. Analytical Hierarchy Process

The Analytic Hierarchy Process (AHP), developed by T. Saaty [15], presents an integration of the additive model [14, 4] with a distinctive determination of the decision matrix, $V_{i,a}$, and criteria weights, w_i, $i = 1,\ldots,n$. Within AHP, a systematic pairwise comparison of alternatives with respect to each criterion is used based on a special ratio scale: for a given criterion, alternative i is preferred to alternative j with the strength of preference given by $a_{ij} = s$, $1 \leq s \leq 9$, correspondingly, $a_{ji} = 1/s$. Then, the same procedure is implemented for $n(n-1)/2$ pairwise comparisons in the same scale for n criteria. The obtained matrices are processed by extracting the eigenvector corresponding to the maximum eigenvalue of the pairwise comparison matrix [3,15] and yield the values $V_{i,a}$ and weights w_i for subsequent use with the model, when preferences are aggregated across different criteria [14].

AHP may thus be considered an MAVT approach with a specific eliciting value function (scoring) and criteria weights (weighting). However, taking into account different assumptions and approaches, proponents of AHP insist that it is not a value function method. Additionally, AHP relies on the supposition that humans are more capable of making relative judgments than absolute judgments. Consequently, the rationality assumption in AHP is more relaxed than in MAVT.

Despite longstanding disputes about the correctness of AHP for analyzing and ranking alternatives (such as the rank reversal problem) [3,15], its transparency and relatively simple pairwise judgments make it a popular decision analysis method.

The MCDA methods for MAVT, AHP, and outranking are the basic MCDA models to be implemented within DECERNS. Some other methods used for analysis of specific problems are described elsewhere [5].

A range of methods for setting criteria weights has been developed, including:

- Ranking method.
- Rating method.
- Pairwise comparison/AHP method.
- Swing methods.

All the methods indicated will be implemented in the MCDA module within DECERNS. Uncertainty analysis is a necessary step within multi-criteria problem analysis [3, 5]. These will also be included in DECERNS.

TABLE 1. List of MCDA tools

Decision Plus	MACBETH	REMBRANDT
Expert Choice	MacModel	RGDB
Decision Lab	M&P	SANNA
DAM	MIIDAS	TOPSIS
ELECCALC	MINORA	UTA Plus
ELECTRE IS	MUSTARD	VIMDA
ELECTRE III-IV	NAIADE	VIP Analysis
Equity	OnBalance	V.I.S.A.
Expert Choice	PREFCALC	Web-HIPRE
HIVIEW	PRIAM	WINPRE
Logical Decisions	PRIME Decisions	

Recently, interest has been growing in the methods and systems which realize *group decision support* approaches. Web-DSS facilitates the realization of group decision support. Group decision making is a process of aggregation of individual preferences of different experts into one group preference. Within a group MCDA method, each expert can have a weight, which corresponds to the competence of the expert. Sometimes each expert is given only one weight, but the expert may be given a separate weight for each criterion, reflecting his/her specific competencies. Group MCDA methods may be divided into elementary/voting and complex methods (which use individual MCDA methods to calculate individual preferences) and methods used under certainty and uncertainty. Group decision support approaches will be implemented in DECERNS.

To prepare for implementation of MCDA tools into DECERNS, a thorough review of existing MCDA software tools including commercial, academic, web, and desktop software solutions was conducted.

This review evaluated software tools to determine their ability to:

- Support different stages of the decision process.
- Model uncertainty.
- Support group decision and collaboration processes.
- Exchange information with external applications and data sources.
- Be used as a part of new software solutions.

The tools reviewed are listed in Table 1. The detailed review focused on the most widely used commercial software tools, such as Decision Lab, Decision Plus, and Expert Choice. These software tools were investigated, described, and evaluated against 10 criteria. It was found that no single software tool was clearly superior to the others. Some excel in documentation and reporting while others excel in the choice of weighting functions and analysis techniques. The results of the review are being used to select the best features to program into the DECERNS MCDA module.

4. DECERNS Software Framework and Architecture

Over the past 10 years, members of the DECERNS team have been developing decision support tools for assessing lands radioactively contaminated due to fallout from the Chernobyl accident in the Bryansk region of Russia [17,18,19]. The Protection and Rehabilitation of Agrosphere after Nuclear Accident (PRANA) system of codes encompasses many of the component functions that will be needed in DECERNS, including GIS modules, radioactive dose modules, cost-benefit modules, and limited MCDA modules. Experience and client-server technology used for development of J-*Prana* (a web-based PRANA version, written in JAVA) will serve as a basis for developing SDSS within the DECERNS project.

J-*Prana* has been developed specifically for addressing problems associated with the rehabilitation of land that was radioactively contaminated due to the Chernobyl accident in the Bryansk region, Russia. The main objectives of J-*Prana* are:

- Radiation and agrochemical monitoring networking support for agriculture.
- Analysis of protective measures and evaluation of different scenarios for agricultural land management.
- Assessing risk values for different scenarios of countermeasure implementation.
- CBA and multi-criteria analysis of different alternatives for contaminated land management.

These objectives will be expanded in DECERNS to cover a more general class of problems that can be handled by MCDA techniques. These problems may include chemical contamination, competing land use alternatives (e.g., redevelopment of contaminated lands), and sediment contamination problems.

DECERNS will be a distributed application utilizing the classical three-tier architecture (information layer, application layer, client layer). The information layer is built on the Interbase Firebird v.1.5 database management system (DBMS). This DBMS was chosen for several reasons; the most important ones are availability, reliability, and simplicity. The database of consists of the following main components:

- The system database contains information about registered users and their privileges as well as information about saved research scenarios and various system settings.
- The geometrical database contains geometric/geographic data for all the objects under consideration in regions of interest as digital maps.

- The attributive database contains information for all the regions of interest in vector maps. This database will be problem specific and could include information on contamination levels and other monitoring data; agricultural data; societal infrastructure data (roads, buildings); and demographic, social, economic, and other data.

Access to the database is realized in session mode, under control of an application server which uses a pool of connections for acceleration of database access and centralized connection management. Any server meeting J2EE specifications can be used as an application server. JBoss v.4 is one of the recommended application servers. It combines such merits as cross-platform availability, accessibility, and acceptable hardware requirements.

The application layer is located on the server. It encompasses a set of program components which realize the base functionality of DECERNS. The key components of the application layer are:

- Model block, which includes models for risk assessment (e.g., external and internal doses to the population, HSI), models for assessing the results of different treatment options, models for evaluating costs/benefits (e.g., enhanced land value after remediation), models for uncertainty and sensitivity analysis, and algorithms for statistical and spatial analysis of input and output data.
- Scenario management block (problem definition and remedial alternatives).
- Block for representation of input/output data as multilayered vector maps, diagrams, and tables.
- Decision-making support block, which implements CBA, CEA, and MCDA methods with uncertainty analysis.

The client layer consists of a set of html pages, java applets, and JavaScript modules which implement the user interface of the system. The structure of the user interface logically corresponds to the following (simplified) scheme:

- Authorization.
- Choice of the object(s) for subsequent analysis.
- Assignment of the evaluation criteria and alternatives.
- Calculation.
- Presentation and analysis of the results (part of decision-making support).

Software quality depends on three factors. First, some general requirements for the system under development should be identified. Usually these include:

- Scalability.
- Security.

- Reliability.
- Portability.
- Interoperability.
- Internationalization (i18n).

Second, appropriate technologies and tools must be selected. This includes both computer programming technologies and simulation models.

Third, it is necessary to organize the software development process, including version control and quality assurance testing. Development of web-based systems requires a range of IT personnel. These include (but are not limited to):

- System administrator.
- Application client developer.
- Web component developer.
- Business component developer.
- Application deployer.
- Application assembler.

Selection of the development platform must take into account general requirements as well as developer experience. One of the leading web platforms that meets all conditions is Java EE. Major advantages of this platform include:

- High development productivity.
- Availability of free (non-commercial) high-quality IDEs.
- Variety of implemented technologies.
- Comprehensive documentation.
- Huge codebase.

An important software problem that will occur with DECERNS involves the integration of existing code developed in a variety of languages (e.g., C++, Visual Basic) into a web-based system. During the planning stage, it is necessary to assess the functional elements of such code and develop an interoperability requirement analysis to evaluate a list of possible approaches to integration and determine which of them can provide the best result. A bridge product architecture that places the bridge function on the client side preserves the existing Java software that resides in the client code. Through proper implementation, bridge architecture fulfills all interoperability requirements. Software products are currently available to facilitate this approach. dotNET-Java bridge software will be used in DECERNS to link between the different codebases (Figure 4). The advantages of this approach include:

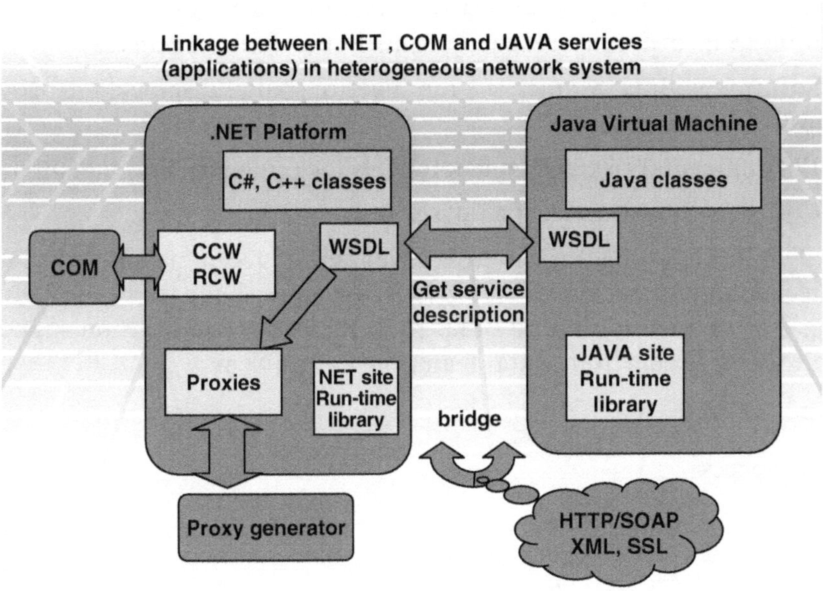

Figure 4. Schematic of bridge between server (.NET Platform) and client (Java Virtual Machine).

- Optimal use of staff by leveraging expertise gained from previously developed software projects.
- Incorporating existing code such as COM components.
- Creation of a stable software system using diverse software technologies.

5. Conclusions

Management of land or coastal estuarine waters and habitats is a complex environmental problem, including economic, managerial, and social elements in addition to technical challenges. The goals and interests of the different stakeholders often conflict and decisions about selecting land use management alternatives must often be viewed as multi-criteria decision problems. Evaluation of multiple criteria associated with each management alternative (such as costs, benefits, and ecological and human health risks) could serve as the basis for effective, transparent, and reproducible policy selection decisions.

The DECERNS software program is being developed to address this issue. The software will offer the following major decision support elements:

1. MCDA approach for defensible and sound decision making, including tools to incorporate value judgments in the decision process.
2. Software with GIS functions for effective spatial display comparison of available options.
3. Environmental risk assessment models for ecological and human health.
4. Cost-benefit or least-cost analysis tools.

The DECERNS program will take advantage of recent advances in computer technology to provide a web-based system enabling users to work with data, models, GIS tools, and other tools for risk-based decision support to resolve complex environmental management problems.

6. References

1. Bardos RP, Mariotti C, Marot F, Sullivan T. Framework for decision support used in contaminated land management in Europe and North America. Land Contamination & Reclamation. 2001; 9(1):149-163.
2. Baverstam U, Fraser G, Kelly GN. Decision making support for off-site emergency management. Radiat. Prot. Dosim. 1997; 73(1-4).
3. Belton V, Steward T. Multiple criteria decision analysis: an integrated approach. Amsterdam: Kluwer Academic Publishers; 2002.
4. Brans JP, Vincke P. A preference ranking organization method: the PROMETHEE method for multiple criteria decision-making. Management Science. 1985; 31:647-656.
5. Figueira J, Greco S, Ehrgott M, editors. Multiple criteria decision analysis: state of the art surveys. New York: Springer; 2005.
6. Kiker GA, Rivers-Moore NA, Kiker MK, Linkov I. QnD: a modeling game system for integrating environmental processes and practical management decisions. In: Morel B, Linkov I, editors. Environmental security and environmental management: the role of risk assessment. Amsterdam: Springer; 2006.
7. Kiker GA, Linkov I. The QnD model/game system: integrating questions and decisions for multiple stressors. In: Arapis G, Goncharova N, Baveye, P, editors. Ecotoxicology, ecological risk assessment and multiple stressors. Dordrecht: Springer; 2006.
8. Kelly GN, Fraser G. Decision making support for off-site emergency management. Radiat. Prot. Dosim. 1993; 50(2-4).
9. Linkov I, Grebenkov A, Andrizhievski A, Loukashevich A, Trifonov A. Risk-trace: software for spatially explicit exposure assessment. In: Kapustka LA, Gilbraith H, Luxon M, Biddinger GR, editors. Landscape ecology and wildlife habitat evaluation: critical information for ecological risk assessment, land-use management activities, and biodiversity enhancement practices, ASTM STP 1458. West Conshohocken: ASTM International; 2004.
10. Linkov I, Grebenkov A, Andrizhievski A, Loukashevich A, Trifonov A, Kapustka L. Incorporating habitat characterization into Risk-trace: software for spatially explicit exposure assessment. In: Linkov I, Ramadan A, editors. Comparative risk assessment and environmental decision making. Amsterdam: Kluwer; 2004.
11. Linkov I, Varghese A, Jamil S, Seager TP, Kiker G, Bridges T. Multi-criteria decision analysis: framework for applications in remedial planning for contaminated sites. In: Linkov I, Ramadan A, editors. Comparative Risk Assessment and Environmental Decision Making. Amsterdam: Kluwer; 2004.
12. Linkov I, Satterstrom K, Kiker G, Batchelor CG, Bridges T. 2006. From comparative risk assessment to multi-criteria decision analysis and adaptive management: recent developments and applications. Environment International. 2006; 32:1072-1093.

13. Malczewski J. GIS and multicriteria decision analysis. New York: John Wiley & Sons Inc.; 1999.
14. Roy B. Multicriteria methodology for decision aiding. Dordrecht: Kluwer Academic Publishers; 1996.
15. Saaty TL. The analytic hierarchy process. New York: McGraw-Hill; 1980.
16. von Winterfeldt D, Edwards W. Decision analysis and behavioral research. Cambridge: Cambridge University Press; 1986.
17. Yatsalo B, Mirzeabassov O, Okhrimenko I, Pichugina I, Kulagin B. *PRANA* - decision support system for assessment of countermeasure strategy in the long-term period of liquidation of the consequences of a nuclear accident. Radiat. Prot. Dosim. 1997; 73(1-4):291-294.
18. Yatsalo B, Bardos P. Decision support on risk based land management and sustainable rehabilitation of radioactive contaminated territories. Radioprotection. 2002; 37(C1):1087-1092.
19. Yatsalo B, Okhrimenko I, Pichugina I, Mirzeabassov O, Didenko V, Okhrimenko D, Golikov V, Bruk G, Shutov V. Decision-making support within risk based land management and sustainable rehabilitation of radioactive contaminated territories. Proc. of the 11 Intern. Congr. IRPA-11, Madrid, Spain, May 23-28, 2004.
20. Yatsalo B, Kiker G, Kim J, Bridges T, Seager T, Gardner K, Satterstrom K, Linkov I. Application of multi-criteria decision analysis tools for management of contaminated sediments. Integrated Environmental Assessment and Management. In press.

PART 5

CASE STUDIES IN RISK MANAGEMENT

CHAPTER 21

A MODEL FOR OPTIMAL INDUSTRIAL WASTEWATER TREATMENT AS A TOOL FOR MANAGING PORT ENVIRONMENTAL SECURITY AND SUSTAINABILITY

N. HARUVY

Netanya Academic College
1 University Street
Netanya, Israel 42365

S. SHALHEVET

Shalhevet Consulting
126 Thorndike Street
Brookline, MA USA 02446

Abstract: One of the major environmental threats to industrial ports is water contamination from nearby industrial plants, causing discharge of toxic metals such as mercury into the ports and harbors. Appropriate industrial wastewater treatment can reduce the environmental damage and save on the very expensive cleanup of contaminated harbors. A socioeconomic-environmental model to determine the optimal level of wastewater treatment of industrial operations should take into account the effect of different wastewater treatment options on the companies' profits, and the short-term and long-term benefit to society. The benefit to society is a function of the environmental aspects, including the short-term and long-term effects of contaminated harbors on human health. An inclusive model takes into account the welfare of all stakeholder groups. The model takes into account the rate of contamination and the cost of cleanup to estimate the economic value of the benefit from wastewater treatment and compare it with the treatment costs in order to reach recommendations for risk managers regarding optimal industrial wastewater treatment in port areas.

1. Introduction

One of the major environmental threats to industrial ports is water contamination from nearby industrial plants, causing discharge of toxic metals such as mercury into the ports and harbors. Appropriate industrial wastewater

I. Linkov et al. (eds.), Managing Critical Infrastructure Risks, 395–399.
© 2007 *Springer*.

treatment can reduce the environmental damage and save on the very expensive cleanup of contaminated harbors.

Several different treatment methods of industrial wastewater have been examined. Treatment with activated carbons is effective and practical for utilization in industrial wastewater treatment for mercury removal [12]. High-rate anaerobic treatment systems are becoming increasingly popular to treat industrial wastewater containing large amounts of organic matter in the form of carbohydrates or proteins [9]. Choosing the optimal treatment depends on the cost-benefit analysis. Panero [8] performed a cost-benefit analysis for a mercury abatement plan for the New York/New Jersey harbor, and developed an abatement cost curve based on a simple cost minimization criterion.

Reuse of industrial wastewater is one of the important strategies for reducing freshwater consumption and wastewater generation [4]. Akinyemi et al. [1] examined industrial waste effluents disposal of five industries in Nigeria, and found that they can all be used as potential water source for irrigation. Wastewater reuse typically requires treatment to meet the required water quality standards, but the cost of the treatment rises exponentially with increasing contaminant removal efficiency [4]. Feng and Chu developed a cost optimization procedure that balances these factors in order to determine the optimal point of wastewater treatment. Carmichael and Strzepek [2] developed a simulation/optimization model of multiple-pollutant wastewater management that includes industrial as well as municipal wastewater treatment.

One possibility is to reuse the wastewater as irrigation water for agriculture. Using small amounts of municipal-industrial wastewater sludge has been found to give crop yields similar to the highest mineral fertilizer dressing. However, a higher rate of applications caused an excessive and harmful N supply, and was associated with possible negative effects on water quality due to increased P availability [7].

1.1. OPTIMAL BEHAVIOR MODEL

A socioeconomic-environmental model to determine the optimal level of wastewater treatment of industrial operations should take into account the companies' profits as well as the short-term and long-term benefit to society. The benefit to society is a function of the environmental aspects, including the short-term and long-term effects of contaminated harbors on human health. In economic terms, the effect is sometimes expressed as the cost of cleaning up the contaminated harbor.

The model was adapted from an urban wastewater treatment model [5]. The individual factory wishes to maximize its net profit, disregarding the pollution effects. This profit is maximized at a given level of pollutants, and is shown as an example for nitrogen pollution, as defined by Equation 1.

$$V = P_q * Q - P_n * N_a - TC \tag{1}$$

Where V is the net profit, P_q is the price of the product, Q is the quantity produced, N_a is the amount of nitrogen used, and TC represents all other costs.

This maximum profit will be reached by the amount of fertilizers N_a, as defined by Equation 2.

$$P_n = P_q * dQ/dN_a \tag{2}$$

The national benefit is given by Equation 3, which includes the cost of the environmental damage caused by increased nitrogen concentration in harbor water, where B is the national net benefit, N is the nitrogen amount added to the harbor, and D is the cost of the pollution change.

$$B = V - D(N) \tag{3}$$

Increasing nitrogen use causes decreasing marginal returns to industry, and increasing marginal cost of harbor cleanup. The optimum national gain takes into account the industrial profits as well as pollution prevention, and is reached at the point where Equation 4 is satisfied:

$$P_q * dQ/dN_a = P_n + dD/dN_a \tag{4}$$

In other words, the optimum point is when the marginal net profit to industry from an additional unit of nitrogen is equal to the marginal net cleanup cost of an additional unit of nitrogen.

1.2. COSTS OF WASTEWATER TREATMENT

Wastewater may be treated and reused for irrigation of some crops, such as field crops, but not for raw-eaten vegetables. This can serve to dispose of industrial wastewater at lower costs than treatment for disposal, provide a reliable and uniform source of water for irrigation, as well as serve as a source of macronutrients for fertilizing [6].

Most of the research on wastewater reuse for agriculture has been done on urban wastewater reuse. However, the cost of industrial wastewater treatment is often higher than the cost of urban wastewater treatment. The U.S.

Department of Defense [20] estimates industrial wastewater treatment costs to vary from \$0.35 to \$1.40 per gallon of untreated water, although newer technologies may bring the range down to \$0.15 to \$0.60 per gallon.

If the wastewater is spilled into the harbor, the treatment cost is replaced by the cleanup cost. The costs of harbor cleanup vary greatly, according to the level of contamination and the quality of the treatment. According to Etkin [3], a simple manual recovery of an oil spill into a port can cost as little as \$0.37 per gallon, while an extensive cleanup, including wildlife rescue and rehabilitation, can increase the cost to \$296 per gallon, and in a few extreme cases even higher. For example, a cleanup of an oil spill in Huntington Beach in California in 1990 cost \$39.96 per gallon.

Reusing wastewater for agriculture enables aquifer recharge, as opposed to disposal to the harbor, which does not affect the aquifer level. The net economic benefit of aquifer recharge has been estimated at \$0.03/m^3 [6]. The cost of water storage and conveyance to the southern region was estimated at \$0.24/m^3 [6]. The estimate of the environmental damage caused by nitrogen leaching was calculated using a linear optimization model, which estimated the maximum profits under a set of constraints on cultivated areas, water supply and nitrogen leaching. The damage was estimated as \$0.10/m^3, which reflects the decrease in farmers' profits as a result of an additional nitrogen leaching restriction of 1 kg/ha [5]. The net cost is then \$0.31/m^3, which is \$0.001/gallon.

2. Conclusion

This analysis shows that the cost of wastewater treatment and reuse in agriculture is about equal, or slightly higher, than the cost of a simple cleanup of a concentrated spill. However, the cost of reuse, including water storage and conveyance, is much lower than most harbor cleanup costs, which generally involve more widespread water pollution. Therefore, for areas where agricultural cultivation is close to a harbor, industrial wastewater should be treated to the level required for agricultural irrigation and diverted to agricultural lands.

3. References

1. Akinyemi JO, Kumapayi OK, Adeokun OP. Potential use of industrial wastewater in irrigation - case study of Ibadan Metropolis, Nigeria. Journal of Environmental Extension. 2003; 4:37-42.
2. Carmichael JJ, Strzepek KM. A multiple-organic pollutant simulation/optimization model of industrial and municipal wastewater loading to a riverine environment. Water Resources Research. 2000; 36(5):1325-32.

3. Etkin DS. Cleanup costs for oil spills in port. In: Port technology international, volume 8. London: ICG Publishing Ltd.; 1998.
4. Feng X, Chu KH. Cost optimization of industrial wastewater reuse systems. Process Safety and Environmental Protection. 2004; 82(3):249-255.
5. Haruvy N, Hadas A, Hadas A. Cost assessment of various means of averting environmental damage and groundwater contamination from nitrate seepage. Agricultural Water Management. 1997; 32:307-320.
6. Haruvy N. Agricultural reuse of wastewater: nation-wide cost-benefit analysis. Agriculture, Ecosystems & Environment. 1997; 66:113-119.
7. Mantovi P, Baldoni G, Toderi G. Reuse of liquid, dewatered, and composted sewage sludge on agricultural land: effects of long-term application on soil and crop. Water Research. 2005; 39(2/3):289-296.
8. Panero MA. The economics of mercury in the New York-New Jersey watershed. Ph.D. dissertation. New York: New School for Social Research; 2005.
9. Ramsay IR, Pullammanappallil PC. Full-scale application of a dynamic model for high-rate anaerobic wastewater treatment systems. Journal of Environmental Engineering. 2005; 131(7):1030-1036.
10. U.S. Department of Defense. ESTCP cost and performance report: ammonium perchlorate biodegradation for industrial wastewater treatment. Arlington: ESTCP; 2000. Available at: http://clu-in.org/download/contaminantfocus/perchlorate/ESTCP1.pdf.
11. U.S. Environmental Protection Agency. Feasibility of onshore ballast water treatment at California ports, a study conducted on behalf of the California Association of Port Authorities (CAPA). 2000. Available at: http://www.epa.gov/owow/mts/capareport/52.html.
12. Zhang FS, Nriagu JO, Itoh H. Mercury removal from water using activated carbons derived from organic sewage sludge. Water Research. 2005; 39(2/3):389-395.

CHAPTER 22

THE SAFE OPERATION OF CONSTANTZA PORT DURING WINTER

M. LAZARESCU

National Research and Development Institute for Environmental Protection

ICIM Bucharest

Spl. Independentei 294, sect 6, cod 060031
Bucharest 78 Romania

Abstract: The port of Constantza is on the western coast of the Black Sea, at about 182 nautical miles (nM) from the Bosporus and about 85 nM from the mouth of Sulina Branch, through which the Danube flows into the sea. Its average coordinates are:

- Latitude 44° 09" N
- Longitude 28° 39" E

It is the main port of the Romanian shore.

The Romanian seaside is generally oriented, 244 km long, with a cliff height of 25–30 m in the range of Constantza port. The continental platform allows safe and easy navigation, because it requires neither excessive turning around navigation obstacles or islands, nor difficult changes of course nor dangerous passages.

The limit of Romania's territorial waters is set at 12 nautical miles (nM) from the shore, beyond which the contiguous zone of 24 nM extends to the open sea. Romania exercises control in this zone in order to prevent and repress infringements of its customs, sanitary, and boundary laws and rules.

The paper will present navigation conditions, safety measures for vessel entry into the port area, and functioning conditions during winter and storms.

1. The History of Constantza Port

The town and the port of Constantza have their origin in the millennial past of the Dobrogea territory. Archaeological sources prove the existence of a pre-colonial settlement in the cape area of the Constantza peninsula.

I. Linkov et al. (eds.), Managing Critical Infrastructure Risks, 401–409.
© 2007 *Springer*.

Archeological finds connect Tomis with important ports from Pontus Euxinus and the Aegean Sea like Thasos, Rhodos, Cridos, Sinope, and the pontic Heraclea.

A significant event for Tomis took place in the year 260 B.C. when the town gained the status of free port. In the second half of the first century B.C. the Roman Empire extended its border on the western coast of the Black Sea. Roman authority coincided with a flourishing period of trade, defense development, and port construction.

The city of Constantza grew larger through the new Constantza district, which later became Constantza. Waves of migratory people conquered the city, destroying it several times. In the 13th century, Genovese and Venetians settled in Tomis, receiving commercial privileges in the Black Sea. Documents dated 1260 certify the existence of Constantza port, where the Genovese and the Venetians practiced a busy barter.

In 1857, a British company—Danube and Black Sea Railways Company Ltd.—obtained a concession to build the Constantza-Cenavodă railway (62 km) and the port of Constantza. The firm was also awarded a 99-year ferry lease. Construction work ended in 1860. At the port's inauguration, there was a 4-ha, 5.2-m-deep basin sheltered by a 200-m-long breakwater. A lighthouse was erected on the cliff in the southeastern corner of the peninsula.

Following the War of Independence (from 1877–1878), Dobrogea and the port of Constantza were reunited with Romania.

In 1882, the Parliament of Romania voted to buy back the port installation and Constantza-Cernavoda railway from the British company. The government sought foreign and Romanian specialists for the design of the modern port of Constantza and in 1895 selected a project proposed by engineers I.B. Cantacuzino and Anghel Saligny.

By 1916, 2,994 m of breakwaters and 4,312 m of wharves were built. The aquatorium was enlarged to 60 ha and the outer port to 13 ha.

By the summer of 1916, the Romanian state had invested more than 2.2 billion lei in the port's construction and fitting works. From 1916–1918, the port of Constantza fell under the occupation of the German, Bulgarian, and Turkish troops. The port installation was destroyed and robbed. After the war and until 1926, the port was reconstructed and extended. Silos, a maritime station, warehouse, a grain stock exchange, and floating dock basin were also built.

Starting with the years 1956–1958, new port systematization studies were conducted. These studies continued until 1968.

The north breakwater was extended to the south by 2,600 m and a new southern breakwater of about 2,700 m was built. The access mouth had a depth of 14.5 m. Inside the port, nine new basins were put into operation.

The total area of the port was 722 ha, of which the water pool was 318 ha. Depth varied from 7.25 m in the basin of the old port to between 11.5 m and 13.50 m in the basin of the new port.

In 1978, a new extension of Constantza port was built to the south. Sheltered by the northern breakwater, the new port, Constantza South extended the northern breakwater by 5 km and included a new southern breakwater for a total length of 5,560 m.

With the new extension, the port precincts encompass an area of 3,626 ha, along about 6.5 km of shoreline. The port was extended by 4 km into to the open sea.

The favorable geographical position of the port allowed the link with the inland navigation network through the Danube-Black Sea Canal, which becomes part of the Danube-Maine-Rhin Corridor.

The port will be able to accommodate and operate vessels of 150,000 dry weight (dwt) to 200,000 dwt; in the river sector, vessels up to 5,000 dwt can be accommodated.

The depth of the new port is 21 m at the port entrance; 16.5 m and 14.5 m in the basins for seagoing vessels depending on the specification of the sector, which calls for a certain vessel size; and 7 m in the basins for the operation, composition, and decomposition of the barge convoys navigating through the canal.

The new port will be almost 3.5 times larger than the existing one on the surface, with an operating capacity four times bigger than that of the north port.

This berth is equipped with special installations for loading and discharging oil products into large vessels (150,000 dwt).

In the Constantza South port, berthing facilities were built for bulk carriers with up to 165,000 dwt capacity. Capacities for general cargo, rolled metals, and cement as well as berths for the operation of Ro-Ro and ferryboat vessels have also been completed. For the transshipment of cargo from seagoing vessels into barges, an access canal from river-maritime basin (depth 7 m) to the maritime basins is provided.

Basin no.	Berth no.	Lengths (m)	No. of Berth	Water Depth (m)	Destination
No. 1	1–4	454	4	11.70	Ro/ro cargo
	1–5	650	5	11.10	General cargo to be stored on platforms or to be directly transshipped.
	6–7	190	2	11.10	General cargo to be stored in warehouses

(continued)

(continued)

Basin no.	Berth no.	Lengths (m)	No. of Berth	Water Depth (m)	Destination
No. 2	11–12	270	2	7.5–7.9	Perishable and general cargo to be stored on platforms or to be directly transshipped
	13–16	470	4	6.8–7.2	General cargo to be stored in warehouses
	17–18	224	2	7.9	Cereals to be stored in silos
	19	114	1	7.7	Edible oil and molasses
	20–21	100	1	8.1	General cargo to be stored in warehouses
	22				Edible oil and fishing vessels
No. 3	23	115	1	7.0	Wooden products and construction items
	24		1	7.7	Cereals to be stored in the silos
No. 6	31–33	682	3	10.9–11.1	Grouts
	36–37	622	3	11.0–11.7	General cargo to be stored in warehouses or to be directly transshipped
	38–39	408	2	11.3–13.0	Vehicles, installation, spare parts for mechanical engineering
No. 7	40	208	1	10.9	Vehicles, installations, spare parts for mechanical engineering
	41–43	622	3	11.3–11.9	General cargo to be stored in warehouses or to be directly transshipped
	44	220	1	11.7	Vehicles, installation, spare parts for mechanical engineering
	45–46	455	2	11.7–11.8	Metallurgical products
	47–48	455	2	11.2–11.5	Construction items, cement, wooden products
No. 8	48–51	688	3	8.8–9.8	Cement, construction items, wooden products
	52	240	1	8.7	Goods in containers
	53	220	1	10.2	Perishable foodstuff
	54–57	930	4	10.7–12.0	Packed chemical products
	58	180	1	11.6	Packed chemicals products
No. 9	59–60	485	2	11.6	Packed chemical products
	61	232	1	10.7	Bulk or bagged fertilizers
	62	232	1	11.7	Phosphates, apatite
	63	220	1	10.5	Bulk or bagged fertilizers
	64–67	852	4	11.0–12.8	Ores, coke, coal
	68	215	1	12.8	Bulk or bagged fertilizers
No. 10	69–70	610	2	11.8	Petroleum and other liquid products

(continued)

(continued)

Basin no.	Berth no.	Lengths (m)	No. of Berth	Water Depth (m)	Destination
No. 11	72–73	610	2	13.5–13.8	Liquid petroleum products
No. 12	75–76	608	2	13.4–13.6	Liquid petroleum products
	78	344	1	14.0	Gas removal, washing, bunkering
No.1 South	79	405	1	18.5	Liquid petroleum products
	80–84	1395	5	12.5–18.5	Ores, coke, coal
	85	355	1	11.5	Ores for vessels with access to the Danube and the canal
River	94–96	600	3	7.0	Ores, coal to barges
Mar-itime	104–114	1747	11	7.0–14.5	General cargo
No. 1	115–119	1195	5	14.5	General cargo-Free zone
	120	210	1	13.5	Ferry Boat
	121	212	1	14.50	Ro/Ro Cargo
	122–125	838	4	14.50	General cargo
No. 2–3	126–137	2822	12	16.50	General cargo

The river-maritime basin includes specialized sectors for ore, coke, and coal and other sectors for bulk chemicals, cement, and rolled metals will be built.

Berths characteristics and utilization are presented in the table below.

2. Navigation Conditions

Low-speed currents do not influence maneuvering and port activity. The general north-south current on the western coast of the Black Sea acts in the roadstead; its speed varies from 0.5–1.5 knots according to the force of the wind from the northern sector.

Weak currents within the port area drive the water southward to the port exit in the case of northwest winds—when the sea water level in the port decreases—or drive the water northward in the case of south-southeast winds—when the water level increases.

Water level variation due to the tide on the western coast of the Black Sea is insignificant (30 cm). Port water levels may drop 50 cm below average (due to lasting north-northwest winds) and increase up to 70 cm above average (due to south-southeast winds). Such situations are extremely rare and do not influence vessel maximum loading capacity. The wind regime is irregular. The prevailing winds are from the north and northeast, especially in autumn and winter, sometimes reaching force 9–10 on the Beaufort Scale.

Under such circumstances, due to the strong waves, sheltering in the roads becomes difficult and requires special safety measures. In spring and summer, the south winds often blow, which usually do not exceed force 6–7 and do not generate high waves in the roads. During all seasons, west and northwest winds blow irregularly, reaching force 7–8 with gusts of up to force 9–10.

The port is generally well sheltered, both by the shore cliff and by breakwaters, so that port activity can be normally performed, although there are few days without wind in the area through the year.

The Constantza port may be considered an ice-free port. Exceptionally, during extremely cold winters, when strong northeast winds blow, temporary floe piling may take place at the port entrance.

3. The Port Roadstead

Situated outside the precincts, the roadstead of Constantza is 4 × 4 km, being limited to the north and south by the 44° 10'5" N and 44° 06'5" N parallels and to the east and west by the 28 49'5" E and 28° 44'00" E meridians.

The meridian of 28° 46'70" E and 28° 48'20" E divide the port roadstead into three zones numbered from west to east:

1. Zone no. 1: anchorage for vessels under 40,000 GRT (except tankers).
2. Zone no. 2: anchorage for vessels over 40,000 GRT (except tankers).
3. Zone no. 3: anchorage for tankers, liquefied gas carriers or vessels with dangerous cargo.

The roadstead is between 25 m and 30 m deep, allowing a safe anchorage for 40–50 large vessels, with the possibility of making a full swing in case of changes in wind or current direction. During bad weather, small vessels may enter the port with the approval of the port authority.

The control, surveillance, and direction of the vessels at the roadstead are performed by Constantza Port Control on VHF channel 67. The orders of Constantza Port Control are compulsory. The anchorage area and the traffic separation zone are under the surveillance of VTS equipment located at Constantza Port Control.

4. Approach to the Port

Navigation to the port of Constantza is performed on courses recommended by the "Notices to Mariners," the passage through the traffic separation zone being compulsory. The traffic separation scheme for vessels entering or leaving the port has the following borders:

- Separation zone 44° 04'42" N-28° 43'22" E; 44° 04'78" N-28° 43'76" E; 44° 00' 19" N-28° 48'96" E; 43° 50'81" N-28° 49'44" E

- Entrance channel into the port having the general direction NW, true course –316 44° 04'74" E; 44° 005'35" E
- Exit channel from the port having the general direction SE, true course −136 44°03'88" N-28° 42' 44" E

5. The Entrance to the Port

The port of Constantza has a single entrance in the southern area, between the edges of the north and south breakwaters, on the courses between 160° and 340°. The entrance has a 1,300 m breadth and a depth of 21 m; the entrance channel has a breadth of 300 m and a depth of 19 m. The entrance to the port is marked by light buoys. Due to the fact than the breakwaters are still under construction, lighthouses on the edges of the north and south breakwaters have not yet been built.

The entrance to Constantza port is marked by a red light to portside and a yellow and green light to starboard. Vessels have to take a 340° course entering the maneuvering basin of the old port.

Vessel navigation in the fairway follows the route marked by the buoys with the green light to starboard the red light to portside. Port entries and exits are allowed day and night.

6. Pilotage

When approaching the port, ships must contact Port Control on VHF channel 67 as well the employed agent on VHF channel 14, or, when this is not available, Constantza Radio Station on channel 16 (VHF) or 500 kHz, call sign YQI, about two hours prior to arrival at anchorage.

Pilotage of vessels in the roadstead and in the port is compulsory for technical port vessels, small trawlers, and pleasure boats. Aquatic sport boat pilotage is performed by the Pilotage Service, permanently, any time of day or night, according to a shifting schedule established by port authorities or in accordance with the vessel requirements (especially in emergencies) but always with the authority's approval.

Vessels asking for a pilot hoist a "G" flag and give three sound signals (two signals and a short one); after the Pilot has boarded the vessel, the "G" flag is substituted by an "H" flag.

The Pilot Boat has the upperwork painted white with the letters "NR" and the hull painted black with the word "Pilot" in white. The Pilot Boat has a radio-telephone (channel 14) and radar.

The roadstead of Constantza port is an open road; therefore, it is necessary that vessels should perform maneuvers in the inner roads during bad weather, to offer shelter for the Pilot when boarding and disembarking.

The Pilot assists the Master in performing maneuvers, giving necessary indications for navigation safety without replacing the Master in vessel command and, therefore, without undertaking responsibilities.

Port Control indicates to each vessel the anchorage, entrance, exit, and maneuvering of vessels as established by the Port Activities Coordination Commission in the ports of Constantza Midia and Mangalia.

To be considered as arrived, the vessels must communicate the dropping of anchor at the established place in the roadstead. As the commercial limits of the port are marked by sheltering breakwaters of the water pool, the "Notice of Readiness" (NOR) is received at the operation berth, after obtaining "Free Pratique."

7. Towage

Towage in the roadstead in the port is performed by a maritime towage company based on a towage contract distributed to the vessels by the vessel agent.

Tug assistance is compulsory for any vessels of 1,000 net register tonnage (NRT) or more; for vessels under 1,000 NRT, towage needs are established based on the weather conditions and the mooring area.

Unpropelled vessels, or those with main engines under repairs or out of order, also have to use the towage service of their NRT.

The tug is allocated depending on NRT, vessel cargo, transfer zone, weather, and other conditions so that maximum navigation safety can be achieved. Usually, tugs offer vessels assistance until berthing, using the vessel towage ropes.

Towage services in the roadstead and in the port are continuously performed, 24 hours a day. Tugs are equipped with high frequency radiotelephone VHF fixed in accordance with tug power, NRT, and duration of towage (assistance).

8. Salvage of Human Lives and Vessels

The management of salvage operations in the open sea, in the roadstead, and inside the port is the responsibility of the Harbor Master's Office in this respect: the vessels and special means of the Salvage group belong to the Civil Navigation Inspectorate.

Depending on the situation , the Harbor Master may use port and roads tugs from current service or specialized vessels and means; for example, rescue and salvage motor tugs, lifeboats for shipwrecked persons , pontoons, lifting pontoons for refloating, coastal rescue installations, and fire appliances. These may belong to other specialized port companies.

Emergency messages (e.g., S.O.S., MAYDAY, N.C.I., and others) established by international norms are received by Constantza Coast Radio Station, which immediately passes them by phone to the Harbor Master's Office, which arranges salvage action.

For the vessels in the roadstead, an emergency message or request for help can be sent by radio-telephone, or channel 16, to Coast Radio Station or directly to the Pilotage Service, who are obliged to inform the Harbor Master's Office immediately. The terms of salvage are according to the usual norms.

9. Conclusions

In Constantza port, all operations and salvages actions follow fixed procedures.

In the winter, if—as happened in the winter of 2006—the water is freezing, approaching the port is forbidden until the weather has improved.

After some strong storms, a breakwater was broken to the north of Constantza port and its repair using stabilopoeds was difficult, requiring time and money.

Generally speaking, the operation of the Constantza port, located on the Romanian Black Sea, is safe and no important accidents have taken place.

CHAPTER 23

ENVIRONMENTAL HEALTH IN PORT AND HARBOR AREAS

Air Humidity Modifying Respiratory Health in Drobeta Turnu Severin, Romania

U. SCHLINK, C. STEINERT, M. RICHTER

Department of Human Exposure Research & Epidemiology, Helmholtz Centre for Environmental Research - UFZ, Leipzig, Germany

C. PETRESCU, O. SUCIU

Department of Hygiene, "Victor Babes" University of Medicine and Pharmacy, Timisoara, Romania

R. IONOVICI

Public Health Direction Mehedinți, Turnu Severin, Romania

O. HERBARTH

Environmental Medicine, Faculty of Medicine, Leipzig University, Leipzig, Germany

Abstract: Given the heterogeneity of individual health responses to air pollution, the severity of health effects experienced by a susceptible subgroup may be much greater than that experienced by the total population. This may hinder adequate protection of an entire population, even when a low threshold value is established.

A particular situation arises when the considered region has special conditions, such as high air humidity, which occurs in the vicinity of large water surfaces. This situation is prevalent for ports and harbors that are also busy industrial regions where many people work and live.

Calculating concentration-risk relationships for respiratory health effects of ambient air quality, we discuss the conditions for which the air humidity modifies the adverse effect of airborne particulate matter. The data used were gathered in Drobeta Turnu Severin, Romania, which is a port at the Danube River just behind the "Iron Gates." Due to the large water surface

I. Linkov et al. (eds.), Managing Critical Infrastructure Risks, 411–422.
© 2007 *Springer*.

retained by the barrage, the studied area experiences strong variations in air humidity that might modify the adverse effect of certain air pollutants. For the study of particulate matter, generalized linear statistical models are widely used and we present and interpret their application to hospital admissions in Turnu Severin.

As a result we find that absolute air humidity significantly weakens the adverse effect of Total Suspended Particles (TSP) on asthma. Humidity has a protective effect on chronic obstructive pulmonary disease (COPD), but this is weakened by the interaction with TSP. The results, presented as concentration-risk relationships, demonstrate that there is no threshold value for the adverse effects of airborne particulate matter.

1. Introduction

Ports and harbors are busy industrial regions where many people work and live. An important natural characteristic of those regions is the vicinity of large surfaces of water, which can increase the humidity in air and, in this way, can influence the respiratory health of the population. Various epidemiological studies have proved that respiratory morbidity is significantly affected by meteorological conditions as well as atmospheric pollutant levels. In some cases the meteorological impact even seems to overwhelm the air pollution effects [5].

Recently, the HEAPSS study analyzed the effects of air pollution on respiratory health in several urban areas throughout Europe [2]. The study was based on health data gathered from medical emergency units and hospital admission data.

The relationship between respiratory morbidity, meteorological factors, and air pollution is modeled in statistical approaches using Poisson regression analysis [4]. Potential confounders such as ambient temperature, humidity, wind velocity and wind direction, and day of the week may adjust the generalized models. A lot of studies used time-series analysis of the air pollution effect on respiratory morbidity, including the meteorological factors as confounders. Some environmental researchers analyzed the effect of meteorological factors themselves on the risk of respiratory diseases under the consideration of airborne pollutants.

The present paper outlines an analysis of the relationship between meteorological data, total suspended particles, and daily hospital admissions of patients with respiratory complaints. For results of a first investigation see Petrescu [1]. The data were gathered in Drobeta Turnu Severin, District Mehedinti, Romania (Figure 1).

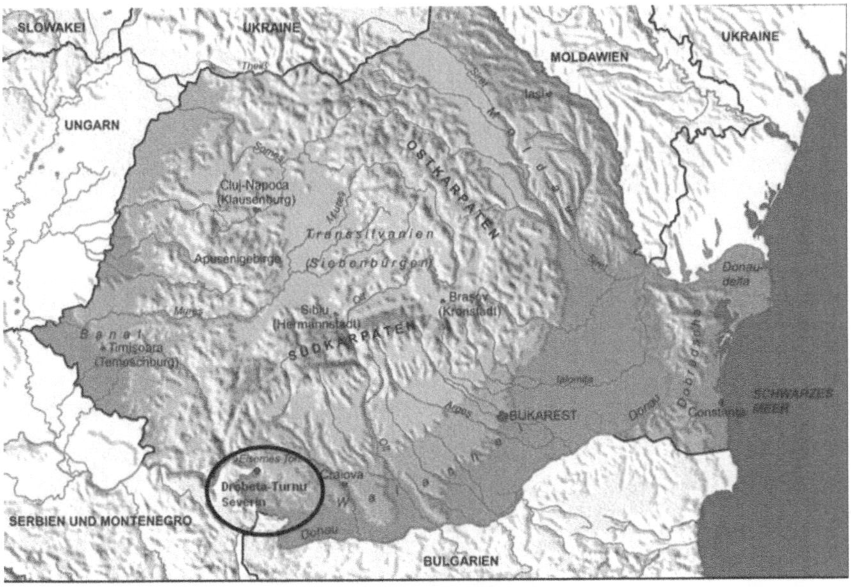

Figure 1. Physical map of Romania indicating the study area.

2. Material

Drobeta Turnu Severin is situated in southwest Romania (Figure 1), close to the river Danube, which is the second largest river in Europe (Figure 2). Drobeta is the capital of the district of Mehedinti and has 103,000 inhabitants [5]. In general, the climate is temperate and continental with cold winters and warm summers. Especially in Drobeta, the climate is influenced by the Danube and the Carpathians.

Meteorological data were provided by the Romanian meteorological service and comprise daily minimum temperature (Tmin in °C), daily average temperature (Tave in °C), daily maximum temperature (Tmax in °C), relative humidity (relHum in %), wind velocity (WV in m/s), and wind direction. The air pollution data (daily average of total suspended particulate matter, TSP in $\mu g/m^3$) were measured by gravimetric method. Between TSP and PM10 mass concentrations a site-specific, but fixed ratio was assumed.

The epidemiological data were gathered from the hospital in Drobeta. Daily admissions were recorded during four years, from 1 January 2000 until 31 December 2003. The diseases listed in Table 1 were especially considered in the present study.

Figure 2. Relevant sites in Drobeta Turnu Severin: (1) Meteo Station, (2) Bread Factory, (3) County Hospital, (4) Candy Factory, (5) Junction T.Vladimirescu - Smardan streets, (6) Junction Smardan - Traian streets, and (7) Paper Plant.

TABLE 1. Epidemiological data considered in the present study

Epidemiological Data	ICD-10 code	Comments
Asthma (A)	345	
Chronic Bronchitis (CB)	341–342	
Chronic Obstructive Pulmonary Disease (COPD)	344	
Lung Emphysema (LE)	343	Very few cases; data were not analyzed.
Total hospital admissions (TOT)	-	
Non-respiratory diseases (OTHER)	-	=TOT −(A+CB+COPD+LE)

3. Methods

The count of daily hospital admissions is a non-negative integer. Therefore it is assumed that the counts follow a Poisson distribution. The probability function for a Poisson distribution is:

$$P(Y = y) = \frac{\lambda^y}{y!}\exp(-\lambda) \qquad y = 0,1,2,... \quad \lambda > 0 \tag{1}$$

Both expected value and variance are equal to:

$$\lambda : E(Y) = \lambda, \qquad Var(Y) = \lambda \tag{2}$$

Calculating environmental effects on the daily counts of hospital admissions, Poisson regression models were used. Poisson regression belongs to the group of generalized linear models (GLM). The regression is linked to the expected value with a logarithmic function:

$$\log(E(Y)) = \log(\lambda) = \beta_0 + \beta_1 x_1 + \beta_2 x_2 + ... + \beta_p x_p \tag{3}$$

Y is the count of hospital admissions on a day. $E(Y)$ is the expected value of hospital admissions on that day. $x_1, x_2, \ldots, x_p$ are the predictors and $\beta_1, \beta_2, \ldots, \beta_p$ are the regression coefficients. Estimates for the regression coefficients $\hat{\beta}_1, \hat{\beta}_2, \ldots, \hat{\beta}_p$ are calculated using maximum likelihood. The freeware R was used for calculations.

One of the measured meteorological parameters is relative humidity. Relative humidity is the per cent ratio between water vapor pressure and saturation vapor pressure. To quantify the total amount of water inhaled during breathing, absolute humidity is needed; that is the content of water vapor in the air, measured in g/m^3.

Saturation vapor pressure $s(T)$ in hPa is calculated using the Magnus formula:

$$s(T) = 6{,}107 \cdot 10^{\frac{7{,}5 \cdot T}{238+T}} \qquad (T = temperature\ in\ {}^\circ C) \tag{4}$$

Then the water vapor pressure e (in Pa) is $e = r \cdot s(T)$ (r = *relative humidity*), and the absolute humidity (absHum) is $a = \frac{e}{R_W T}$ (in g/m^3), with the gas constant of water vapor R_W=461J/kgK.

The frequency of the wind directions was calculated as follows: If there were more than one wind direction per day, the direction was specified as "variable." If the wind velocity, measured in m/s, was smaller than or equal to 1 m/s, it was taken as calm. Missing TSP data are imputed by means of a special Kalman filter, which acts as a low pass for periods longer than 100 days.

TABLE 2. Descriptive data analysis[1]

	Minimum	25 perc	Mean	Median	75 perc	Maximum	Variance	Std Err.
T min	−14.00	1.00	7.78	8.20	14.70	25.80	67.59	8.22
T ave	−9.20	5.10	12.78	12.80	20.80	31.70	88.72	9.42
T max	−6.00	9.60	18.61	19.40	27.60	42.60	118.86	10.90
TSP	18.00	75.00	111.42	105.00	142.00	289.00	2245.85	47.39
rel Hum [%]	29.00	58.00	67.96	68.00	79.00	100.00	211.68	14.55
WV	0.00	3.00	5.64	5.00	8.00	16.00	10.14	3.18
m Chronic bronchitis	0.00	0.00	0.20	0.00	0.00	3.00	0.22	0.47
f Chronic bronchitis	0.00	0.00	0.16	0.00	0.00	3.00	0.16	0.40
m Lung Emphysema	0.00	0.00	0.01	0.00	0.00	1.00	0.01	0.08
f Lung Emphysema	0.00	0.00	0.01	0.00	0.00	2.00	0.01	0.10
m Asthma	0.00	0.00	0.23	0.00	0.00	4.00	0.27	0.52
f Asthma	0.00	0.00	0.46	0.00	1.00	5.00	0.66	0.81
m COPD	0.00	0.00	0.89	1.00	1.00	7.00	1.38	1.18
f COPD	0.00	0.00	0.44	0.00	1.00	4.00	0.50	0.70
total COPD	0.00	0.00	1.32	1.00	2.00	10.00	2.22	1.49

4. Results and Discussion

4.1. DESCRIPTIVE STATISTICS

Table 2 presents descriptive statistics for all data. Note that there are only 820 measured values for TSP; i.e., there are 641 missing values. The measuring instrument for TSP mostly failed on weekends. The longest breakdown resulted in a data gap of 38 days. All missing data are imputed as is explained in the previous section.

4.2. HEALTH EFFECTS

4.2.1. Asthma (A)

In our study we analyzed hospital admissions for the diseases listed in Table 1. The considered diseases are differently influenced by the environment. For asthma, TSP has a significant adverse effect (positive and significant coefficient in Table 3). The relative risk due to an increase in 10μgTSP/m^3 is RR = exp(0.0167) ≈ 1.02.

[1] (f female, m male); n = 1461 except for TSP: n = 820.

Interest ingly, the absolute humidity (absHum) modifies the adverse effect of TSP and lowers the frequency of hospital admissions (statistically significant negative coefficient for the interaction absHum:TSP in Table 3). The effect modification is summarized by the relative risk RR = exp(−0.00222) ≈ 0.998 for a fixed TSP level and an increase in 1gH2O/m^3 that indicates the protective effect. The significance of SW winds is just weak.

Additionally, we analyzed the relative risk for TSP pollution exceeding predefined thresholds. These thresholds are 0, 10, 20, 30, 40, ... µgTSP/m^3 and the risks are plotted in Figure 3 as a concentration-risk relationship. The grey plot represents the risk due to TSP alone, which yields in adverse effects. As results from the estimated coefficients of the model (Table 3), the interaction of absolute humidity and TSP is protective and the risk attributable to this effect modification is plotted in black in Figure 3. The relative risk attributable to both the direct and the modified effect of TSP is the product of the relative risks.

For example, at 60µgTSP/m^3 we have RR(TSP) = 1.018 and RR(absHum: TSP) = 0.998. That means the relative risk for TSP values exceeding 60µg/m^3 is 1.018 × 0.998 = 1.016 per increase of 10µgTSP/m^3 (this is indicated by a black cross in Figure 3).

4.2.2. Chronic Bronchitis (CB)

For chronic bronchitis (CB) a statistically significant effect of TSP was not detected (Table 4, for TSP $p > 0.05$).

However, the absolute humidity in air has a protective effect, lowering the risk by a factor of exp(−0.0567) ≈ 0.95 for an increase of 1gH2O/m^3 (Figure 4). As situations with a > 6 g/m^3 occur rarely, there is a low sample size and this results in nonsignificant estimates of the relative risk.

TABLE 3. Estimated parameters of the Poisson model for asthma; confounding by the day of the week was included

Variable	coeff	s.e.	p-value
Intercept	0.3169	0.1053	0.0026
Tuesday	−0.3226	0.0911	0.0004
Wednesday	−0.5215	0.0972	<0.0001
Thursday	−0.5972	0.0993	<0.0001
Friday	−1.1167	0.1194	<0.0001
Saturday	−1.7127	0.1507	<0.0001
Sunday	−1.7060	0.1507	<0.0001
TSP	0.0167	0.00836	0.0453
SW	0.5157	0.3041	0.0899
absHum:TSP	−0.0022	0.00073	0.0023

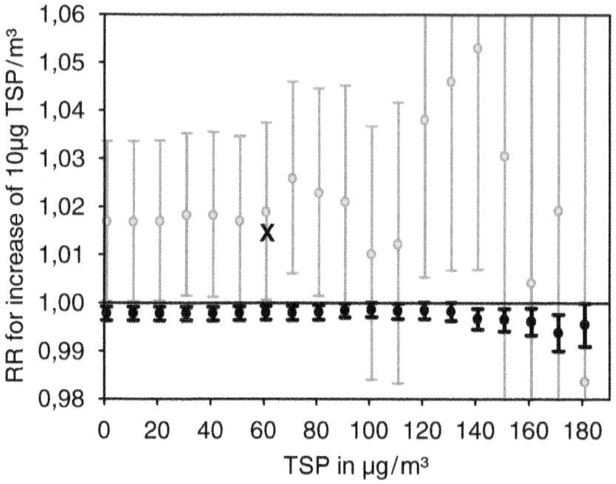

Figure 3. Concentration-risk relationship for the effect of TSP on asthma (grey) and the effect modification of absolute humidity (black); the cross indicates the total relative risk at 60 μgTSP/m^3.

TABLE 4. Poisson model estimates for CB; note that the effect of TSP is not significant

Variable	coeff	s.e.	p-value
Intercept	−0.2344	0.1893	0.2155
Tuesday	−0.0258	0.1294	0.8413
Wednesday	−0.2507	0.1387	0.0707
Thursday	−0.3463	0.1422	0.0148
Friday	−0.7924	0.1637	<0.0001
Saturday	−1.7480	0.2365	<0.0001
Sunday	−1.3205	0.1989	<0.0001
TSP	0.0106	0.0103	0.3006
absHum	−0.0567	0.0124	<0.0001

4.2.3. Chronic Obstructive Pulmonary Disease (COPD)

A protective effect of absolute humidity was observed for COPD (statistically significant negative coefficient in Table 5). Here RR = exp(−0.0347) ≈ 0.965 for an increase of 1gH2O/m^3. Different from asthma, for COPD the protective effect of humidity is modified by TSP. The relative risk of this modification is rather small and so is the statistical significance: RR = exp(0.001) ≈ 1.001, $p = 0.08$.

Figure 5 plots the pure and the modified effect for different threshold values of the absolute humidity. While the pure effect of humidity to COPD resembles the protective effect of humidity on CB, for COPD patients the

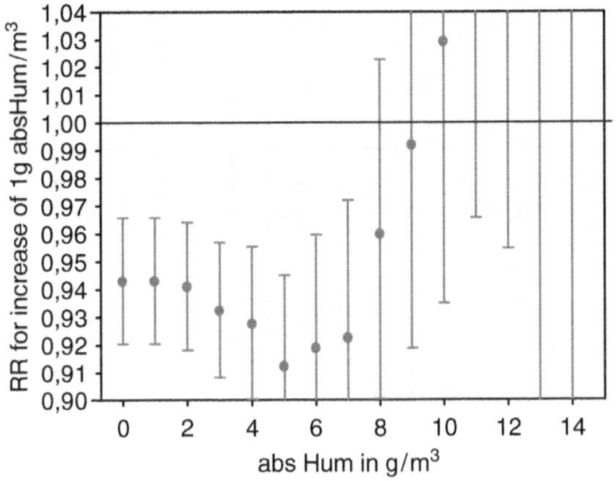

Figure 4. Concentration-risk relationship for the effect of absolute humidity on chronic bronchitis; the effect of TSP was insignificant and, therefore, removed from the model.

TABLE 5. Estimates for COPD; the significance of the effect modification is weak

Variable	coeff	s.e.	p-value
Intercept	1.187	0.0640	<0.0001
Tuesday	−0.4496	0.0670	<0.0001
Wednesday	−0.5425	0.0693	<0.0001
Thursday	−0.7590	0.0743	<0.0001
Friday	−0.9780	0.0803	<0.0001
Saturday	−1.741	0.1084	<0.0001
Sunday	−1.844	0.1134	<0.0001
absHum	−0.0347	0.0088	<0.0001
absHum:TSP	0.0011	0.0006	0.0833

exposure to TSP modifies this effect and increases the risk. For example, see the cross-marked total relative risk in Figure 5 for a level of 5 gH2O/m^3.

The modifying effect of TSP is further explored in Figure 6, where low and high relative risks for admission due to COPD are presented by light and dark areas, respectively. At low TSP levels, there is a protective action of humidity that decreases the risk. For TSP > 60 µg/m^3, however, the risk remains nearly independent of humidity and the protective effect is suppressed.

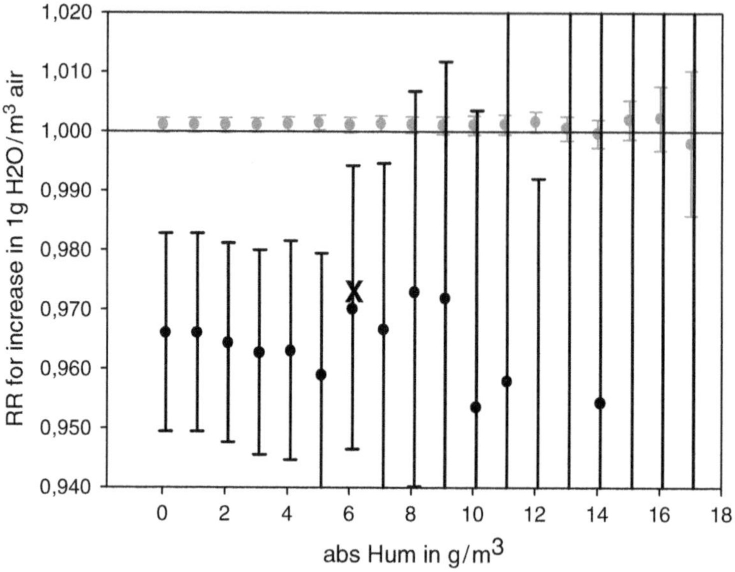

Figure 5. Concentration-risk relationship for the effect of absolute humidity on COPD (black) and the effect modification of TSP (gray).

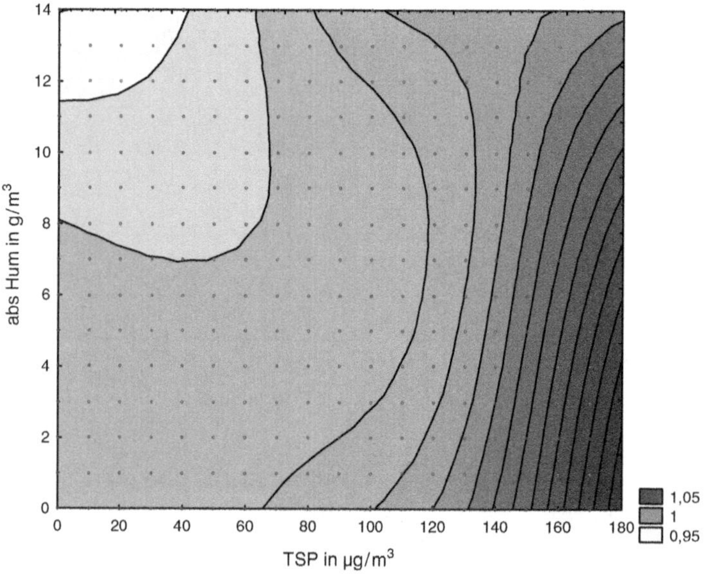

Figure 6. Concentration-risk relationship for the combined effect of TSP and absolute humidity on the relative risk of COPD; risk values are calculated at the grid points and interpolated by splines.

4.2.4. Hospital Admissions Due to OTHER Complaints

Considering total hospital admissions due to OTHER diseases—i.e., OTHER = TOT – (A+CB+COPD+LE)—absolute humidity has a protective effect and TSP has an adverse effect with a RR = exp(0,005156) ≈ 1,0051.

This result suggests that, besides the considered respiratory diseases, other complaints, comprising acute respiratory diseases, lung carcinoma, silicosis, and even cardiovascular diseases, which are included in the total hospital admissions, may be associated with TSP. Future studies would be necessary, because for OTHER there is also a statistically significant effect of the daily average temperature. This might indicate an annual pattern in the frequency of hospital admissions (Table 6).

5. Conclusions

The analysis of hospital admission data in Drobeta Turnu Severin confirms the adverse effect of TSP on asthma and gives evidence for a modification of this effect by absolute air humidity. While air pollution, such as TSP, is a trigger for asthma exacerbations, the airways of individuals suffering from COPD are obviously protected by the humidity contained in the air. That means high air humidity appears to be beneficial to the respiratory system and attenuates any complaints.

TABLE 6. Poisson model estimated for total hospital admissions without the respiratory diseases

Variable	coeff	s.e.	p-value
Intercept	5.0290	0.02170	<0.0001
Tuesday	−0.3890	0.00906	<0.0001
Wednesday	−0.4908	0.00943	<0.0001
Thursday	−0.5975	0.00970	<0.0001
Friday	−0.8371	0.01052	<0.0001
Saturday	−1.5120	0.01356	<0.0001
Sunday	−1.4710	0.01335	<0.0001
TSP	0.0052	0.00161	0.0014
absHum	−0.0292	0.00274	<0.0001
Tave	0.0066	0.00070	<0.0001
NNW	−0.2169	0.04780	<0.0001
TSP:absHum	0.0007	0.00019	0.0001
NNW:WV	0.0159	0.00756	0.0355

As there is a site-specific but fixed ratio between TSP and PM10, the former can be considered a proxy for PM10.

All study results demonstrate that there is no threshold value for the adverse effects of particulate matter. In contrast, Schlink et al. [3] observed a threshold-like health effect for ground-level ozone pollution. Further investigations might apply generalized additive models to the detection of a possible threshold value. Additionally, future studies have to solve the drawback that, because of limited data availability, the achieved results could not be adjusted for further confounding of air pollution factors.

6. Acknowledgements

The financial support of "Internationales Büro" of the Ministry of Education and Research (BMBF, Germany), project MOE04/R70, and the support of the Ministry of Education & Research (Romania) are gratefully acknowledged. The meteorological data were provided by the Regional Meteorological Centre Craiova and hospital admissions data by the County Hospital Turnu Severin, Romania. The authors are grateful for the data support of their partners. This research was financially partly supported by the European Commission (project NoMiracle, FP6 Contract No 003956).

7. References

1. Petrescu C. Some aspects of air pollution and its effects on health in population from Turnu Severin. Solness Publishing House; 2000.
2. Paatero P, Aalto P, Picciotto S, et al. Estimating time series of aerosol particle number concentrations in the five HEAPSS cities on the basis of measured air pollution and meteorological variables. Atmospheric Environment. 2005; 39(12):2261-2273.
3. Schlink U, Fritz GJ, Herbarth O, Richter M. Longitudinal modelling of respiratory symptoms in children. International Journal of Biometeorology. 2002; 47:35-48.
4. Tenias JM, Ballester F, Perez-Hoyos S, Rivera ML. Air pollution and hospital emergency room admissions for chronic obstructive pulmonary disease in Valencia, Spain. Archives of Environmental Health. 2002; 57(1):41-47.
5. Wilson AM, Salloway JC, Wake CP, Kelly T. Air pollution and the demand for hospital services: a review. Environment International. 2004; 30:1109-1118.

CHAPTER 24

ENVIRONMENTAL SECURITY IN URBAN AREAS

Health-Relevant VOC Exposure in the Greater Cairo Area, Egypt

U. SCHLINK, M. REHWAGEN, M. RICHTER, O. HERBARTH

Department of Human Exposure Research and Epidemiology, Helmholtz Centre for Environmental Research - UFZ, Leipzig, Germany

A.B. RAMADAN

National Centre for Nuclear Safety, Atomic Energy Authority, Cairo, Egypt

Abstract: Air pollution of volatile organic compounds (VOCs) is a potential mediator of adverse health effects. VOCs are emitted mainly from industrial complexes, such as port facilities and refineries, any sort of traffic, and also in closed rooms during redecoration and other behavior-linked activities, such as smoking and cooking, from building materials and furniture. Assessing and managing the health risk in terms of VOCs, the spatiotemporal variation of VOC concentrations has to be taken into account.

The paper presents an analysis of the spatial and temporal variation of VOCs in Cairo, Egypt, and, secondly, an approach to the temporal variation of VOCs in closed rooms in Leipzig, Germany. During a period of one month we exposed passive samplers at measuring sites that were selected randomly and that are representative of the surrounding area.

Cairo is one of the most traffic-polluted cities of the world and VOCs were observed at 10 sites. Comparing the concentrations with Leipzig, we found differences in aromatics, in particular in benzene pollution. Interestingly, though outdoor concentrations are up to six times higher in Cairo than in Leipzig, indoor concentrations are generally high and at similar levels in both cities. The latter, however, differ in composition of organic compounds.

I. Linkov et al. (eds.), Managing Critical Infrastructure Risks, 423–434.
© 2007 *Springer*.

In both cities, the group of aromatic hydrocarbons constitutes the principal part with more than 56%. Their main sources are traffic emissions. The indoor VOC concentrations are in Cairo 1.5 times higher than in outdoor air. In Leipzig the indoor VOC load is 6 times as high as outdoors and follows a seasonal cycle. For both cities, the composition of indoor air in apartments contrasts with the composition of outdoor air in the content of alkanes, terpenes, and aromatics. For the greater Cairo area we found that indoor air pollution exceeds outdoor pollution, in particular for alkanes. Sites characterized by high traffic have higher benzene concentrations compared to low-traffic regions.

The annual cycle for total VOC concentrations observed in Leipzig proved to be the most dominant pattern. The highest concentrations occurred during the winter months and amount to approximately three to four times the summer burden. As seasonality is very important for risk management, we fit a model to the observed concentrations and develop a procedure for seasonal adjustment, which enables estimation of the annual peak concentration utilizing an observation taken in an arbitrary month.

1. Introduction

Environmental health investigators realized the significant role of air pollution in urban areas. An interesting issue is the importance of the duration of the indoor stay in different geographical regions. Both indoor and outdoor concentrations determine the effective personal exposure. Two large representative population-based studies, the USA TEAM studies [19] and the European EXPOLIS study [7], pioneered the assessment of personal exposure to air pollutants with particular consideration of the workplace and indoor air pollution.

In outdoor environments, volatile organic compounds (VOCs) are released from industrial complexes, in particular by chemical industries, refineries, power plants, and—most important— traffic. In indoor air, VOCs are emitted mainly from building materials and furniture, during redecoration and other behavior-linked activities, such as smoking and cooking.

Under the conditions of industrialized countries at moderate climate, a particular problem arises as a result of the decreasing air-change rate due to energy-saving measures, which causes the accumulation of pollutants in indoor air. As a consequence, in regions of moderate climate, indoor concentrations of VOC generally may exceed the outdoor burden [11,17,1]. This is not necessarily valid for the north of Egypt, which is characterized by a Mediterranean climate.

Due to the lack of legal regulations, an evaluation of indoor air quality can be based just on established reference values (e.g., given by the *Umweltbundesamt*, the federal environmental agency in Germany). These reference values are specified for the total VOC concentration (TVOC) observed within apartments intended for a longer residence [16]. They represent a target value, which if attained, would most probably have no effects upon health. It was not, however, taken into consideration, whether the sampling within the apartments took place in summer or in winter. If the VOC concentrations in apartments follow a seasonal pattern, this would require a seasonal adjustment of the indoor measurements, however. A similar problem was considered by Wallace and colleagues [18], who developed a method for estimating long-term distributions of exposure based on repeated short-term measurements.

On the basis of measurements, which were taken in and in front of apartments in Cairo, as well as in the course of several epidemiological field studies in Germany, the purpose of the present paper is twofold:

1. Analyze the differences in VOC burden in outdoor and indoor air between Mediterranean and moderate climate; i.e., between Cairo and Leipzig.
2. Suggest a procedure that can be used for a seasonal adjustment of the VOC concentration.

2. Methods

2.1. FIELD STUDIES

The Greater Cairo area has about 12 million inhabitants and is located on the banks and islands of the Nile River, immediately south of the point where the river leaves its desert-bound valley and breaks into two branches into the low-lying Nile Delta region. For the Greater Cairo area, the statistics of the Egyptian government reported 1,735,724 vehicles in April 2003; about 47% of them were motor vehicles. The number of vehicles has increased by 10% per annum and electricity consumption by 8.7% per annum during the 10-year period from 1990 to 2000. About 64% of industry is concentrated in Greater Cairo. Thus, this metropolitan area is one of the largest in the world, with an estimated population of 20 million (about 27% of the Egyptian population). From the above statistics, it is clear that the population of Greater Cairo is exposed to an alarming level of air pollution. Owing to these facts, and Greater Cairo's latitude at 30°N with more than 340 days of sunshine, the formation of high levels of photochemical smog are to be expected; these levels are expected to dramatically increase in the near future.

VOC samples were collected between 28 August 2004 and 23 November 2004 at 10 sampling sites in different parts of the Greater Cairo area (see Table 1 and Figure 1). The study area is bounded by El Haram (Pyramids) at the west, El Mokatam Hill at the east, Helwan in the south, and Shoubra and El Obour in the north and northeast. These sites also represent different micro environments (roads, industrial areas, urban areas, and residential areas). All air monitoring sites are classified according to the predominant land use in the surrounding area (Table 1). Though all sites are in residential areas, some of them are affected by other land uses, such

TABLE 1. Sampling sites in the Greater Cairo area

Site #	Site name	Description
1	El Tahrir	Residential / traffic
2	El Haram (Pyramids)	Residential / traffic
3	Heliopolice	Residential / traffic
4	Nasr city	Residential
5	El Maadi	Residential / traffic
6	Helwan	Residential / industrial
7	El Attaba	Residential / traffic
8	Shoubra	Residential / industrial
9	El Mokatam Hill	Residential / background
10	El Obour	Residential

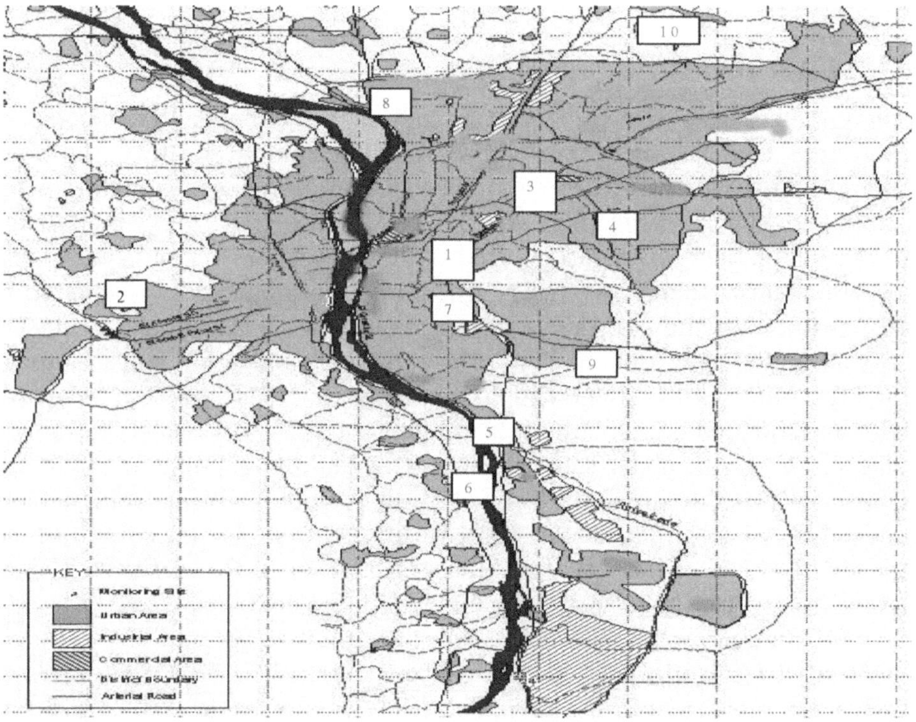

Figure 1. Map of sampling sites in Greater Cairo Area (high traffic sites 1, 2, 3, 5 and 7, low traffic sites 4, 6, 8, 9 and 10).

as industrial or traffic. One site can be considered as a background monitor (El Mokatam).

Site 6 was selected in Helwan City, which is one of the most important industrial regions, located in the southern part of Cairo. It covers an area of 35 km^2, is the domicile of 563,000 inhabitants and is bounded to the north by El Massara, to the south by Kafr El Ellow, to the east by the El-Autostorade highway, and to the west by the Nile River. Representing 17% of the total industrial activity in Cairo, Helwan contains 27 industrial facilities, including factories for the production of cement, metallurgy, steel, textiles, and cars, which contribute directly to urban pollution. The air in Helwan is characterized by high turbidity, due to the surrounding traffic and the heavy industry. To the northeast and north of Helwan, the cement industry is found in Toura and often there is advection of the pollutants to Helwan. Moreover, in the south there are iron, steel, and cement industries, the pollutants of which are transported to Helwan during winter. The major sources of pollution at Helwan are three types of industries:

1. Cement factories distributed from Toura (in the north) to El-Tebeen (in the south) at a distance of about 30 km.
2. Engineering industries (car factory, pipes and tubes factory).
3. Iron and steel factories.

In Leipzig, VOC measurements were carried out between 1994 and 2001. The data were gathered for use in several epidemiological studies designed to assess the impact of VOC exposures in environmental levels on the development of respiratory illnesses and allergies in children of different age groups [5,3,15].

Leipzig has about 500,000 inhabitants and is located in the center of the eastern part of Germany in a plains region. Though chemical and iron industries as well as power plants are in and around the city, the urban area is mainly affected by traffic pollution. The climate is moderate and the weather is characterized by passing low-pressure areas that contribute to the exchange of air in the urban area. As energy-saving measures are very modern, the ventilation of rooms is often a problem associated with indoor VOC pollution.

For the present analysis, we used indoor VOC measurements taken in randomly selected apartments, which was a necessary procedure for the epidemiological studies. We did not make any exclusion on grounds of the inhabitants' activities, such as smoking or recent renovation. Each of the apartments was sampled only once. Sampling was started close to the birthday of the child that was recruited for the epidemiological study. According to their structure and composition, the sampled VOC components belong to the following groups: alkanes, cycloalkanes, aromatic compounds, halogenated hydrocarbons, and terpenes.

2.2. SAMPLING

For passive sampling of VOCs, we used 3M monitors of the type OVM 3500 from the company 3M GmbH in Neuss. We selected a sampling interval of four weeks, which can be assumed to describe the personal exposure much better (and integrative) than a short-term measurement. During the four-week period the monitors were exposed to the indoor air of the rooms where the study participants spent most of their time: mostly the child's bedroom. The passive collectors were placed in the middle of the room at 1.5–2 m height with a minimum distance of 50 cm from furniture. Outdoor air measurements were carried out parallel to the indoor air measurements. The monitor was placed at a rain-protected position directly on the outer side of one of the apartment's windows. The outdoor samplers were also exposed for four weeks.

All our measurements refer to normal conditions of 25°C and 760 mmHg. Further parameters characterizing the climate of the rooms were not registered because of the multitude of measurements. The authors are aware of the influence of indoor air environmental conditions modifying the diffusion of VOC onto the sampler. Assessing the impact of parameters such as humidity, temperature, and linear velocity on the sampler, we concluded from a review of the literature [6,14, 8] that unknown conditions to the VOC monitor may result in an error of ≤13%. At the end of the sampling period, each monitor was tightly closed with a cap belonging to it and stored by room temperature until analysis.

After exposure the VOCs were desorbed from the adsorption layers (charcoal pads) by means of 1.5 ml carbon disulfide (with low benzene, from Merck) containing 1% methanol. During the extraction period of 30 minutes, the monitors were shaken every 10 minutes and then the solution was decanted into GC vials for analysis.

2.3. GASCHROMATOGRAPHIC ANALYSIS

VOC analysis was performed on the Perkin Elmer GC-MS system Turbomass equipped with an RTX-1 column (60 m × 0.32 mm I.D., 1.0 μm film thickness; Restek). The oven temperature was held at 43°C for 5 minutes, and then programmed to 200°C at a rate of 2.5°C/min. The flow-rate of carrier gas (helium) was held at 1.1 ml/min. The injector and transfer line temperatures were held at 250°C. For ionization we used electron energy of 70 eV. The source temperature was held at 200°C. A sample volume of 1 μl was splitless injected. Integrated areas of selected fragment ions from each of the 30 VOCs were obtained with the software Turbomass, Version 4.4 (Perkin Elmer).

2.4. CALIBRATION AND CALCULATION

A stock solution was prepared, diluting original reagents in highest available purity with methanol (it can be stored frozen for later use). For calibration, we freshly prepared a set of five levels in the 0.8–4 µg/ml range by diluting the stock solution with carbon disulfide, which contained an internal standard (cyclododecan, and since 1999 benzene d6). For each component, a calibration function was determined based on the integrated area of the chromatogram (for details see Rehwagen et al. [12]).

Recovery coefficients were determined by direct injection of a known amount of the standard into a 3M sampler and subsequent extraction with carbon disulfide containing 1% methanol. The recovery was between 98% and 102%. For the studied VOCs, the background amounts contained in the monitors and in the carbon disulfide were determined for each charge of samplers and chemicals and subtracted from the sample results.

The detection limits for the components were estimated as the threefold standard deviation of five replicate measurements of monitor blanks. For components with blank values too low to be registered it is usual practice to use the threefold standard deviation of replicate measurements of a low-level standard solution [2]. The detection limits of the GC/FID/ECD system were observed to be between 0.01 µg/m^3 and 1.0 µg/m^3 regarding a sampling interval of four weeks. For the GCMS system we found detection limits from 0.01 µg/m^3 to 0.05 µg/m^3 converted to a sampling interval of four weeks.

We tested the storage stability of five passive samplers at room temperature for a five-month period and observed no significant loss, which corresponds with observations of Shields and colleagues [17] and Elke et al. [4]. The precision was evaluated as the percent relative standard deviation (% RSD). The % RSD was typically below 10%.

2.5. STATISTICAL METHODS

For statistical processing, GenStat 6.1 was applied [10]. To calculate the rank values $q_{u,l}$ of upper and lower confidence limits for the median of monthly subgroups we used the normal approximation $q_{u,l} \approx E[q] \pm u_{1-0.05/2} \cdot \sqrt{\mathrm{var}[q]}$ with $u_{1-0.05/2} = 1.96$ representing the 95th-percentile of the standard normal distribution. This approximation is valid for sample sizes $N \geq 10$ [9].

As percentiles follow a binomial distribution [9], for the median (which is the 50th-percentile; i.e., $q = \frac{1}{2}$) we can specify the expectation value $E[q] = Nq = \frac{N}{2}$ and the variance $\mathrm{var}[q] = Nq(1-q) = \frac{N}{4}$.

As the data did not follow a normal distribution function, we used the Mann-Whitney U-test and the H-test of Kruskal-Wallis to detect significant differences between groups of data. These tests refer only to the statistics of the data and do not include the measurement errors.

To assess the amplitude of the observed annual cycle of indoor VOCs and to enable a conversion between concentrations observed in different months, we fitted a seasonal model to the logarithms of the monthly medians of concentrations. The predictors are harmonic functions of the time (m in months) and the coefficients have been estimated by means of regression analysis.

3. Results and Discussion

3.1. VOC COMPOSITION IN CAIRO

In Cairo, all VOC group totals are higher indoors than outdoors (Figure 2), but only the concentrations of alkanes (1) and terpenes (5) are higher indoors by a statistically significant amount ($p=0.003$ and $p=0.001$, Man-Whitney-U-test). This result is plausible because there are many indoor sources of these VOCs, especially for alkanes that are typically emitted with renovation activities such as repapering and painting walls, and for terpenes that are emitted from wooden furniture and flooring. This result clearly demonstrates that in Cairo's apartments there are indoor sources of VOCs.

Besides indoor sources, another important source of VOCs is traffic. Comparing regions with different traffic levels, at sites with high traffic burdens the outdoor pollution was statistically significant higher for the total concentration of aromatic hydrocarbons, comprising 10 individual

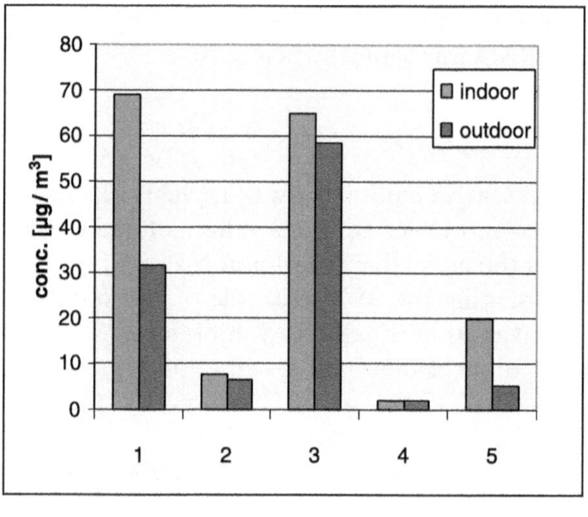

Figure 2. In Cairo indoors and outdoors observed sum-concentrations for alkanes (1), cycloalkanes (2), aromatics (3), halogenated hydrocarbons (4), and terpenes (5).

compounds. For eight of them there was a significant difference in regions without elevated traffic (Table 2).

For apartments located in regions of high traffic (6.7 ± 5.3 µg/m³) we found significantly elevated indoor-air benzene compared with low-traffic sites (3.1 ± 1.0 µg/m³); both groups are significantly different ($p = 0.014$; U-test).

3.2. COMPARISON OF VOCS IN CAIRO AND LEIPZIG

The measurements in the greater Cairo area were compared with exposition data of Leipzig dwellings. Figure 3 presents the arithmetic mean of the sum of the VOC concentrations in indoor and outdoor air of both cities and, within the columns, the composition of the substance groups.

TABLE 2. Mean outdoor concentrations of aromatic hydrocarbons (in µg/m³) and significance level of their difference in regions with high and low traffic (Man-Whitney-U-test)

Substance	high traffic	low traffic	p-value
benzene	7.4	3.5	0.009
toluene	40.7	12.8	0.028
ethylbenzene	4.5	1.7	0.028
m+p-xylene	16.1	6.1	0.028
styrene	0.8	0.3	0.252
o-xylene	4.9	1.9	0.028
4-ethyltoluene	6.7	2.7	0.028
3-ethyltoluene	2.9	1.2	0.028
2-ethyltoluene	1.8	0.8	0.028
naphthalene	0.8	0.2	0.175

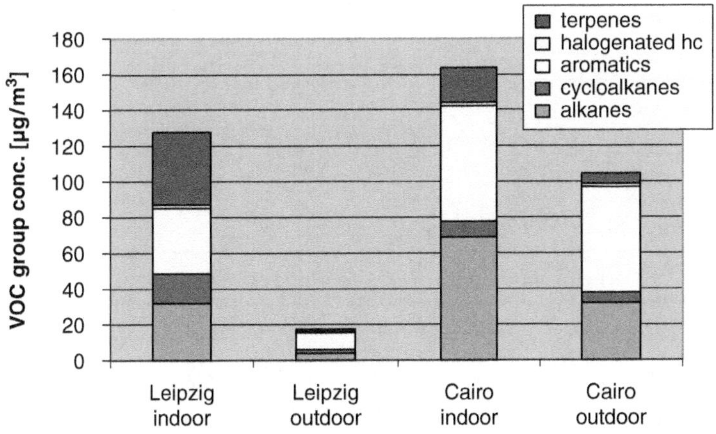

Figure 3. Concentrations observed indoors and outdoors in Cairo and Leipzig.

Though the total outdoor pollution level is higher in Cairo than in Leipzig, the percentage composition is very similar. That means there is resemblance in the pollution sources of both cities. We conclude from Figure 4 that the exchange between outdoor and indoor air is more intense in Cairo than in Leipzig.

Generally, in both cities, the VOC load is higher indoors than outdoors. We find that the indoor-outdoor ratio is 7.5:1 for Leipzig, but 1.5:1 for Cairo. Outdoor air in the greater Cairo area is six times more polluted with VOCs than outdoor air in Leipzig. Aromatic hydrocarbon concentrations are nearly identical indoors and outdoors in Cairo; in Leipzig, indoor concentrations are higher.

Considering the fractions of the VOC groups in the outdoor air, in Table 3 the aromatic hydrocarbons, whose main source is traffic, represent the principal part with >56% in both cities. The main fraction in Leipzig dwellings represents the group of terpenes, which is due to furniture and flooring made from pure wood. In Cairo the group of alkanes dominates.

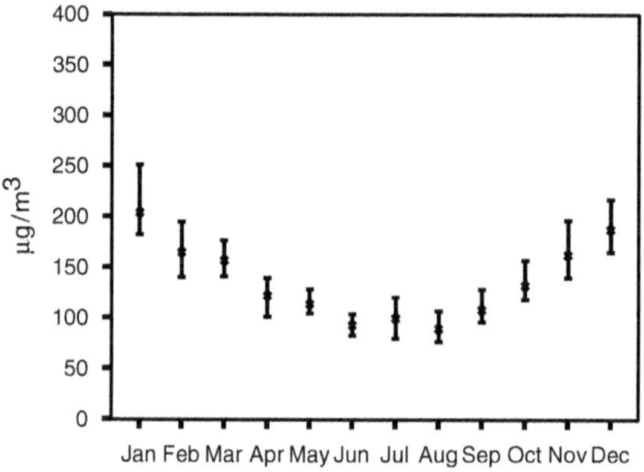

Figure 4. Seasonal cycle of the sum of VOCs observed in apartments in Leipzig; median with 95% CI.

TABLE 3. Fractions of VOC groups [%]

	Leipzig outdoor	Cairo outdoor	Leipzig indoor	Cairo indoor
Alkanes	23.4	30.5	24.9	42.3
Cycloalkanes	9.2	6.2	13.1	4.8
Aromatics	58.1	56.3	28.8	39.6
Halogenated hc	1.9	2.0	0.9	1.3
Terpenes	7.4	5.1	32.3	12.0

3.3. TEMPORAL VARIATION OF VOCS IN LEIPZIG

Classifying the VOC observations as to the months when the measurement was made, we observe a clear annual pattern for the indoor pollution in apartments in Leipzig (Figure 4). Fitting a harmonic function to the measurements, we calculated adjustment factors that enable calculation of the annual maximum and annual mean burden of the residents [13].

4. Conclusions

Apartments in regions of high traffic have high benzene pollution indoors and outdoors. Benzene is a carcinogenic substance and, therefore, of the highest relevance to health. It should be useful to analyze, in further studies, the average fraction of daytime that a person usually spends indoors in greater Cairo area.

Causes for the different chemical composition of indoor air are certainly different life style factors. In addition, the very small sample size for Cairo, sampled all in one month, represents only a small part of the year. The Leipzig exposure data were sampled over a total year. Due to our results about the seasonal variation of VOCs in indoor air of Leipzig and other German cities, with approximately three times higher concentrations in the winter compared to summer (Figure 2), we are very interested to examine whether in the greater Cairo area this annual variation is also observable. This is a subject of future research cooperation.

5. Acknowledgements

The help and financial support of Dr. Manthey and Ms. Wolfgram, International Office of the German Ministry of Education and Research, is gratefully acknowledged.

6. References

1. Begerow J. Passive sampling for volatile organic compounds (VOC) in air at environmentally relevant concentration levels. Fresenius' Journal of Analytical Chemistry. 1995; 351:549-554.
2. Begerow J, Jermann E, Türkan K, Dunemann L. Performance of two different types of passive samplers for the GC/ECD-FID determination of environmental VOC levels in air. Fresenius' Journal of Analytical Chemistry. 1999; 363:399-403.
3. Diez U, Kroessner T, Rehwagen M, Richter M, Schlink U, Krumbiegel P, Fritz G, Schulz R, Wetzig H, Metzner G, Herbarth O, Borte M. Effect of indoor chemicals exposure on

the development of infections of upper respiratory tract and atopy in the first weeks of live. Pediatrics and Related Topics. 1999; 37:469-485.
4. Elke K, Jermann J, Begerow L, Dunemann L. Determination of benzene toluene, ethylbenzene and xylenes in indoor air at environmental levels using diffusive samplers in combination with headspace solid-phase microextraction and high resolution gas chromatography-flame ionization detection. Journal of Chromatography. 1998; 826:191-200.
5. Fritz GJ, Herbarth O, Schlink U, Richter M, Rehwagen M. Urban area pollution profiles and patterns of respiratory disease and allergies. Epidemiology. 1998; 9(4):105.
6. Hendricks W. Development of a protocol for laboratory testing of diffusive samplers. Salt Lake City: OSHA Salt Lake Technical Center; 1996.
7. Jantunen MJ, Hanninen O, Katsouyanni K, Knoppel H, Kuenzli N, Lebret E, Maroni M, Saarela K, Sram R, Zmirou D. Air pollution exposure in European cities: the "EXPOLIS" study. Journal of Exposure Analysis and Environmental Epidemiology. 1998; 8(4):495-518.
8. Larsen D. Storage characteristics of exposed activated carbon diffusion monitors. AIHCE 1996 abstracts, abstract 413. 1996.
9. Lienert GA. Verteilungsfreie methoden in der biostatistik. Meisenheim am Glan: Verlag Anton Hain; 1978.
10. Payne RW. The guide to GenStat. Rothamsted: Lawes Agricultural Trust; 2000.
11. Rehwagen M, Schlink U, Herbarth O, Fritz GJ. Pollution profiles at different kindergarten sites in Leipzig, Germany. Environmental Toxicology. 1999; 14:321-327.
12. Rehwagen M, Schlink U, Herbarth O. Seasonal cycle of VOCs in apartments. Indoor Air. 2003; 13:283-291.
13. Schlink U, Rehwagen M, Damm M, Richter M, Borte M, Herbarth O. Seasonal cycle of indoor-VOCs: comparison of apartments and cities. Atmospheric Environment. 2004; 38:1181-1190.
14. Schultz G. How pressure and temperature affect sampling with passive monitors. Salt Lake City: OSHA Salt Lake Technical Center; 1995.
15. Schulz R, Borte M, Diez U, Herbarth O, Lehmann I, Jakob B, Heinrich J, Wichmann HE. LISA-studie: einfluβ von lebensbedingungen und verhaltensweisen auf die entwicklung von immunsystem und allergien im Ost-West-Vergleich. Allergo Journal. 1999; 8:25.
16. Seifert B. Richtwerte für die innenraumluft. Bundesgesundheitsblatt-Gesundheitsforschung – Gesundheitsschutz. 1999; 42:270-277.
17. Shields HC, Fleischer DM, Weschler CJ. Comparisons among VOCs measured in three types of U.S. commercial buildings with different occupant densities. Indoor Air. 1996; 6:2-17.
18. Wallace LA, Duan N, Ziegenfus R. Can long-term exposure distributions be predicted from short-term measurements? Risk Analysis. 1994; 14(1):75-85.
19. Wallace LA, Pellizzari ED, Hartwell TD, Sparacino C, Whitmore R, Sheldon I, Zelon H, Perritt R. The TEAM (total exposure assessment methodology) study: personal exposures to toxic substances in air, drinking water, and breath of 400 residents of New Jersey, North Carolina, and North Dakota. Environmental Research. 1987; 43(2):290-307.

CHAPTER 25

POTENTIAL RISK AND CONTROL OF CONTAMINATION IN THE GULF OF AQABA-JORDAN

N. AL-HALASAH

Royal Scientific Society, Amman 11941 Jordan

B.Y. AMMARY

Balqa Applied University, Huson College, Huson 21510 Jordan

Abstract: The Gulf of Aqaba (GOA) has a semi-enclosed nature that has led to its unique ecosystem and biodiversity. This nature, which results in low water circulation and renewal, along with the rapid development of industry, tourism, and transport in the area, has made it susceptible to marine pollution and ecosystem degradation. The present paper discusses the potential risks to the GOA ecosystem resulting from a number of activities. These include coral reef damage caused by tourism, oil spills, waste oil contamination and air pollution associated with land transportation, solid waste disposal, phosphate dust deposition from ship loading activities, chemical and thermal pollution from coastal industries, and sewage discharges into the marine and terrestrial environment. Management and control of these potential risks requires regional cooperation with the riparian countries.

1. Introduction

The Gulf of Aqaba (GOA) is one of two northward extensions of the Red Sea. It is surrounded by Egypt in the east, Saudi Arabia to the west, and Jordan and Israel to the north (Figure 1). It has a width that ranges from 5 km at its northern tip to a maximum of about 25 km and then narrows back to 6 km at the straits of Tiran, where the GOA is connected to the Red Sea. The GOA has an average water depth of 800 m. Its maximum depth is about 1800 m. The straits of Tiran have a maximum depth of 250 m. The length of the GOA on the Jordanian borders is about 26.5 km. The semi-enclosed nature of the GOA, along with the semi-enclosed nature of the Red Sea, has given it a unique ecosystem with unique biological diversity. In fact, the Red Sea supports higher levels of species diversity than any eastern Indian Ocean region [7]. At the same time this semi-enclosed nature has limited the

I. Linkov et al. (eds.), Managing Critical Infrastructure Risks, 435–444.
© 2007 *Springer*.

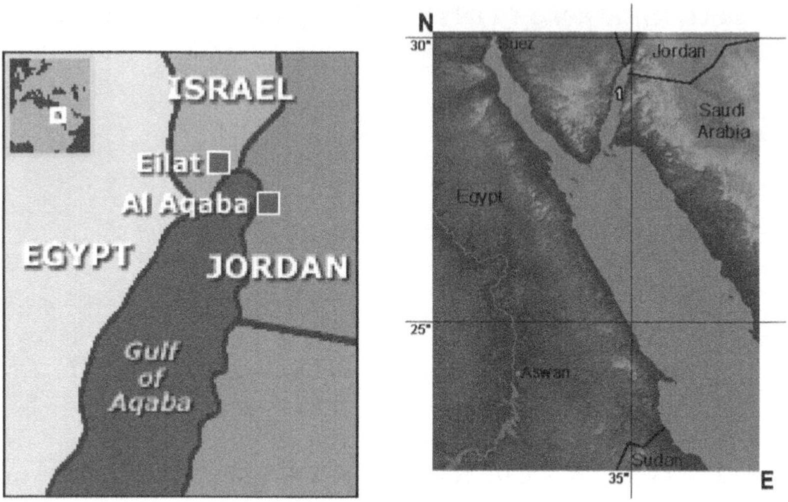

Figure 1. Gulf of Aqaba and Red Sea areas.

rate of exchange of its water with the Red Sea and the Indian Ocean. The mean residence time of water in the GOA ranges from one to three years [20] with an average of about two years. This—along with the relatively small volume of the Gulf—results in the accumulation of pollutants, which is one of the most important concerns about the sustainability of inland and offshore activities in the Gulf. This is particularly true in the Jordanian and Israeli parts of the GOA, which are the furthest from the Red Sea and have relatively shallow waters.

The city of Aqaba, which lies on the Jordanian side of the GOA, is the only city in Jordan that has an access to the sea, and therefore has its sole port. The city has a population of about 75,000. In 2001, the government of Jordan has declared the city of Aqaba along with the surrounding area a free zone called Aqaba Special Economic Zone (ASEZ). The purpose of such a move is to make the zone a haven for commerce, tourism, and technology. This is expected to increase the total population, tourists, industrial and commercial activities, and wastewater discharge, to a point that the environmental richness of the Gulf becomes threatened unless special steps are taken.

In order to preserve the marine environment in ASEZ, regulations call for the preparation of a comprehensive Environmental Impact Assessment (EIA) for major projects like oil refineries; thermal power stations; land reclamation for the purpose of industry or recreation; construction of major hotels and entertainment complexes; and construction of highways, railways, airports, harbors, or marine ports [6]. Comprehensive EIAs are also required for various industrial activities, waste disposal activities, and fish breeding farms in the sea. Other projects need either preliminary evaluation or are exempted from EIA. Environmental auditing is an important part of the regulation to monitor, prevent, and mitigate any environmental threats. The regulation also has provision for air protection, solid and hospital waste, and used mineral oil. In order to protect the marine environment, the regulations prohibit the disposal of oils, oil mixtures, garbage, litter, sewage, and ballast waters from ships frequenting the port without prior approval and in designated areas. The captain of the ship is required to maintain an oil record book recording all processes conducted related to transferring, loading, unloading, leakage, oil discharging, and so forth. Ships visiting the GOA should be equipped with pollution minimizing equipment. Fines are specified in the regulation as a means of deterring any person from polluting the environment.

2. Sea Water Quality in the Gulf of Aqaba

Typical values of a number of water quality parameters in the GOA are shown in Table 1.

TABLE 1. Typical water quality parameters in the GOA

Parameter	Value or Range	Reference
Temperature (°C)	20.8-27.5	[17]
Water Transparency (m)	22-35	[17]
Chlorophyll a (μ g/L)	0.02-2.01	[17]
	0.14-0.23	[13]
Ammonia (μ M/L)	0.43-0.58	[5]
Nitrate	0.37	[5]
Inorganic Nitrogen (μ M/L)	0.98	[5]
	0.13-0.58	[13]
Phosphate (μ M/L)	0.175-0.35	[5]
	0.02-0.07	[13]
Silica (μ M/L)	1.36	[5]
	0.70-1.43	[13]
Organic carbon (mg/L)	0.04-3.79	[17]
Suspended Matter (mg/L)	0.71-3.44	[17]
Heavy Metals (μ g/L)		[14]
Cd, Pb, Cu, Zn, Fe, Ni, Al	Negligible	

The table shows that water temperature is high, within a narrow range, and there is high water transparency. This may be due to low levels of suspended matter, low phytoplankton biomass, and the trapping of sediments in deep water [17]. The low chlorophyll a concentration (to assess phytoplankton biomass) in Table 1 also displays a clear temporal trend. These low levels concur with the low phytoplankton biomass in Aqaba, which partly explains the high water transparency. During summer, the water column in the GOA is stratified and surface layers are depleted of nutrients. In winter, by contrast, the thermocline weakens and deep convective mixing persists for several months, bringing up relatively nutrient-rich water to the surface. Inorganic nutrient concentrations in the GOA waters are within the typical range reported for oligotrophic tropical oceans [5].

3. Potential Sources of Pollution

In the GOA, the major threats to coral reefs stem from urbanization and tourism, oil transport, industrial pollution, sewage discharges [7], and heavy souvenir collection. Development in the GOA has only occurred during the last three decades, producing major industrial facilities, shipping activities, and rapidly expanding tourism. The following is a discussion of the major environmental threats that may affect the quality of the marine ecosystem in the GOA and possible mitigation and preventive measures.

3.1. TOURISM

The GOA is a highly valued area as a national, regional, and international tourism center. It has amazingly clear, deep, calm, and warm water with spectacular underwater life. There are about 1,000 fish species, 110 species of reef-building coral, and 120 species of soft corals. This has made the GOA a popular tourist destination [7]. At the present time, the number of visitors to Aqaba is estimated at 600,000 [16], and is expected to increase sharply in the future as a result of the major development and investment plans in the area.

In many areas of the world, a large influx of tourists has brought with it environmental degradation that threatened the same components of the ecosystem that caused the influx in the first place. The spectacular coral reefs of the Gulf of Aqaba are an example of these threatened components. High numbers of swimmers, divers, snorkelers, and recreational boats have caused severe damage to coral reef structures in these waters. Tourists tend to step on living corals or anchor their boats in coral reef areas, thus breaking off pieces of corals. Unless educated and regulated, higher numbers of tourists in the Gulf of Aqaba would severely degrade the quality of the marine ecosystem. Along with tourists, the construction of infrastructure and resort hotels produces heavy sedimentation on the corals, which is detrimental to their well-being.

It should be noted that successful tourism and environmental protection of the GOA are interdependent. The lack of environmental control will severely restrict Jordan's ability to attract private investments and international tourism. Successful tourism will depend on the availability of clean, non-polluted water and healthy coral reef ecosystems. If beaches are frequently not swimmable and coral reef areas display a declining diversity of marine life, tourists are likely to choose other travel destinations. Tourism should provide the countries bordering the GOA with an economic incentive to preserve their underwater ecological system. The time may come in the GOA when environmental improvement projects would be more economically feasible than building hotels and resorts. Therefore, EIA studies should be implemented for tourism development projects in the GOA [15].

In 1994, Jordan and Israel signed a peace treaty between the two countries. As a result, the Red Sea Marine Peace Park (RSMPP) in the GOA was established. The RSMPP encompasses the Coral Reef Reserve in Eilat, extending along approximately 2 km of Israeli coastline, and the Aqaba Marine Park, extending along approximately 7 km of Jordanian coastline [8]. The RSMPP calls upon Jordan and Israel to collaborate in research efforts and to implement comparable policies and regulations designed to protect the coral reefs as a tourist attraction. Long- term monitoring and research programs along with educational and awareness programs are planned as part of the RSMPP [7].

3.2. OIL SPILLS

An oil spill is probably the most devastating and visually compelling form of pollution. Oil pollution can damage the reproductive system of corals, interfere with the production of larvae, and inhibit normal settling. Oil can also induce feeding and behavioral changes among fauna. Experiments indicate a range of responses to oil, including abnormal mouth opening and feeding behavior, mucus secretion, decreased growth rate, and increased rate of tissue death [18]. Factors that reduce light penetration, such as continuous thick films of oil, interrupt the photosynthetic cycle of specific coral reefs.

Numerous shipping and industrial activities can cause an oil spill in the Gulf of Aqaba. Spills from ship traffic range from small to large oil spills. Accidental or intentional releases of oil-contaminated bilge or ballast water from freighters are examples of small spills. Spills from oily ballast water or bunker oil releases or from defective equipment or procedures at an oil terminal or pipeline facility are examples of minor to medium oil spills.

Rupture of a bunker oil tank, or a tanker collision causing the rupture of a single oil tank; and disastrous spills caused by oil tanker wreckage are examples of major oil spills [20].

During 1994 and 1995, Egypt, Israel, and Jordan prepared an Upper Gulf of Aqaba Oil Spill Contingency Plan. The plan established emergency response centers in each country. These centers are supposed to coordinate efforts in the event of a spill requiring a regional response [20]. Large spills, however, may exceed the capacity of these centers, causing devastating impacts on coral reefs and marine life located mostly in shallow waters.

In order to reduce the possibilities of large spills and reduce their effects, Jordan, Egypt, and Israel have to allocate responsibilities and financial burdens, conduct joint training exercises, and set mandatory design and operational standards for cargo vessels and oil tankers operating within the Gulf (requiring double-skin oil tankers and new navigational safety provisions, for example) [20].

It should be noted that small oil spills in the Gulf of Aqaba are inevitable. However, the impacts of these spills can be minimized if proper regulations are implemented and enforced.

3.3. WASTEWATER DISCHARGES

Wastewater discharge into marine areas has a negative impact on the sustainability of underwater ecosystems. High suspended solids concentrations in the effluent can accumulate on coral reefs which prove to be detrimental to their well-being. High suspended solids concentrations can also reduce light penetration to the Gulf's bottom, which is essential to the growth of corals. Reduced light penetration interrupts the photosynthetic cycle of specific coral reefs. Corals need sunlight because they depend for their survival on tiny algae that live in their tissues. The photosynthetic algae provide the polyps with carbon, and benefit in return from nitrates and phosphates produced as waste by the polyps [19].

In addition, nutrients like nitrogen and phosphorous enhance the growth of algae in sea waters, thus reducing light penetration. When there are enough nutrients in the water, the polyps algae no longer depend on their host for these materials. Although the algae proliferate in the host cells, they begin to withhold the products of photosynthesis from the polyp, which starves. Eventually the algae either leave or are expelled by the host [19]. Heavy metals, coliform bacteria, and oil and grease are among many other wastewater characteristics with potential harmful effects on both the ecosystem and tourist population. The total area of the Gulf of Aqaba is comparatively small with high concentrations of economic activities that produce wastewater along the coast. Stringent regulations, monitoring, and enforcement are required to preserve such an ecosystem for future generations.

In 1987 the city of Aqaba started its first wastewater treatment plant using waste stabilization ponds, and the effluent was used for irrigation. Recently the plant has been converted to an extended aeration plant, thus improving the treatment efficiency. Therefore no wastewater discharges to the Gulf of Aqaba have occurred from the Jordanian portion on the Gulf since 1987. However, until 1996, municipal wastewater from the City of Eilat/Israel was discharged into the shallow coastal waters of the Gulf of Aqaba after only minimal treatment, causing frequent beach closures and coral reef decline. Eilat municipality has stopped discharging its sewage into the Red Sea and diverts its effluents through a 40-kilometer long pipeline to the north of the bay where the low quality reclaimed wastewater is used to irrigate date fields. A new treatment facility is in the advanced planning stages.

Development in ASEZ, which is expected to increase its population to about 250,000 will dramatically increase water demand and wastewater generation. A ban on wastewater discharges into the Red Sea should be carefully investigated. Wastewater reuse in areas near the Gulf of Aqaba with poorly treated wastewater may not prevent it from reaching the Gulf's water through runoff or groundwater. Advanced wastewater treatment or irrigation projects far from the coastline may be the solution to this problem. New wastewater reuse applications should also be investigated.

3.4. URBANIZATION AND INDUSTRIAL POLLUTION

Urban development and industrial pollution impose high environmental pressure on the Gulf of Aqaba [3,1]. The last three decades have altered 30 to 40% of the Jordanian coastline from a pristine natural environment to a heavily used port and industrial area. With the declaration of ASEZ, a rapid increase in port and industrial activities is to be expected, causing degradation of coral reefs and leading to coral death and loss of biological diversity [10].

Urbanization affects the environmental conditions of the Gulf in various ways. One of the most serious aspects of expanding urbanization is the strong tendency to develop the coast in a strip-like fashion, which inevitably degrades the environment along the coast [11]. Development projects should be directed away from the coast. This would preserve the marine ecosystem and shore for recreational activities and tourism. Near-coast infilling and construction increases sediment loading onto the seabed which can not be tolerated by coral reefs. Industrial discharges into the Gulf also contain suspended particles and heavy metals. Sources of metals include land-based operations such as klinker production and fertilizer manufacture in Aqaba and seawater desalination in Eilat.

Industrial activities have put the fringing reefs under pressure [9] and increased coral diseases by tenfold [4]. Values of phosphate concentrations, heavy metals [3], and algal cover are higher at or near industrial sites. In addition, a reduction of 50% in fish abundance and changes in the relative abundance of different fish communities were observed [10]. It was suggested that future industrial development should be restricted to already industrialized areas, in order to preserve the remaining coral reefs and seagrass meadows [10].

The exportation of phosphate in the port of Aqaba from Jordanian industries is a major environmental problem. Approximately 1% of phosphate was lost into the atmosphere during the loading process before 1993. This has been reportedly to be reduced by 80 to 85% [20]. The dust settles in the Gulf of Aqaba, increasing the waterborne phosphate concentration, reducing water clarity, and increasing sediment load. Phosphate dust deposition was suggested as a factor contributing to depressed coral growth rates in Jordanian coastal waters [20].

Thermal pollution is also a consequence of industrial development in the Gulf of Aqaba. At present heated cooling water is discharged mainly by power stations and fertilizer production. A consequence of the discharge into the Gulf is that the water in these areas is about 5 to 8 degrees Celsius higher than ambient water temperature [20]. Considering that the Gulf of Aqaba marine organisms live within a few degrees of their upper thermal limits, an increase in temperature of a few degrees Celsius can have profound effects [18]. With the planned new power plant and other industrial facilities in the future, careful monitoring should be practiced in the coming years.

3.5. OTHER SOURCES OF POLLUTION

Recreational activities are a major source of solid waste and litter. This poses a constant, high-visibility environmental problem. Jordan's Gulf of Aqaba beaches and near-shore reef and sea grass areas are heavily impacted by discarded plastic and other synthetic materials [20, 2]. Solid waste and litter have negative ecological effects on sea fauna. The effects include strangulation and entanglement, starvation, and death from ingestion of plastics and concentration of synthetic chemicals up the food chain [2]. The Port of Aqaba has solid waste receptacles in port areas and additionally provides a daily garbage collection service via barge to ships anchored offshore. No systematic recycling is currently practiced in Aqaba. The problem of solid waste and litter should be solved on a regional scale, as much of the litter found on the Jordanian side of the Gulf of Aqaba has regional origin [2]. A solid waste strategy that integrates land-based solid waste management issues with those associated with ship and boat-generated marine litter should be developed.

In addition, sound solid waste management practices in Aqaba town, Aqaba ports, and in the whole region of the Gulf of Aqaba and the Red Sea should be promoted [2].

Environmental hazards associated with land transportation result from the high truck volume to and from the Aqaba Port and the industrial facilities. Inadequate provisions for waste oil collection and recovery pose a hazard to the Aqaba marine environment. The use of waste oil as a dust suppressant in unpaved truck repair yards is a further environmental hazard. Air pollution—especially from leaded gasoline, which is still the main gasoline type used in the Aqaba region—is an additional environmental hazard. In addition, the absence of a clear hazardous waste management policy has caused used asbestos brake linings, exhausted automotive batteries, and worn tires to be discarded in a haphazard and, in some cases, health-endangering manner [20].

All the above problems are shared with the Israeli and Egyptian sides of the Gulf of Aqaba. A new and unique source of pollution, however, comes from the Israeli side: pollution from fish farms. They are increasingly expanding and cover a large area. Limits should be imposed on these farms in order not to harm the corals and sea creatures in the Gulf.

4. References

1. Abelson A, Shteinman B, Fine M, Kaganovsky S. Mass transport from pollution sources to remote coral reefs in Eilat (Gulf of Aqaba, Red Sea). Mar Pollut Bull. 1999; 38:25-29.
2. Abu-Hilal AH, Al-Najjar T. Litter pollution on the Jordanian shores of the Gulf of Aqaba (Red Sea). Marine Environmental Research. 2004; 58:39-63.
3. Abu-Hilal AH, Badran MM. Effect of pollution source on metal concentration in sediment cores from the Gulf of Aqaba. Mar Pollut Bull. 1990; 21:190-197.
4. Al-Moghrabi. Unusual black band disease (BBD) outbreak in the northern tip of the Gulf of Aqaba (Jordan). Coral Reefs. 2001; 19:330-331.
5. Al-Rousan S, Rasheed M, Badran M. Nutrient diffusive fluxes from sediments in the northern Gulf of Aqaba, Red Sea. Scientia Marina. 2004; 68(4):483-490.
6. Aqaba Special Economic Zone Authority. Regulation for the protection of the environment in the Aqaba Special Economic Zone. Regulation No. (21) for the year 2001. 2001.
7. Crosby MP, Abu-Hilal A, Al-Homoud A, Erez J, Ortal R. Interactions among scientists, managers, and the public in defining research priorities and management strategies for marine and coastal resources: is the Red Sea Marine Peace Park a new paradigm. Water, Air, and Soil Pollution. 2000; 123:581-594.
8. Crosby MP, Al-Bashir B, Badran M, Dweiri S, Ortal R, Ottolenghi M, Perevolotsky A. 2002. The Red Sea Marine Peace Park: early lessons learned from a unique trans-boundary cooperative research, monitoring and management program. Proceedings of IUCN/WCPA-EA-4 Taipei Conference. 2002.
9. Gladstone W, Tawfiq N, Nasr D, Andersen I, Cheung C, Drammeh H, Krupp F, Lintner S. Sustainable use of renewable resources and conservation in the Red Sea and Gulf of Aden: issues, needs and strategic actions. Ocean Coast Manag. 1999; 42:671-697.

10. Khalaf MA, Kochzius M. Changes in trophic community structure of shore fishes at an industrial site in the Gulf of Aqaba, Red Sea. Marine Ecology Progress Series. 2002; 239:287-299.
11. Mancy KH. Gulf of Aqaba ecological overview and call to action. In: Sandler D, Adly E, Al-Khoshman MA, editors. Protecting the Gulf of Aqaba: a regional environmental challenge. Washington, DC: Environmental Law Institute; 1993.
12. Maragos JE, Crosby MP, McManus J. Coral reefs and biodiversity: a critical and threatened relationship. Oceanography. 1996; 9:83-99.
13. Rasheed M, Badran MI, Richter C, Huettel M. Effect of reef framework and bottom sediment on nutrient enrichment in a coral reef of the Gulf of Aqaba, Red Sea. Marine Ecology Progress Series. 2002; 239:277-285.
14. Royal Scientific Society. National Project for Water Resources Monitoring in Jordan. 1994.
15. Sultan F. Tourism Development: An Egyptian Perspective. In: Sandler D, Adly E, Al-Khoshman MA, editors. Protecting the Gulf of Aqaba: a regional environmental challenge. Washington, DC: Environmental Law Institute; 1993.
16. UNEP. Overview of the socio-economic aspects related to the management of municipal wastewater in west Asia (including all countries bordering the Red Sea and Gulf of Aden). Workshop on Municipal Wastewater Management in West Asia, Bahrain. 2001.
17. Wahbah MI, Zughul MB. Temporal distribution of chlorophyll *a*, suspended matter, and the vertical flux of particles in Aqaba (Jordan). Hydrobiologia. 2001; 459:147-156.
18. Wahbeh M. Agenda for scientific research in the Gulf of Aqaba. In: Sandler D, Adly E, Al-Khoshman MA, editors. Protecting the Gulf of Aqaba: a regional environmental challenge. Washington, DC: Environmental Law Institute; 1993.
19. Watzman H. Red Sea pays the price of peace. New Scientist. 1995 February 18; 1965.
20. World Bank. The Hashemite Kingdom of Jordan Gulf of Aqaba environmental action plan. 1996.

CHAPTER 26

SUSTAINABILITY AND VULNERABILITY ANALYSIS OF CRITICAL UNDERGROUND INFRASTRUCTURE

N. BOBYLEV

United Nations University
53–70, Jingumae 5-chome, Shibuya-ku
Tokyo 150–8925 Japan

Abstract: The paper discusses sustainability and vulnerability issues related to critical underground infrastructure (CUI). CUI is defined on the basis of its provision of vital services for a community. Most CUI can be found in urban areas. This analysis is based on understanding the natural characteristics of underground space and the benefits and difficulties of its utilization. Sustainability analysis considers how CUI is sustainable itself and what its impact on the environment is. Similarly, vulnerability analysis breaks down external and internal threats to CUI, and CUI impact on vulnerability of the physical and functional environment. The multiple-criteria decision-making analytic hierarchy process is used for aggregation impacts and integral assessment of CUI. In conclusion, the importance of CUI planning is highlighted.

1. Critical Underground Infrastructure

1.1. IDENTIFYING CRITICAL UNDERGROUND INFRASTRUCTURE

Underground spaces have been offering protection to humans since the dawn of civilization. During the last centuries, underground space has been increasingly utilized for an expanding number of purposes. In many areas, a set of underground facilities creates infrastructure that is of a high importance or critical for the well-being of the area. In recent decades it has been revealed that new, emerging threats, such as terrorism, put underground infrastructure at high risk and underground spaces have to be protected.

Infrastructure has been defined as "the basic facilities, services, and installations needed for the functioning of a community or society…" [1]; "facilities with high fixed costs, long economic lives, strong links to economic development…" [15]; and "the framework of interdependent

I. Linkov et al. (eds.), Managing Critical Infrastructure Risks, 445–469.
© 2007 *Springer*.

networks…" [8]. A key characteristic of infrastructure is the interdependence of facilities or structures that together provide some services. Infrastructure is also characterized by sizable capital investment in constructed facilities and equipment and substantial commitments of land and other natural resources.

The US Critical Infrastructures Protection Act of 2001 defines "critical infrastructure" as systems and assets, whether physical or virtual, so vital to the United States that the incapacity or destruction of such systems and assets would have a debilitating impact on security, national economic security, national public health or safety, or any combination of those matters. The notion of "critical" related to facilities, infrastructure, or system elements means that any disability of those would cause disability of many other systems, facilities, or services.

Underground space encompasses structures of various functional purposes: storage (e.g., food, water, oil, industrial goods, and waste, including hazardous and radioactive waste); industry (e.g., power plants); transport (e.g., railways, motor roads, and pedestrian tunnels); utilities and communications (e.g., water, sewerage, gas, and electric cables); military (e.g., rocket launches and command centers); public (e.g., shopping centers, hospitals, and civil defense structures); and private and personal (e.g., car garage).

Underground infrastructure can be defined as a set of below-surface-level structures, interconnected physically or functionally. Critical underground infrastructure (CUI) is an underground structure—or a set of underground structures—which is functionally vital for a community. Functional interconnection often means physical connection as well; e.g., a sewage treatment plant is connected with city sewerage through a collector. But physical connectivity in itself is a very important property that characterizes CUI. Facilities with different functional purposes may be located in the same underground structure, or in proximity, forming a particular soil stress-strain condition, so that damage or destruction of one facility can cause the failure of others.

1.2. CRITICAL UNDERGROUND INFRASTRUCTURE IN URBAN AREAS

Most CUI can be found in urban areas, and this infrastructure is heavily used on a daily basis by the vast majority of urban inhabitants. Most traditional underground networks that can be classified as CUI are water supply and wastewater treatment, transport, and electric cable tunnels. Some important elements and structures include sewer collectors and pumping stations, water supply pipes, cable tunnels, rail and motor tunnels, and ventilation shafts.

CUI includes all civil defense and emergency response structures, such as shelters, food storage, and emergency evacuation routes. Highly important are CUI facilities that have dual functionality: normal and emergency. For example, some mass transit infrastructure facilities, like underground metro stations, can be equipped with airtight gates, so that these underground spaces can be used as a civil defense shelters. Underground civil defense shelters can be successfully used during major military and terror attacks and during most natural disasters.

Underground space in urban areas has been extensively utilized [4]. In many cities underground space at the depth of 5–10 m is very congested: it is a mixture of old—in many cases unused—service tunnels, forgotten cables, and recently installed utilities. Often CUI is found in a densely utilized underground space, which makes its maintenance, modernization, renovation, and redevelopment difficult.

The urban agglomeration of Tokyo, with a population of 35 million people [16,17] has a significant volume of developed underground space. Until recent times, land property rights included all the underground space beneath privately owned land. Now, since adoption of the deep underground utilization law in 2001, the Japanese government has the right to use underground space below 30 m for public purposes including the space below privately owned land. Unfortunately, this legislation constrains the development of CUI. As a result of this constraint, most of the public underground infrastructure in Tokyo, including CUI, is found on scarce public land. Many underground facilities are not located in places which would be most beneficial according to their functional purposes. Winding metro lines in Tokyo illustrate these underground space use constraints, which have resulted in the accumulation of all types of underground infrastructure, including CUI, beneath publicly owned roads (see Table 1).

Table 2 shows typical distribution by depth of different types of underground infrastructure. These infrastructure elements can be potentially considered CUI, depending upon their particular location and the size of the area to which services are provided.

TABLE 1. Length of utility networks under major roads in central Tokyo [7]

Utility	Total Length, km	Length per 1 km of Road, km
Telecommunication	2,737	17.0
Electricity	1,498	9.3
Gas	326	2.0
Water network	367	2.3
Drainage	311	1.9

TABLE 2. Distribution of CUI in a typical street cross-section in Tokyo (revised data provided by Tokyo Metropolitan Government, 2005)

Depth, m	CUI
0–3	Electricity cables; telecommunication cables; gas lines; water supply pipes
3–5	Motorways; metro stations; car pickings and shopping malls that include security and disaster response facilities; electricity cables; telecommunication cables; gas lines; water supply pipes
6–10	Motorways; metro stations and train tracks; shopping malls that include security and disaster response facilities; sewerage collectors; electricity cables; telecommunication cables
10–30	Motorways; metro stations and train tracks; sewerage collectors, electricity, and telecommunication cables tunnels

1.3. CRITICAL UNDERGROUND INFRASTRUCTURE IN PORT AREAS

CUI is found predominantly in industrial sea and river ports rather than in passenger ports. CUI of industrial ports is represented by oil, gas, and chemical underground storage and pipelines.

A specific feature of port CUI is a significant potential threat to groundwater due to leakages from or into reservoirs and pipelines. In port areas, the groundwater level is high due to its proximity to surface water. Groundwater flow is likely to have significant gradients due to proximity to the discharge zone—surface water boundary. High groundwater levels and gradients directed towards surface water reservoirs facilitate potential contamination of surface waters. It is especially dangerous in the case of toxic chemical storage and leakage of groundwater into and subsequently out of toxic storage.

Ports in regions prone to extreme weather conditions such as hurricanes may benefit from locating critical or potentially dangerous facilities in underground space.

2. Characteristics of Underground Spaces

Underground space has certain natural peculiarities, or characteristics, which are summarized in Table 3.

From this summary, one can draw an overall conclusion: underground spaces are resilient to impacts from the outside and vulnerable to impacts that emerge inside. Terrorism threats are an excellent illustration of this. It is difficult to penetrate a protected underground facility and difficult to damage it by external explosion. But if terror actions are performed inside the underground space, damage can be immense. Even relatively small explosions, fires, air pollutants, or simply stampedes in public underground facilities can cause many casualties and lead to the collapse of structural elements. In case of fire or explosion, many

TABLE 3. Major characteristics of underground spaces

Characteristic	Description	Relation to CUI, Examples
Isolation	Underground spaces are less susceptible to impacts from outside, and their impact on the external environment is less than aboveground facilities (e.g., noise)	Energy supply facilities
Temperature stability	Less need for heating or cooling; in many cases underground facilities do not require any temperature adjustment at all.	Storage, civil defense shelters
Protection	Underground spaces have limited areas of connection with outside, and flow or movements through these connection areas are easy to control.	Military, civil defense, valuable or hazardous goods storage
Vulnerability to floods	Inundation can cause severe and unpredictable damage to underground structures, e.g., floor structure of upper level can collapse under the weight of water.	Measures to manage flooding: waterproof and semi-waterproof doors, emergency drainage tunnels and reservoirs, pumping stations.
Resilience during earthquakes	Deep underground structures suffer significantly less damage during earthquakes than aboveground structures.	Storage, civil defense shelters, emergency response centers
High cost of construction	Construction cost is highly dependent on ground conditions.	Underground structures are planned for many years of operation (usually 100 – 300 years for public infrastructure)
Low cost of operation	No maintenance for outer walls is needed, low cost for heating/cooling. Some extra costs for ventilation and emergency prevention and response systems.	Vast range of public structures, particularly in regions with severe climate
Possibility to locate in proximity	In urban areas very often surface space is already occupied by valuable developments, and underground space is the only option to locate a new facility in the required area.	Transport and emergency response infrastructure.

components of damaging impacts are aggravated: blast waves are stronger and smoke cannot be removed easily. Fires are the most common and damaging problem. High temperatures during burning are retained longer in closed spaces, although the absolute temperature of burning can be less than in open spaces due to limited oxygen access (refer to section 4.2.2). High temperatures may cause the collapse of reinforced concrete floor beams and tunnel linings.

3. Benefits and Drawbacks of Underground Facilities

The previous section describes the utilization of underground space and design of underground facilities. This section describes the main benefits and drawbacks of locating facilities underground. Significant attention is given to urban areas, due to their vast and diverse utilization of underground facilities.

3.1. BENEFITS

3.1.1. Rational Land Use

Given significant continuous global population growth and urbanization, rational land use becomes an increasingly important issue from many points of view:

- Property rights.
- Energy efficiency.
- Ecological footprint.

The negative effects of urban expansion, including urban sprawl, propel city planners to find solutions for more sustainable management of urban growth. Placing facilities underground allows the creation of a compact city, with high concentrations of build stock. Compact cities help to save not only land resources, but also energy and commuting time. The latter has many positive environmental implications: less air pollution, less goods consumption, and less municipal waste. A densely built and populated homogeneous city might be an unpleasant place to live. Alternatively, a compact city encompasses activity centers, where public services are concentrated, and vast green areas, which are public amenities and support ecological diversity.

Underground facilities are an indispensable part of compact cities [6]. Underground infrastructure can be located in activity centers and used for a vast array of purposes, including critical functions. Transport infrastructure can also be located in underground space, avoiding impacts on green areas.

Underground facilities help to rationalize land use. They play an important role in spatial planning, creating compact urban areas with high environmental quality.

3.1.2. Reliability, Safety, Security

Underground facilities are very reliable structures from an engineering point of view. Thanks to the natural properties of soil, artificial underground structures can persist for centuries with minimum maintenance.

Underground infrastructure contributes to the safety of urban areas by providing:

1. Spatial partition of traffic streams, decreasing the number of accidents.
2. Shelter during natural and artificial disasters.
3. Continuous operation of emergency response systems in proximity to important and potentially vulnerable locations.

Underground facilities naturally provide excellent security for critical functions. The security of underground facilities should be guaranteed by border control, since underground places are mainly vulnerable to impacts from the inside.

3.1.3. Environment

An extensive list of the environmental benefits of underground facilities includes:

- Energy efficiency.
- Enhancing green areas.
- Noise reduction.
- Reducing air pollution.

Underground facilities are energy efficient and they also help to save energy in city cycles, allowing the reduction of urban ecological footprints and reducing pressure on the global environment. For instance, the reduction of distances required for transporting goods makes urban economies more efficient. Compact urban areas hold promise for sustainable development because of their ability to support a large number of people while limiting their per capita impact on the natural environment [22].

Lack of open and green spaces is a common characteristic of urban areas. The worst situation occurs when city development exceeds previously approved master plans, and green and open spaces are replaced by buildings or—more often—by transport infrastructure. Increased built-stock density affects aesthetics, creates an unfriendly landscape, and reduces area amenities. Underground infrastructure helps not only to install vital new facilities with minimum deterioration of landscape while preserving green and open spaces, but also offers opportunities to rectify previous mistakes in aboveground infrastructure arrangements.

Air pollution in cities has strong impacts on residents' health as well as on vegetation and soils at a considerable distance from city borders. Urban transport is a major contributor to air pollution. Projections for 2010 show that 70% of the urban population is still likely to be exposed to particulate matter levels in excess of threshold values [22]. Locating motorways underground allows centralized treatment of emissions; e.g., in the ventilation shafts of motor tunnels. Underground infrastructure also contributes indirectly to reduction of atmospheric emissions:

1. Reducing the need to travel by transport.
2. Reducing congestion in transport infrastructure and increasing speed of transit.

This apparently correlates with two major factors that according to recent estimations have contributed to the increase in urban air pollution:

1. Increase in the number of motor vehicles.
2. Increase in travel time due to road congestion [22].

Noise is an important factor to be considered in urban environments. In Europe, more than 30% of citizens who reside in urban communities live in dwellings with significant exposure to road noise. To confront noise pollution, noise protection screens and fences are erected but these protection measures have negative impact on landscape quality. Underground space offers good isolation and noise adsorption.

Waste management can be made more cost effective by utilizing a widespread, well-planned underground infrastructure. Solid waste underground transportation systems can significantly reduce the costs of collecting and transporting municipal waste.

3.2. DRAWBACKS

3.2.1. Challenges During Construction

Major difficulties and risks related to underground facilities occur during their construction. Two main factors define the cost of construction and whether installation of an underground structure is possible:

- Ground conditions.
- Existing built environment.

Underground construction has encountered significant growth during the last decades. This growth is associated with modern construction technologies such as bored piles, diaphragm walls, and automated tunneling, including microtunneling. These technologies are expensive, and the cost of locating a facility underground can be dozens of times higher than locating one aboveground.

Historic parts of cities may represent the most challenging and expensive areas for underground construction. This is because of difficult ground conditions and the proximity of valuable built stock. Underground construction work also exposes existing CUI to the risk of damage due to uneven settlement. Tunnels outside densely built areas may also require risk management for uneven ground settlement.

Other drawbacks and risks during underground construction in urban areas include:

- Transport disruption due to construction work.
- Increased air pollution due to traffic congestion.
- Water pollution due to construction chemical leakage.
- Nearby structures.
- Pipeline disruption.

Undertaking underground construction in urban areas, especially when the open-cut method is used, requires recalculation of emergency response times for the vicinity.

3.2.2. *Psychological Discomfort*

Individuals have different sensitivities to being in closed spaces without daylight access. Psychological discomfort can be felt by people continuously staying in underground spaces due to their work requirements. Dwellings should be located underground only in exceptional cases.

3.2.3. *Difficulties with Renovation and Replacement*

Underground space is a resource that cannot be replaced; i.e., a cavern—once excavated—can never be refilled with soil containing the same properties. Also underground structures cannot be demolished. Replacement of an underground structure would require installation of another—larger—structure in its place. Therefore, careful planning of underground installations is important.

4. Sustainability and Vulnerability Analysis of Critical Underground Infrastructure

Sustainability and vulnerability analysis of CUI considers the following questions:

- How do external and internal impacts affect CUI?
- How can CUI impact its environment?

These considerations are formalized in Table 4 and analyzed below.

TABLE 4. Questions for analysis of sustainability and vulnerability of CUI

Considered Question	Considered Object	Nature of Impact	Some Issues to Consider
Sustainability	CUI	Internal	• Energy consumption • Waste generation
	Physical and functional environment of CIU	External—from functioning of CUI	• Uninterrupted supply of services (e.g., transport) • Energy efficiency • Environmental quality of impacted area
Vulnerability	CUI	External—from the environment	• Terrorism threat • Natural and manmade disasters
	CUI	Internal	• Terrorism threat • Threat of equipment and structural failures
	Physical and functional environment of CIU	External—from functioning or malfunctioning of CUI	• Failure of vital services (e.g., water supply) • Increased risk of terrorism attractiveness

4.1. SUSTAINABILITY

Sustainable development is development that meets the needs of the present generation without compromising the ability of future generations to meet their needs [25]. Underground infrastructure and CUI contributes to sustainability in many ways:

- Saving natural resources, including land, water, and biodiversity.
- Reducing air pollution (mainly in the transport sector).
- Creating opportunities for less energy use and waste generation (compact city).
- Enhancing overall landscape and environmental quality.

Underground infrastructure is needed to reduce the amount of land covered by manmade structures. Urban sprawl consists of low- and medium-density residential areas around urban centers. It is known that urban sprawl has a damaging effect on biodiversity. Urban sprawl is inherent to developed countries, where citizens can afford and prefer to use private motorcars for commuting. In the urban areas of Western Europe, half of all car trips are less than 6 km and 10% are less than 1 km [22]. The last figure suggests the car trip involves moving within downtown to obtain different services. This means that the city is not compact enough to offer diverse services within walking distance. Underground infrastructure can tackle this problem and combat urban sprawl, maintaining compactness of the urban community, preserving natural landscapes, and protecting biodiversity.

In urban areas, extensively developed underground infrastructure creates an independent spatial layer of communication and services, including critical facilities that enhance the city's coherence and resilience. It is known that systems with many alternative connections are more stable (sustainable) than those without such alternative connections.

The internal sustainability of CUI is characterized mainly by energy efficiency and rational land resource use.

Reliable provision of services and performance at high environmental standards are the most important contributions of CUI to sustainability.

4.2. VULNERABILITY

4.2.1. External Threats to CUI

A major issue in the analysis of CUI vulnerability is external threats. Those threats include natural disasters, accidents, and military and terror attacks. For CUI vulnerability analysis it would be improper to consider threats according to their causes, but consideration according to actual impact on CUI is important. For example, CUI could be inundated due to retainer wall, sewer, or levee failure. This structural failure can be caused by natural disasters, including flood, earthquake, or storm; or artificial ones, including accidents, or military or terror attack.

External threats to CUI can be considered in three categories according to their origin:

- Above/on the ground (explosions, air pollution, wind).
- In underground space (earthquakes, soil conditions alteration).
- Water (ingress and leakages).

Underground structures are scarcely affected by aboveground-originated impacts, like extreme wind, temperature, and explosions. However, ventilation shafts and exits and entrances to underground facilities are vulnerable. Normally, air is conveyed to public underground facilities through ventilation shafts and is conveyed out of underground facilities through exits and entrances due to pressure difference. Ventilation shafts are vulnerable to air pollution, which can be caused by fire or terror attack. A common and underestimated threat in urban areas is routine air pollution from motor transport that can create discomfort in underground facilities. Impacts can be aggravated when a combination of the following factors is unfavorable:

- Traffic congestion.
- Wind direction.
- Temperature.

Temporary suspension of ventilation is the best solution for ensuring stable air quality inside an underground facility. Ventilation shafts of CUI should be equipped with chemical detectors and automatic systems for air supply management. CUI that has been designed as a civil defense shelter (also may be multifunctional) has hermetic doors and air purification systems. At an advanced level, these systems are also capable of providing refuge during radioactive pollution.

External impacts that originate in underground space can represent a serious threat to underground facilities. These impacts include earthquakes and impacts on soil properties from other structures or construction work.

Underground structures are more resilient to earthquakes than aboveground ones (Figure 1). Upper levels and exits and entrances of an underground structure experience the most damage during earthquakes. Evacuation routes from underground can be blocked, so having emergency supplies in the facility would be an important preventive response measure.

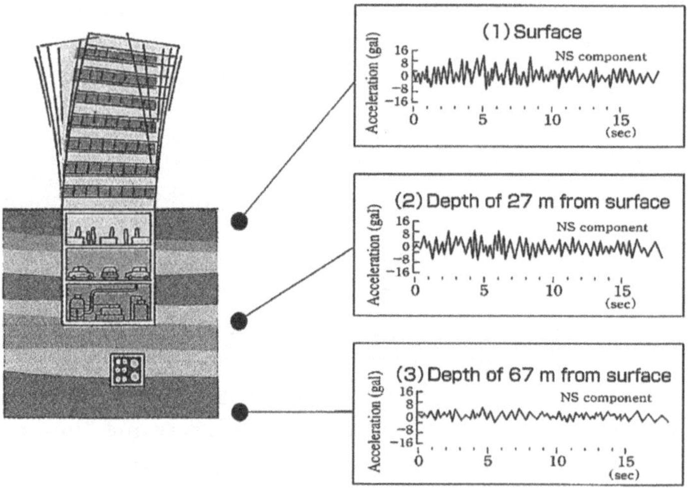

Figure 1. Impact during an earthquake at different depths [7].

Damage to structural elements of an underground facility can be caused by changes in surrounding soil stress-strain conditions initiated by construction work (excavation and tunneling) and suffosion (a process of washing out the small particles of sand and limestone by underground waters, which causes soil destruction and loss of strength). Suffosion can be caused by construction work, leakages from or into conduits and caverns, or increased groundwater flow. The process of suffosion is difficult to detect, and very often it is revealed after damage to structures (e.g., cracks and uneven settlement) has already occurred. CUI—especially sewer collectors and cable tunnels—is vulnerable to suffosion that can cause serious disruption of critical services if not detected in time.

Inundation of underground structures can be caused by:

- Rain.
- Floods.
- Failure of surface water structures.
- Leakage from underground water containers and conduits.
- Groundwater ingress.

Regardless of the causes and origins of the inundation emergency, response can pursue two strategies:

- Preventing water from coming into the facility and preventing water movement in the facility.
- Moving water to the lower levels of facility and subsequently pumping it out.

The first strategy includes engineering solutions such as hermetic doors inside the underground facility and spatial design that prevents easy water flow underground (drainage gutters, elevated entrances). These options are good for small, rather isolated facilities, such as parking garages or security control centers. The second strategy involves design solutions that allow free water flow in wide spaces to prevent high water levels inside facility premises. There should be also a water storage cavern that can accommodate a significant amount of water and pumping equipment. This strategy is adequate for CUI, widespread underground networks, and public facilities. Free water flow into the structure avoids the problem of blocking exits and entrances, which might be unacceptable for public facilities.

Combining the two inundation-combating strategies (preventing and allowing water flow underground) can be beneficial for CUI that include public facilities and facilities that functionally require high security and safety. Traffic control facilities at metro stations and a security center at a multifunctional mall are examples where a combination of the two strategies should be used. Special equipment, such as semi-waterproof doors and small pumps, can be used. A semi-waterproof door has a movable lower part that allows water flow in one direction. There are also doors that allow free water movement but prevent air (smoke) movement. Writing about floods and CUI would be incomplete without mentioning how underground space can help to combat floods (refer to section 4.2.3.).

4.2.2. *Internal Threats to CUI*

It has been argued (sections 2 and 4.2.1.) that CUI is more susceptible to internal impacts than external ones. Internal threats can be potentially very damaging, especially explosions and air pollution, due to limited air volume in underground spaces. Terror attacks would be at the top of the list while considering possible causes of these disasters. The major risk during fires in underground spaces is smoke. Due to limited oxygen access, burning is generally not intense, and heat effects in most cases are not high enough to damage reinforced concrete.

Equipment failures in CUI are important to consider. Ventilation fans and water pumps are vital equipment. Vertical transport equipment (lifts, escalators) are critical for public places, especially mass transit systems, where their failure can create a stampede.

4.2.3. *Impact of CUI on Vulnerability of the Environment*

Underground infrastructure, especially CUI, can increase or decrease the vulnerability of its physical and functional environment. From the functional point of view, CUI should provide uninterrupted service; ensuring that the underground location of a facility makes it less vulnerable to many potential threats.

In urban areas CUI can decrease a city's vulnerability by:

1. Creating an extra layer of facilities.
2. Doubling functions of—for example—transport systems.
3. Enabling emergency response services by placing them in proximity to potential accident zones.
4. Providing civil defense shelters.

The G-Cans project in Tokyo, Japan, is a remarkable example of how CUI can make urban areas less vulnerable. The G-Cans project is an underground infrastructure for flood prevention during rain and typhoon seasons. The infrastructure consists of five concrete containment silos

Figure 2. G-Cans project. Tokyo, Japan. Photo: Ministry of Land, Infrastructure and Transport Kanto area maintenance bureau Edogawa river office tailrace section.

with a height of 65 m and a diameter of 32 m, connected by 6.4 km of tunnels, 50 m beneath the surface, as well as a large water tank with a height of 25.4 m, a length of 177 m, a width of 78 m, and 59 massive pillars connected to a number of 14,000 horsepower (10 MW) turbines that can pump up to 200 tons of water per second into the Edogawa river [24] (Figure 2). This project also illustrates the difference between vulnerability and sustainability with regard to CUI. The G-Cans project decreases the vulnerability of the environment to natural disasters, but cannot be considered sustainable. The project consumes significant land and energy resources while addressing only the consequences of flooding; not its roots, which are deforestation and impermeable urban surfaces that prevent rainwater infiltration.

Another interesting project that is designed to make urban areas less vulnerable is the stormwater management and road tunnel to be completed in 2006 in Kuala Lumpur, Malaysia. The stormwater tunnel—9.7 km long and 13.2 m in outer diameter [21]—will function as a double-decker motorway and a stormwater tunnel to combat flash floods.

4.2.4. Vulnerability of CUI Components

Table 5 identifies the main critical components of CUI and summarizes potential impacts.

TABLE 5. Vulnerability of CUI components

Component of CUI	Potentially Disabled Function	Possible Reasons for Disruption	Consequences of the Function Failure	Emergency Response to Restore Function or Prevent Damage
Structural elements	Load- bearing capacity	• External explosion • Earthquake • Flood • Ground settlement • Internal fire or explosion	Structure collapse	• Evacuation
Ventilation system	Air supply	• External air contamination • Ventilation shaft damage • Fan equipment failure	Poor air quality	• Temporary suspension of air pumping • Emergency oxygen supply • Evacuation
Exits/entrances	Ability to evacuate to or out of the facility	• External explosion • Earthquake • Internal explosion • Blockage due to stampede	Isolation of facility users	• Provision of emergency supplies inside the structure
Vertical transport equipment	Ability to move within or outside the facility	• Earthquake • Internal explosion • Blockage due to stampede • Equipment failure	Isolation of facility users	• Provision of emergency supplies inside the structure

5. Assessment Tool: Multiple-criteria Decision Making

Sustainability and vulnerability analysis of CUI can be undertaken while developing plans for new infrastructure installation, upgrading and renovating existing CUI, performing integral analysis of urban sustainability and vulnerability, and planning a disaster response. Sustainability and vulnerability assessment of CUI involves analysis of a significant number of issues that are diverse in nature and degree of potential impact. Descriptive evaluation of each issue, or criteria, is indispensable, but not sufficient to get a holistic view of the overall impact. Assessment tools, such as multiple-criteria decision making (MCDM), can aggregate miscellaneous issues under consideration, compare and prioritize them, and eventually arrive at an overall conclusion.

Multiple-criteria analysis (MCA), or multiple-criteria decision analysis (MCDA), is a method of decision analysis that involves the use of scoring and weighting systems based on criteria in order to test and compare the impacts of alternatives. MCDA is applicable to a very broad range of interdisciplinary issues, and is widely utilized in economics, social science, and environmental studies. MCDA addresses situations where decisions are needed to handle complex problems (e.g., sustainability or vulnerability) involving alternative options (e.g., strategies for upgrading CUI) that are evaluated against several conflicting criteria. MCDA aggregates miscellaneous criteria and makes a tradeoff between positive and negative impacts to and from CUI.

MCDA embraces an array of different methods. These methods can be classified according to an aggregation rule [11], operative approaches [18], and a finite or infinite number of alternatives [23]. A comprehensive comparison of MCDA methods, addressing their benefits and drawbacks, can be found in Getzner et al. [10]. The most-used MCDA methods are listed below:

- Multiple attribute value theory.
- Analytic hierarchy process.
- Evaluation matrix.
- Electre.
- Regime.
- Novel approach to imprecise assessment and decision environments, multi-objective programming.
- Goal programming.

These MCDA methods can be viewed as competing ones; the method to be used in a particular assessment case is selected on the basis of assessor expertise, experience, and preference. If time and resources are available, it is an advantage to process the same input information using different methods, which will increase the reliability of results.

The strongest argument for using MCDA is its liberal requirements for initial information input, or criteria values, which can be estimated qualitatively as well as quantitatively. This has crucial importance to sustainability and vulnerability assessment of CUI, since the whole array of interdisciplinary criteria should be taken into account.

5.1. ANALYTIC HIERARCHY PROCESS

Analytic hierarchy process (AHP) is an MCDA method that was developed by Tomas Saati and encompasses three main steps:

- Arrangement of a hierarchy structure.
- Pairwise comparisons between elements of a hierarchy structure to establish priorities.
- Checking the logical consistency of pairwise comparisons [19, 20].

The first step is based on findings that when elaborating information, the human mind recognizes objects and concepts, and identifies relations existing between them. Because the human mind is not able to perceive simultaneously all factors affected by an action and its connections, it helps to break down complex systems into simple structures: this simplification is made possible by means of a logical process which aims at the construction of suitable hierarchies. Hence, hierarchies can be useful in helping the human mind to make decisions by constructing a framework of different and numerous elements, which are separated and connected at the same time. Saati defines a hierarchy as a particular type of system, which is based on the assumption that the entities can be grouped into disjoined sets, with the entities of one group influencing the entities of only one other group, and being influenced by the entities of only one group [19].

5.1.1. Stages of AHP

Table 6 summarizes stages of AHP and associated assessor actions.

5.1.2. Assessment Goals and Alternatives Generation

Identification of an assessment goal of a hierarchy does not require extensive analysis, but rather clear vision of opportunities and limitations in the assessment process. The epistemology of AHP, or consideration of what new information it is possible to get as a result of an assessment process, is based on

TABLE 6. Stages of MCDA AHP for sustainability and vulnerability assessment of CUI

MCDA stage	Questions to Be Addressed During CUI Assessment
Establishment of the decision context	How would MCDM AHP fit into overall sustainability and vulnerability analysis of a particular CUI? How can the overall assessment problem be broken down? How many AHP would be needed?
Identification of an assessment goal	AHP epistemology: What do we want to know? Can we know? Examples of assessment goals: CUI internal sustainability, CUI contribution to environmental quality, CUI vulnerability to external impacts.
Identification of the options to be assessed and/or generating an alternatives	Identification of alternative policies, programs, plans, projects or design solutions. Are there any options? How many options should be compared in an AHP? Which options should be compared with each other? In case of only one existing option, generation of alternatives or formulating baseline (option/no action).
Identification of available data for assessment	Is there enough data? Can criteria be identified and formulated?
Elicitation of a hierarchy structure	Selecting an approach to start elicitation: top-down or bottom-up. Identification of criteria and elaboration of nodes.
Weighting – assigning weights for the hierarchy structure elements	Identification of priorities at all the levels of the hierarchy structure; i.e., what is the assessment focus? Which hierarchy elements are more or less important than others?
Scoring – assessment of performance of each option against criteria	Pairwise comparisons of lower-hierarchy-level criteria performance under different options.
Analysis of the assessment results	Sensitivity analysis. Understanding performance of different criteria groups under different options. What is the impact of CUI on sustainability or vulnerability? How and why it can be changed?

clear formulation of the assessment goal. Formulation of the assessment goal should not be governed by available input data, or criteria, but by the ultimate decision at which one should arrive through the assessment process.

Table 7 gives some examples of assessment goals, options that may exist prior to the MCDM process, and possible alternatives.

Options, or alternatives, in a simple assessment case would be presented by alternative policies, programs, plans, projects, or design solutions of a particular element or structure of CUI. The term "option" can be used if several different CUI design solutions are fed in advance to the MCDM

TABLE 7. Assessment goals and alternatives generation

Main Theme	Example of an Assessment Goal	Example of a Baseline	Example of a Given Option	Example of a Generated Alternative
Sustainability	CUI internal sustainability	Conditions in existing CUI	Different design solutions for construction or reconstruction of CUI	CUI replacement with another type of infrastructure that provides the same functions
	CUI external sustainability	Existing environmental conditions	Different design solutions for construction or upgrading CUI	Different urban development policies
	CUI contribution to environmental quality of the impacted area	Existing environmental conditions	Different policies, programs, plans, projects, design solutions	CUI replacement with another type of infrastructure that provides the same functions
Vulnerability	CUI vulnerability to external impacts	Present state	Different design solutions for reconstruction and upgrading CUI	Actions to reduce threat/impact of natural disasters
	CUI vulnerability to internal impacts	Present state	Different design solutions for upgrading CUI	Actions to reduce terrorism threat in the area
	Vulnerability of the CUI environment	Existing environmental conditions	Different policies, programs, plans, projects, design solutions	Different location of CUI

process. The term "alternative" can identify some other solution, or option that is not known before conducting the MCDM process. This situation is common at a policy analysis level. In this case, alternative(s) should be generated. The basic option that always exists is "do nothing – no action"; that is, the environmental baseline: the environmental conditions that would exist if no CUI (re)development were undertaken, or if CUI did not exist. Alternatives facilitate a broader view on a problem. However, alternatives should be of high quality: focused and relevant to achieving the proposed action (e.g., CUI development) goals (e.g., providing public transportation). Normally, the total number of options and alternatives is between three and five.

5.1.3. Elicitation of a Hierarchy Structure

Elaboration of a hierarchy structure for assessment is the basic and important step in MCDA AHP. The whole concept and philosophy of assessment depends upon the way in which criteria are composed into the hierarchy structure. Composition of a hierarchy structure is governed by a CUI assessment goal; e.g., sustainability or vulnerability. Elaboration of a hierarchy can be based on either a bottom-up or top-down approach.

Hierarchy structure arrangement using a bottom-up approach starts with obtaining criteria that would constitute the lowest hierarchy level. These criteria can be identified based on descriptive analysis of CUI sustainability and vulnerability. It is also possible to use criteria databases, like the set of 78 interdisciplinary criteria for environmental assessment of urban underground infrastructure [5]. The main task in the bottom-up approach is to aggregate these criteria into groups, or nodes. The number of hierarchy levels will depend upon the nature of the criteria and aggregation tactics. The bottom-up approach is effective when a number of criteria are clear at the beginning of the aggregation process and criteria are well defined.

The top-down hierarchy approach starts with defining groups of criteria at the highest level of the hierarchy. Elaboration of the hierarchy elements in this case is governed by previous experience in considering sustainability or vulnerability based on similar CUI assessment case studies or broader considerations from relevant areas of knowledge. The top-down hierarchy approach is used when there is no detailed, clear data on criteria. This is often the case at the initial stages of an assessment study, during policy elaboration, or during feasibility studies.

Both approaches can be successfully utilized for CUI assessment. In environmental assessment processes, such as environmental impact assessments, strategic environmental assessments, and sustainability appraisals, it is possible to compose a single hierarchy structure. This hierarchy structure integrates all the assessment criteria under a singular assessment goal;

e.g., sustainability or environmental quality. In the case of CUI, analysis for sustainability and vulnerability assessments appears to be rather complex, involving conflicting issues such as external or internal vulnerability. Analysis using a single hierarchy structure might be difficult in this case, and elaboration of several hierarchies with particular assessment goals is recommended. Assessment goals may include CUI internal sustainability, CUI contribution to environmental quality, and CUI vulnerability to external impacts (see Table 7). Different hierarchies can utilize the same criteria, but the composition of nodes can vary. Figure 3 illustrates an example of a hierarchy structure. Overall judgement on a particular action should be based on a holistic interdisciplinary view, and might go beyond utilizing MCDM tools.

5.1.4. *Weighting Hierarchy Structure Elements and Scoring Criteria*

After a hierarchy structure has been arranged, the next important AHP step is pairwise comparisons between elements of a hierarchy structure to:

1. Assign weights of their relative importance to establish priorities.

2. Compare criteria performance in different options.

All elements are compared in their primary group; then groups are compared with other groups in the same node. Finally, nodes at the highest hierarchy level are compared with each other. When weights are assigned, the relative contribution of each criterion at the lowest hierarchy level (e.g., provision of

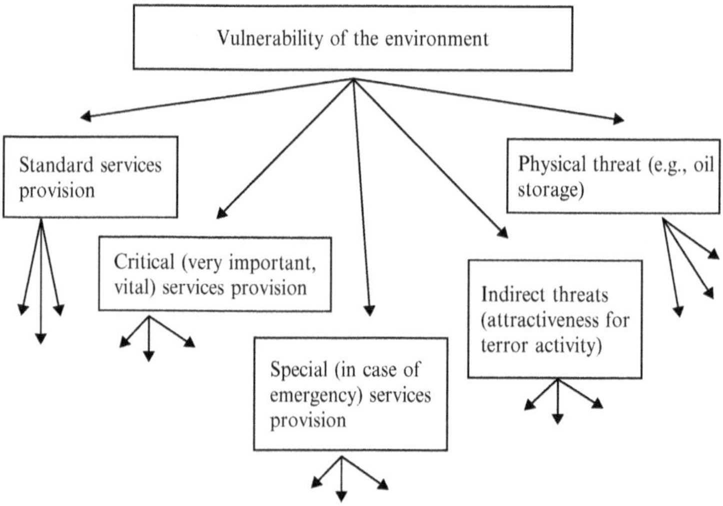

Figure 3. Example of a hierarchy structure of CUI impacts with a reverse assessment goal of "vulnerability of the environment."

some particular function) to the overall assessment goal (sustainability or vulnerability) becomes fixed.

Scoring assesses the performance of each criterion in different project options. Criteria can be assessed on a qualitative (expert judgments) or quantitative (direct input with normalization) basis.

5.1.5. Sensitivity Analysis and Discussion of the Results

The result of AHP is the numerical rating of alternatives against their contribution to the assessment goal. Sensitivity analysis is the last step of AHP; it facilitates checking the logical consistency of decisions and the overall accuracy of the assessment. Sensitivity analysis also helps to thoroughly understand assessment results and draw conclusions beyond figures.

Results of MCDM AHP help to analyze CUI sustainability or vulnerability by:

- Rating alternatives.
- Showing how priorities that were assigned affect the assessment results.
- Grouping alternatives according their impacts.
- Understanding which groups of criteria are the most significant ones.
- Understanding tradeoffs between groups of criteria.
- Providing ideas for alternative solutions.
- Optimizing CUI performance.

AHP can be reiterated several times, by making changes in groupings of element and reassigning weights to the elements, but normally conducting sensitivity analysis is sufficient to understand the impact of different criteria on the assessment goal.

6. Conclusion: Planning of CUI

CUI is planned to operate for hundreds of years. It requires significant initial investment, and consumes a nonrenewable resource: underground space. Thus, planning of CUI is a very important issue, and sustainability and vulnerability analysis is an inherent component of a planning and decision-making process.

Results of sustainability and vulnerability analysis of CUI can be used for planning and development of new cities, industries, and facilities. Analysis of existing CUI is important for its upgrading and renovation.

MCDM tools—AHP in particular—can be successfully utilized for optimization of such CUI parameters as location, function, design, and overall management policies and strategies.

Although sustainability and vulnerability are different characteristics of CUI and require independent analysis, combined consideration of these two issues can provide synergies and clearer results about the overall role and needs of CUI.

It is important to mention that infrastructure can be planned as critical or may acquire critical importance due to changes in its function or in the environment. Observations of urban growth indicate that most underground infrastructure elements in city centers have become CUI due to changing environments and high demand. Similarly, in relation to increasing terrorism threats, more structures can be classified as critical because of their vulnerability and the potential extent of environmental damage.

7. Acknowledgements

The author would like to acknowledge the contribution of the following organizations, which provided opportunities for conducting this research: Japan Society for Promotion of Science, United Nations University, and University of Tokyo.

8. References

1. American Heritage dictionary of the English language, fourth edition. Boston: Houghton Mifflin Company; 2002.
2. Andrews CJ, Valverde ALJJR. Analysis in support of the environmental decision-making. In: Linkov I, Ramadan A., editors. Comparative risk assessment and environmental decision making. Dordrecht: Kluwer Academic Publishers; 2004.
3. Belluck DA, Hull RN, Benjamin SL, Linkov I. Environmental security, critical infrastructure and risk assessment: definitions and current trends. In: Linkov I, Kiker G, editors. Environmental security and risk assessment. Dordrecht: Kluwer Academic Publishers; 2005.
4. Bobylev N. Environmental assessment of urban underground infrastructure. In: Pan T, Meguro K, editors. New technologies for urban safety of mega cities in Asia. Singapore: Nanyang Technological University; 2005.
5. Bobylev N. Report on environmental assessment of urban underground infrastructure. University of Tokyo, United Nations University. 2006.
6. Bobylev, N. Strategic environmental assessment of urban underground infrastructure development policies. Tunnelling and Underground Space Technology. 2006; 21:469-479.
7. Deep Underground Utilisation Planning Office. Space of new potential. Planning Division, Metropolitan Areas Development Bureau, National Land Agency of Japan. 2001.
8. Executive order 13010, critical infrastructure protection. Federal Register. 1996; 61(138):37347-37350.
9. G-Cans Project, Tokyo. Available at: http://www.g-cans.jp/. Accessed on June 1, 2006.
10. Getzner M, Spash CL, Stagl S. Alternatives for environmental valuation. Abingdon: Routledge; 2005.

11. Holland M et al. Tools for sustainability: development and application of an integrated framework. 2004. Available at: http://www.arirabl.com/papers/%20Final%20Report%20SusTools.pdf.
12. Little RG, Pattak PB, Schroeder WA, editors. Use of underground facilities to protect critical infrastructures: summary of a workshop, National Research Council. Washington, DC: National Academy Press; 1998.
13. Matsumoto R. Deep underground utilization law and its benefits. Proceedings of the World Tunnel Congress, Istanbul. 2005.
14. Moteff J, Parfomak P. Critical infrastructure and key assets: definition and identification, CRS report for Congress. Congressional Research Service; 2004. Available at: http://www.fas.org/sgp/crs/RL32631.pdf.
15. National Council on Public Works Improvement. Fragile foundations: a report on America's public works, final report to the President and Congress. Washington, D.C. 1988.
16. Population Division of the Department of Economic and Social Affairs of the United Nations Secretariat. World population prospects: the 2004 revision. Available at: http://esa.un.org/unpp/. Accessed on June 1, 2006.
17. Population Division of the Department of Economic and Social Affairs of the United Nations Secretariat. World urbanization prospects: the 2005 revision. Available at: http://esa.un.org/unup/. Accessed on June 1, 2006.
18. Roy B. Multiple criteria methodology for decision auditing. Dordrecht: Kluwer Academic Press; 1996.
19. Saati TL. The analytic hierarchy process for decision in a complex world. Pittsburgh: RWS Publications; 1980.
20. Saati TL. Fundamentals of decision making and priority theory with the analytical hierarchy process. Pittsburgh: RWS Publications; 1994.
21. Stormwater Management and Road Tunnel (SMART) Project. Available at: http://www.smarttunnel.com.my/. Accessed on June 1, 2006.
22. United Nations Environment Programme. Global Environment Outlook Geo-3. 2002.
23. Vooged H. Multiple criteria evaluation of urban and regional planning. London: Pion; 1983.
24. Wikipedia, the free encyclopedia. Available at: http://en.wikipedia.org/. Accessed on June 1, 2006.
25. World Commission on Environment and Development. Commission. Our common future. 1987; Oxford: Oxford University Press.

CHAPTER 27

APPLICATION OF RAPID IMPACT ASSESSMENT MATRIX (RIAM) METHOD FOR WASTE DISPOSAL SITE

Case Study - Canakkale

A. BABA

Department of Geological Engineering, Canakkale Onsekiz Mart University, Canakkale, Turkey

Abstract: Environmental impact assessment (EIA) work is becoming more and more extensive in the world. The environmental assessment process has been defined differently in different countries. In fact, it appears that no two countries have defined it in exactly the same way. It is an aid to decision making and to the minimization or elimination of environmental impacts at an early planning stage. The EIA process is potentially a basis for negotiations between the developer, public interest groups, and the planning regulator.

The Rapid Impact Assessment Matrix (RIAM) method is based on a standard definition of the important assessment criteria as well as the means by which semi-quantitative values for each of these criteria can be collected to provide an accurate and independent score for each condition. The impact of project activities is evaluated against the environmental components and for each component a score is determined, which provides a measure of the impact expected from the component. The RIAM method makes it possible to carry out a global analysis of the results based on the individual environmental score for each component, which is classified in ranges and so can be compared to each other.

In this study, the RIAM method was used for the new waste disposal site in Canakkale City where one of the most important archaeological, historical, and natural parks of Turkey is located. Compared with other Turkish cities, a great number of EIA reports have been prepared based on numerous projects within the border of Canakkale Province, Turkey. The result of the RIAM method shows that the new waste disposal site will cause positive impacts resulting from economic-operational changes. But local people have a negative reaction to the new waste disposal site. Most impacts are of class –A (impact) for the waste disposal site, because of the waste, biological-ecological, and physicochemical effects prominent in the study area.

I. Linkov et al. (eds.), Managing Critical Infrastructure Risks, 471–481.
© 2007 *Springer.*

1. Introduction

Disposing of waste in a landfill is the most traditional method of waste disposal, and it remains a common practice in most countries. In the United States, approximately 50% or 128.3 million tons of the municipal solid waste generated is buried in landfills [1]. Historically, landfills were often established in disused quarries, mining voids, or borrow pits. Running a landfill that minimizes environmental problems can be a hygienic and relatively inexpensive method of disposing of waste materials.

Older or poorly managed landfills can create a number of adverse environmental impacts, including wind-blown litter, attraction of vermin (rats and mice), and soluble pollutants such as leachate metals and other trace elements that are present in municipal waste landfill leachates in considerable concentrations [2]. The long-term mobility of landfill leachate into ground or surface waters may locally be of environmental concern [3]. Landfills represent a widespread and significant threat to groundwater quality, human health, and even some ecosystems due to their nature of operation and abundance. In communal language, landfill means waste disposal on land [4]. Following the infiltration of rainwater through these waste materials, a water-based byproduct known as landfill leachate is produced. These leachates are responsible for the mobilization of a complex mixture of organic and inorganic constituents [5, 6]. Another product of landfills containing putrescible wastes is landfill gas, mostly composed of methane and carbon dioxide, which is produced as the waste breaks down. Municipal waste landfills are large heterogeneous areas in which organic wastes undergo degradation in anaerobic, acidic environments resulting in a high generation rate of landfill gas [7, 8]. Emissions from municipal landfill sites can potentially be detrimental to both local and global air quality [9]. Landfill gas consists of up to 65% methane and 35% carbon dioxide, both of which are considered to be greenhouse gases contributing to global climate change [9]. Trace volatile organic compounds represent less than 1% of landfill gas. However, these compounds are often odorous [10]. With increased levels of urbanization and consequent location of landfills in close proximity to highly populated areas, there has been an increasing level of intolerance to odor, which is now arguably the greatest nuisance associated with landfill sites [10]. Consequently, odor control has become an increasingly important aspect of regulation of landfills. A reduction in odorous emissions is one of the several objectives associated with the use of daily cover material at landfill sites [11]. Conventionally, material suitable for use as daily cover has been sourced from waste generated by the construction and demolition industry. However, with a global decline of the industry, coupled with the introduction of landfill taxes and increasing pressure to recycle and reuse resources, the amount of suitable material for daily cover being sent to landfills is declining [12].

There is a need to identify suitable alternatives to traditional daily cover materials. Materials investigated to date include paper mill sludge, fly ash, mulched wood material, and foams [10, 13, 14, 15, 26]. With the increasing drive towards sustainable waste management, use of a waste-derived product such as compost, widely used in biofilters, appears favorable. The ability of compost to remove chlorinated hydrocarbons and sulfur compounds has been reported by Muntoni and Cossu [17].

For determining the location of waste disposal sites, many methods, including geographical information systems, have been used. Canakkale City is the location of one of the most important archaeological, historical, and natural parks of Turkey. Compared with other Turkish cities, a great number of environmental impact assessment (EIA) reports have been prepared based on numerous projects within the border of Canakkale Province. In this study, the RIAM method was used for a new waste disposal site in Canakkale.

2. Landfill of Canakkale City

The Canakkale waste disposal site was in the valley of a small creek, located 3 km south of the city center (Figure 1).

The surface area of the waste disposal site was around 10,000 square meters. This waste disposal site has been used by the municipality of Canakkale since 1990. The major risks of the dump are its closeness to the Canakkale Onsekiz Mart University campus area to the south, the airport to the north, and the merging residential areas towards the dump site. The nuisance caused by odor is high. The waste pile emplaced at the head of the

Figure 1. Location map of waste disposal site in Cannakale.

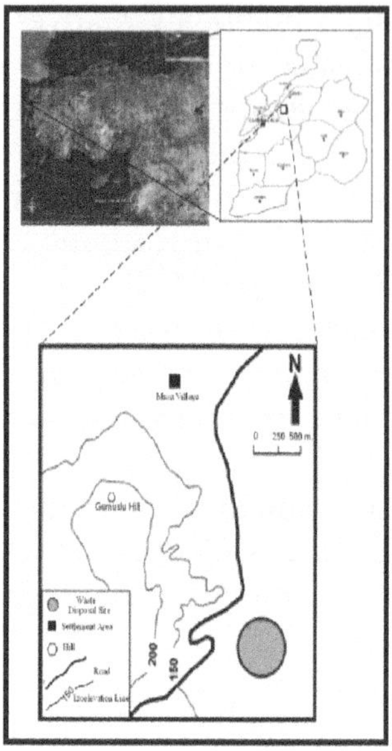

Figure 2. Location map of the new waste disposal site.

tributary stream has caused channeling of the leachate to the creek, which is eventually being washed towards agricultural fields [18,19].

The waste disposal site is located in a developed area. Therefore a new waste disposal site, 17 km northeast of Canakkale City, was found. The new waste disposal site is located near Canakkale-Lapseki highway and north-northwest of Dolluk Hill (Figure 2).

The new waste disposal site was used for mining activity by Canakkale Cement Factory. The site is about 1.8 km and 2.3 km from Kemel village and Musakoy, respectively.

This site was favorably accepted by the local authorities and government due to the following features:

- The location is an old mining pit.
- It is far enough from developed areas.
- Meteorological conditions and size are suitable to meet the needs of Canakkale's future population.

3. Geological Properties of Study Area

Volcanic rock such as andesite and tuff outcrop around the new waste disposal site (Figure 3).

Figure 3. View from new waste disposal site.

Because of the mining activity, some parts of the area have high slopes. Also the elevation of this area changes from 30 m to 35 m (Figure 4).

The new waste disposal site will be deposited on tuff rock, which is about 20 m thick in the study area. Density discontinuities and alteration can be seen in the area. The tuff is overlain by alluvium in the study area (Figure 5). The catchments portion of the study area is about 750 m^2.

4. Rapid Impact Assessment Matrix

The historical development of EIA shows that a number of attempts have been made to improve the quality of the EIA analysis by seeking to improve the accuracy of the judgment, resulting in a number of formats being developed for EIA analysis [20, 21]. The RIAM method is based on a standard definition of the important assessment criteria as well as the means by which semi-quantitative values for each of these criteria can be collected to provide an accurate and independent score for each condition [22]. These methods were used on various projects in different countries [23–28].

Figure 4. View of new waste disposal site.

Figure 5. View of the study area.

An analysis using the RIAM was performed for the Canakkale waste disposal site in 2004 and 2005 (Canakkale-Turkey). The impact of project activities is evaluated against the environmental components and for each component a score is determined, which provides a measure of the impact expected from the component [22, 29]. The important assessment criteria fall into two groups:

- *Group A:* Criteria that are of importance to the condition, and which can individually change the score obtained.
- *Group B*: Criteria that are of value to the situation, but individually should not be capable of changing the score obtained.

The value described to each of these groups of criteria is determined by the use of a series of simple formulae. These formulae allow the scores for the individual components to be determined on a defined basis. The process can be expressed:

$$
\begin{aligned}
&(A1) * (A2) = AT \\
&(B1) + (B2) + (B3) = BT \qquad (1) \\
&(AT) * (BT) = ES \quad \text{is the assessment score for the condition}
\end{aligned}
$$

Environmental Components: The RIAM methods require a specific evaluation of the components to be defined through the process of "scoping" and these environmental components will be in one of four categories that are described as follows:

- Physical-Chemical (PC)
- Biological-Ecological (BE)
- Sociological-Cultural (SC)
- Economic-Operational (EO)

4.1. APPLICATION OF RAPID IMPACT ASSESSMENT MATRIX (RIAM) IN NEW LANDFILL IN CANAKKALE CITY

The rapid impact assessment matrix for new landfill for Canakkale is shown in Table 1.

Most impacts are of class –A (impact) for the waste disposal site (Figure 6). Because of the waste, biological-ecological and physicochemical effects are prominent in the study area. This waste disposal site will cause positive impacts resulting from the economic-operational changes due to waste disposal.

TABLE 1. Physical-chemical, biological-ecological, sociological-cultural, and economical-operational components

Sociological and cultural components (SC)		ES	RB	A1	A2	B1	B2	B3
SC1	Natural protection	−6	−A	1	−1	2	2	2
SC2	Science	0	N	0	1	1	1	1
SC3	Education	0	N	1	0	1	1	1
SC4	Archaeological	−7	−A	1	−1	3	2	2

(continued)

TABLE 1. (continued)

Physical and chemical components (PC)		ES	RB	A1	A2	B1	B2	B3
PC1	Land use change	−7	−A	1	−1	3	2	2
PC2	Noise	0	N	0	−1	2	1	2
PC3	Groundwater	−6	−A	1	−1	3	1	2
PC4	Surface water	−12	−B	1	−2	3	1	2
PC5	Soil	−7	−A	1	−1	3	2	2
PC6	Agriculture	−7	−A	1	−1	3	2	2
PC7	Air	−6	−A	1	−1	3	2	1
PC8	Landscape	−6	−A	1	−1	2	2	2
PC9	Geology	0	N	1	0	1	1	1
PC10	Tectonic	0	N	1	0	1	1	1

Biological and ecological components (BE)		ES	RB	A1	A2	B1	B2	B3
BE1	Vegetation	−6	−A	1	−1	2	2	2
BE2	Fauna	−6	−A	1	−1	2	2	2
BE3	Flora	−6	−A	1	−1	2	2	2
BE4	Habitat loss	−6	−A	1	−1	2	2	2
BE5	Recreation	−6	−A	1	−1	2	2	2
BE6	Health	−8	−A	1	−1	3	3	2

Economical and operational components (EO)		ES	RB	A1	A2	B1	B2	B3
EO1	Transport	−6	−A	1	−1	2	2	2
EO2	Services	0	N	1	0	2	2	2
EO3	Work	7	A	1	1	3	2	2
EO4	Tourism opportunity	0	N	0	−1	3	2	1
EO5	Access roads improvement	6	A	1	1	2	2	2
EO6	Employment opportunity	6	A	1	1	2	2	2

ES: Environmental score B1: Permanence
RB: Range value B2: Permanence
A1: Importance of condition B3: Cumulative
A2: Magnitude of change-effect

Range	−108	−71	−35	−18	−9	0	1	10	19	36	72
Score	−72	−36	−19	−10	−1	0	9	18	35	71	108
Class	−E	−D	−C	−B	−A	N	A	B	C	D	E
PC	0	0	0	1	6	3	0	0	0	0	0
BE	0	0	0	0	6	0	0	0	0	0	0
SC	0	0	0	0	2	2	0	0	0	0	0
EO	0	0	0	0	1	2	3	0	0	0	0
Total	0	0	0	1	15	7	3	0	0	0	0

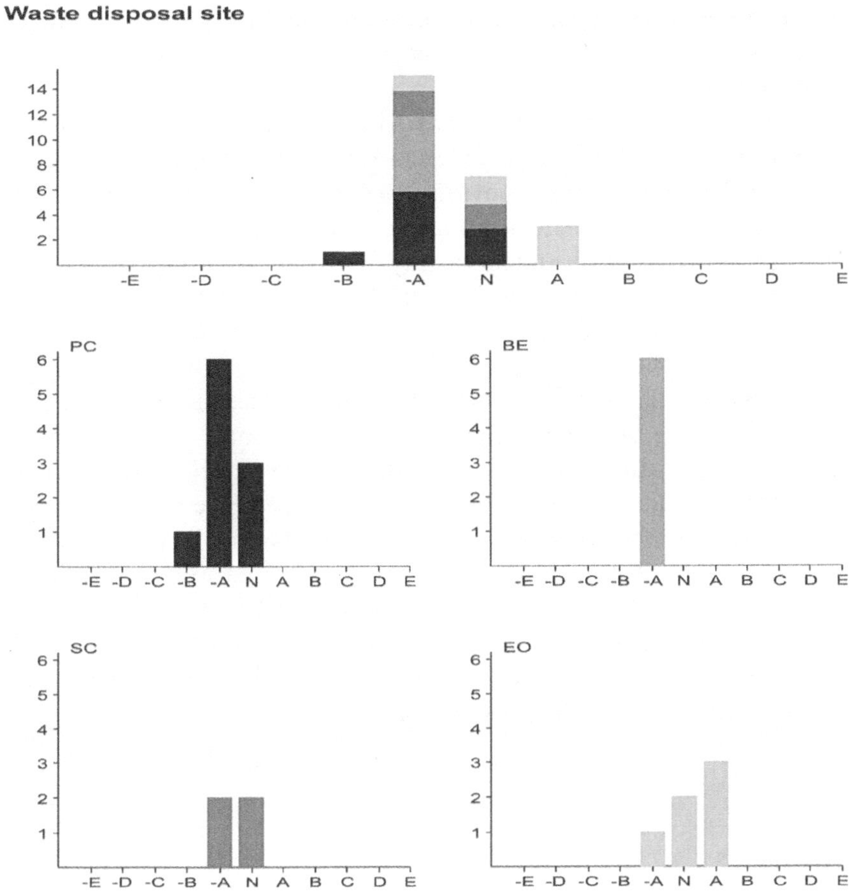

Figure 6. RIAM results for the four components, and the overall assessment.

5. Results and Discussion

Canakkale is a city that has dynamic economic development and rapid population growth. Environmental problems are increasing in those parts of the city with growing populations. People have a negative reaction to the new waste disposal site because the municipal waste is not disposed regularly in Canakkale.

The results visualized in Figure 6 show that negative impact is found especially for physicochemical, biological-ecological components and sociological-cultural components. Generally negative (−A) and neutral (N) effects are obtained for two main reasons. First, many people who live in Canakkale do not have enough knowledge about the application of the current waste disposal site. However, people who have enough information about the waste disposal site in operation at this time are not proud of the government because it has obvious negative effects on people health, soil, and groundwater.

The new waste disposal site is located near Canakkale-Lapseki road and north-northwest of Dolluk Hill, between Kemel village and Musakoy. People who live in these villages have obvious concerns and negative reactions about the selection of the waste site. For many years, residents of these villages have used the proposed location and its vicinity for agriculture and grassland for animals. Villagers worry about their fields and animals such as bees, because they have seen the negative effects of the old waste disposal site in Canakkale.

Also, one of the mining companies wanted an open mine in this area. At the beginning of the mining activity, many villages did not want an open mine in this area, but the company convinced people to allow the mine. The company said that it rehabilitated this area very well. But the company did not rehabilitate this area. Therefore people who live in this area have a negative attitude. They do not want see another bad facility in this area.

Local and government authorities are planning very well organized panels and informative meetings about well structured waste disposal site examples. Residents of these villages are strongly encouraged to attend these panels and meetings by the local authority. This will certainly help to soften residents' intolerance of the proposed landfill.

6. References

1. USEPA. Municipal solid waste in the United States: 2000 facts and figures executive summary, EPA-530-5-02-001. Washington, DC: Office of Solid Waste and Emergency Response; 2002.
2. Baun DL, Christensen TH. Speciation of heavy metals in landfill leachate: a review. Waste Manage. Res. 2004; 22:3-23.
3. Ettler V, Matura M, Mihaljevič M, Šebek O, Bezdička P. Heavy metal scavenging mechanisms in leachate-polluted waters: a case study of an old municipal landfill in Prague, Czech Republic. In: Christensen TH, Cossu R, Stegman R, editors. Sardinia Ninth International Waste Management and Landfill Symposium, Cagliari, Italy. 2003.
4. Yenigül NB, Amro MME, Johannes CG, van den Cees A, André TH, Dekking FM. Reliability assessment of groundwater monitoring networks at landfill sites. Journal of Hydrology 308(1-4):1-17; 2005.
5. Bozkurt S, Moreno L, Neretnieks I. Long-term processes in waste deposits, Sci. Total Environ. 2000; 250:101-121.
6. Kjeldsen P, Barlaz MA, Rooker AP, Baun A, Ledin A, Christensen TH. Present and long-term composition of MSW landfill leachate: a review. Crit. Rev. Environ. Sci. Technol. 2002; 32(4):297-336.
7. Davoli E, Gangai ML, Morselli L, Tonelli D. Characterisation of odorants emissions from landfills by SPME and GC/MS. Chemosphere. 2003; 51:357-368.
8. Schmidt CE, Wilsey SD, Hasek T. Technical approach for the assessment of air emissions from municipal landfills using the US EPA flux chamber and dispersion modelling to

predict off-site impact potential. Proceedings of the 91st Air & Water Management Association Annual Meeting, San Diego, California, 14-18 June 1998.
9. Allen MR, Braithwaite A, Hills CC. Trace organic compounds in landfill gas at seven UK waste disposal sites. Environmental Science Technology. 1997; 31:1054-1061.
10. Bradley AD, Cook DJ, Edwards JS, Johnston AG, Linforth RS, Taylor AJ. The control and measurement of landfill odours. Proceedings of Eighth International Landfill Symposium, Sardinia. 2001.
11. USEPA. Review and guidance on the use of landfill cover materials, version 3.7. 2000.
12 Greedy DR. Impact of daily cover on landfill operations. Proceedings of Fifth International Landfill Symposium, Sardinia. 1995.
13. Bracci G, Paci B, Giardi M. Paper mill sludge used a daily cover in landfills. Proceedings of Fifth International Landfill Symposium, Sardinia. 1995.
14. Carson DA. Municipal solid waste landfill daily cover alternatives. Geotextiles and Geomembranes. 1992; 11:629-635.
15. Hancock JS, Newman SJ, Chadwick BR. Mulched woody material mixtures: an improved performance daily cover. Proceedings of Seventh International Landfill Symposium, Sardinia. 1999.
16. Shimaoka T, Miyawaki K, Hanashima M, Tsuji H, Ito H. Impact of daily cover soil on the stabilisation of a landfill. Proceedings of Sixth International Landfill Symposium, Sardinia. 1997.
17. Muntoni A, Cossu R. Influence of compost covers on landfill gas surface emissions. Proceedings of Sixth International Landfill Symposium, Sardinia. 1997.
18. Baba A, Deniz O. Effect of waste disposal site on ground and surface water and investigation of water quality of Canakkale city. COMU, Research project (AFP) no. 2001-b/04. 2001.
19. Baba A, Deniz O, Kavdir Y. The impact of open waste disposal site on groundwater. International Journal of Environment and Pollution. 2004; 22(6):676-687.
20. Bisset R. Developments in EIA methods. In Walthern P, editor. Environmental impact assessment- theory and practise. UK: Unwin Hyman; 1988.
21. Wathern P. An introductory guide to EIA. In Walthern P, editor. Environmental impact assessment- theory and practise. UK: Unwin Hyman; 1988.
22. Pastakia CMR. The rapid impact assessment matrix (RIAM) - a new tool for environmental impact assessment. In K Jensen, editor. Environmental impact assessment using the rapid impact assessment matrix (RIAM). Fredensborg: Olsen & Olsen; 1998.
23. Baba A. Geothermal environmental impact assessment with special reference to the Tuzla, geothermal area, Canakkale Turkey. Geothermal Training Programme. Iceland: UN University; 2003.
24. Baba A. Environmental impact of the utilization of a geothermal area. Journal of Istanbul Kültür University. 2004; 1:33-38.
25. Jensen A. Environmental Impact Assessment of Halong city sanitation project, Vietnam. In: K. Jensen, editor. Environmental Impact Assessment using the Rapid Impact Assessment Matrix (RIAM). Fredensborg: Olsen & Olsen; 1998.
26. Hagebro C. Flood damage assessment Dac La Province, Vietnam. In: K. Jensen, editor. Environmental impact assessment using the rapid impact assessment matrix (RIAM). Fredensborg: Olsen & Olsen; 1998.
27. Jensen K, Bach HK, Knudsen CH, Foster TM, Mangor K, Haslov DB. Initial impact evaluation of a tourism development project in Malaysia. In: K. Jensen, editor. Environmental impact assessment using the rapid impact assessment matrix (RIAM). Fredensborg: Olsen & Olsen; 1998.
28. Birklund J. Application of the RIAM on the Oresund. In: Jensen K, editor. Environmental impact assessment using the rapid impact assessment matrix (RIAM). Fredensborg: Olsen & Olsen; 1998.
29. Baba A. Rapid impact assessment matrix (RIAM) method for Tuzla Geothermal and Çan Thermal Power Plant in Çanakkale, Turkey. Fresenius Environmental Bulletin. 2005; 14(2):113-118.

LIST OF PARTICIPANTS

Al-Halasah, Nizar	Royal Scientific Society P.O. Box 1438 Al-Jubeiha Amman 11941 Jordan	T: +30-2310-995702 F: +30-2310-995740 dangelid@civil.auth.gr
Apitz, Sabine E.	SEA Environmental Decisions Ltd. 1 South Cottages The Ford Little Hadham Hertfordshire SG 11 2AT United Kingdom	T: +44-0-1279-771890 F: +44-0-1279-771403 drsea@cvrl.org
Atoyev, Konstantyn	Institute of Cybernetics 40, Glushkov pr. Kiev 03022 Ukraine	038-044-2583125 038-044-526-1558 fax atoe@isofts.kiev.ua
Baba, Alper	Canakkale Onsekiz Mart University Department of Geological Engineering Terzioglu Campus Canakale 17100 Turkey	T: +9 0286218 0018/1212 F: +9 0286213 0155 alperbaba@comu.edu.tr
Bennett, Steven	NBACC Program Office 7435 New Technology, Ste. A Frederick, MD 21703 USA	301-846-2159 301-682-3662 fax steve.bennett@dhs.gov
Bridges, Todd S.	U.S. Army Engineer Research and Development Center Waterways Experiment Station EP-R, 3909 Halls Ferry Rd. Vicksburg, MS 39180 USA	T: +1 601-634-3626 F: +1 601-634-3120 fax todd.s.bridges@erdc.army.mil
Bronfman, Roman	The Democratic Choice Faction The Knesset Israel	T: +9 7226753832 dan.goldenblatt@gmail.com
Citron, Marta	Direzione Sicurezza & Ambiente Safety & Environment Dept. Zattere, 1401 Venezia 30123 Italy	+39-041-5334291 +39-041-5334254 fax marta.citron@port.venice.it
Del Pup, Michele	Comunita Porti Adriatici S. Marco 1092 Venezia 30100 Italy	39-041-523-7940 39-041-522-6063 fax delpup@tin.it

Della Sala, Stefano	Direzione Sicurezza & Ambiente Safety & Environment Dept. Zattere, 1401 Venezia 30123 Italy	+39-041-5334249 +39-041-5334254 fax stefano.dellasala@port.venice.it
Ditmer, Renae	US Defense Threat Reduction Agency 8725 John J. Kingman Hwy. Ft. Belvoir, VA 22060 USA	703-989-6750 or 703-767-3384 703-767-1892 fax Renae.Ditmer_Contractor@DTRA.mil
El Sheltawi, Shakinaz	Cairo University Cairo Egypt	202-0106044605 202-5728232 fax Chakina@hotmail.com
Elliott, Katherine	US Defense Threat Reduction Agency 47-701 Hui Kelu Street, #6 Kaneohe, HI 96744 USA	808-258-2090 808-356-1983 fax kate.m.elliott@gmail.com
Figueira, Jose Rui	Instituto Superior Tecnico Technical University of Lisbon DEG, Tagus Park, Av. Cavaco Silva Porto Salvo 2780-990 Portugal	351-21-423-35-07 351-21-423-35-68 fax figueira@ist.utl.pt
Fontaine, Francois	INERIS Institute National de Industrial et des Risques Parc Alata - BP n 2 Verneuil en Halatte 60550 France	33-3-44-55-68-17 33-3-44-55-66-99 fax francois.fontaine@ineris.fr
Ganoulis, Jacques	Aristotle Univeristy of Thessaloniki Department of Civil Engineering Thessaloniki 54124 Greece	T: +30-2310-99-5682 F: +30-2310-99-5681 iganouli@civil.auth.gr
Geleel, Mohamed Abdel	National Centre for Nuclear Safety & Radiation Control 3 Ahmed El Zomor Street Nasr City 11762 Egypt	202-7575223 202-2740238 fax magelee12000@yahoo.com
Haruvy, Nava	Netanya Academic College One University Street Netanya 42100 Israel	T: +972-8-946-3189 F: +972-8-936-5345 navaharu@netvision.net.il
Kapustka, Lawrence	Golder Associates Ltd. 1000, 940-6th Avenue, SW Calgary T2P 3T1 Canada	403-260-2265 403-299-5606 fax Larry_Kapustka@golder.com

Kiker, Gregory A.	University of Florida Dept. Agricultural & Biological Engineering P.O. Box 110570 Gainesville, FL 32611-0570 USA	T: +1 352-392-1864 gkiker@ufl.edu
Lambert, James H.	University of Virginia P.O. Box 400747 Charlottesville, VA 22904 USA	434-982-2072 434-924-0865 fax lambert@virginia.edu
Lazarescu, Mihaela	National R&D Institute for Environmental Spl. Independentel 294, sector 6 Bucharest 78 060031 Romania	++-401-3182074 ++-401-23182009 fax mihaelalazarescu@yahoo.com
Linkov, Igor	INTERTOX, Inc. 83 Winchester Street, Suite 1, Brookline, MA 02446 USA	T: +1 617-225-0812 ilinkov@intertox.com
Menoni, Scira	Politecnico di Milano DIAP Via Bonardi 3 Milano 20133 Italy	0039/02/23995457 0039/02/23995435 fax menoni@mail.polimi.it
Morgan, Jeff	US Coast Guard International Port Security Program 2100 2nd Street, SW Washington, DC 20593 USA	202-267-4339 jmorgan@comdt.uscg.mil
Parnell, Gregory	US Military Academy Mahan Hall, Room 432 West Point, NY 10996 USA	914-720-3989 845-938-5919 fax gregory.parnell@usma.edu
Peruzzo, Francesca	Direzione Sicurezza & Ambiente Safety & Environment Dept. Zattere, 1401 Venezia 30123 Italy	+39-041-5334249 +39-041-5334254 fax francesca.peruzzo@port.venice.it
Ramadan, Abou Bakr	Atomic Energy Authority Nasr City P.O. Box 7551 Cairo, Egypt	T: 202 27 48 787 F: 202 27 40 238 ramadan58@yahoo.com
Rix, Glenn	Georgia Institute of Technology 790 Atlantic Drive Atlanta, GA 30332 USA	404-894-2292 404-894-2281 glen.rix@ce.gatech.edu

Rowan, Steve	ENVIRON UK Port of Liverpool Building, 3rd Floor Pier Head Liverpool L3 1BY United Kingdom	44-151-227-1989 srowan@uk.environcorp.com
Rutjes, Rob	Environ Netherlands Zeisteroever 17 Zeist 3704 GB The Netherlands	313-0698-6215 313-0698-6239 fax rrutjes@environcorp.nl
Salvi, Olivier	INERIS National Institute for Industrial Risks & Environment Protection BP 2 – Parc Alata Verneuil en Halatte 60550 France	33-3-44-55-61-01 33-3-44-55-62-95 olivier.salvi@ineris.fr
Sargsyan, Vardan	Yerevan State Institute of Economy 128 Nalbandyan Str. Yerevan 75025 Armenia	T: 376-91-32-92-58 vsarg@freenet.am
Schlink, Uwe	UFZ - Centre for Environmental Research P.O. Box 5001235 Leipzig D-04301 Germany	++49-341-235-2742 ++49-341-235-2288 uwe.schlink@ufz.de
Schnelle, Debra	DTRA/CBT 4704 Marymead Drive Fairfax, VA 22030 USA	703-631-4577 703-767-1892 fax DebraSchnelleContractor@ DTRA.mil
Stockwell, David T.	US DoD DTRA 8725 John J. Kingman Road MSC 6201 Ft. Belvoir, VA 22060 USA	703-767-3384 703-767-1892 david.stockwell@dtra.mil
Valverde, Jr., L. James	Insurance Information Institute 110 William Street New York, NY 10038 USA	212-346-5522 212-732-1916 fax jamesv@iii.org
van de Voort, Maarten	RAND Europe Newtonweg 1 Leiden 2333 CP The Netherlands	31-71-524-5180 31-71-524-5191 fax voort@rand.org
Wenning, Richard J.	ENVIRON 6001 Shellmound Street, Ste. 700 Emeryville, CA 94608 USA	510-420-2556 510-658-1191 fax rjwenning@Environcorp.com

AUTHOR INDEX

GPSR Compliance

The European Union's (EU) General Product Safety Regulation (GPSR) is a set of rules that requires consumer products to be safe and our obligations to ensure this.

If you have any concerns about our products, you can contact us on ProductSafety@springernature.com

In case Publisher is established outside the EU, the EU authorized representative is:

Springer Nature Customer Service Center GmbH
Europaplatz 3
69115 Heidelberg, Germany

Batch number: 10809439

Printed by Printforce, the Netherlands